T0250174

Chevrolet
Impala SS &
Caprice
Buick
Roadmaster
Automotive
Repair
Manual

by Jeff Kibler
and John H Haynes
Member of the Guild of Motoring Writers

Models covered:
Chevrolet Impala SS, Caprice and
Buick Roadmaster models
1991 through 1996
Does not include information specific to V6 models

(24046-1R5)

Haynes Group Limited
Haynes North America, Inc.

www.haynes.com

Disclaimer

There are risks associated with automotive repairs. The ability to make repairs depends on the individual's skill, experience and proper tools. Individuals should act with due care and acknowledge and assume the risk of performing automotive repairs.

The purpose of this manual is to provide comprehensive, useful and accessible automotive repair information, to help you get the best value from your vehicle. However, this manual is not a substitute for a professional certified technician or mechanic.

This repair manual is produced by a third party and is not associated with an individual vehicle manufacturer. If there is any doubt or discrepancy between this manual and the owner's manual or the factory service manual, please refer to the factory service manual or seek assistance from a professional certified technician or mechanic.

Even though we have prepared this manual with extreme care and every attempt is made to ensure that the information in this manual is correct, neither the publisher nor the author can accept responsibility for loss, damage or injury caused by any errors in, or omissions from, the information given.

About this manual

Its purpose

The purpose of this manual is to help you get the best value from your vehicle. It can do so in several ways. It can help you decide what work must be done, even if you choose to have it done by a dealer service department or a repair shop; it provides information and procedures for routine maintenance and servicing; and it offers diagnostic and repair procedures to follow when trouble occurs.

We hope you use the manual to tackle the work yourself. For many simpler jobs, doing it yourself may be quicker than arranging an appointment to get the vehicle into a shop and making the trips to leave it and pick it up. More importantly, a lot of money can be saved by avoiding the expense the shop must pass on to you to cover its labor and overhead costs. An added benefit is the sense of satisfaction and accomplishment that you feel after doing the job yourself.

Using the manual

The manual is divided into Chapters. Each Chapter is divided into numbered Sections, which are headed in bold type between horizontal lines. Each Section consists of consecutively numbered paragraphs.

At the beginning of each numbered Section you will be referred to any illustrations which apply to the procedures in that Section. The reference numbers used in illustration captions pinpoint the pertinent Section and the Step within that Section. That is, illustration 3.2 means the illustration refers to Section 3 and Step (or paragraph) 2 within that Section.

Procedures, once described in the text, are not normally repeated. When it's necessary to refer to another Chapter, the reference will be given as Chapter and Section number. Cross references given without use of the word "Chapter" apply to Sections and/or paragraphs in the same Chapter. For example, "see Section 8" means in the same Chapter.

References to the left or right side of the vehicle assume you are sitting in the driver's seat, facing forward.

Even though we have prepared this manual with extreme care, neither the publisher nor the author can accept responsibility for any errors in, or omissions from, the information given.

NOTE

A **Note** provides information necessary to properly complete a procedure or information which will make the procedure easier to understand.

CAUTION

A **Caution** provides a special procedure or special steps which must be taken while completing the procedure where the Caution is found. Not heeding a Caution can result in damage to the assembly being worked on.

WARNING

A **Warning** provides a special procedure or special steps which must be taken while completing the procedure where the Warning is found. Not heeding a Warning can result in personal injury.

Acknowledgements

Technical writers who contributed to this project include Rob Maddox and Mike Stubblefield. Wiring diagrams originated exclusively for Haynes North America, Inc. by Valley Forge Technical Communications.

© **Haynes North America, Inc. 1997, 1999**
With permission from Haynes Group Limited

A book in the Haynes Automotive Repair Manual Series

All rights reserved. No part of this book may be reproduced or transmitted in any form or by any means, electronic or mechanical, including photocopying, recording or by any information storage or retrieval system, without permission in writing from the copyright holder.

ISBN-10: 1-56392-249-5

ISBN-13: 978-1-56392-249-7

Library of Congress Catalog Card Number 97-71414

While every attempt is made to ensure that the information in this manual is correct, no liability can be accepted by the authors or publishers for loss, damage or injury caused by any errors in, or omissions from, the information given.

Contents

Haynes photographer, mechanic and author with 1996 Chevrolet Caprice

Introduction to the Chevrolet Impala SS, Caprice and Buick Roadmaster

The models covered by this manual are available in four-door sedan or four-door station wagon body styles.

V8 engines used in these vehicles include the 4.3 liter, 5.0 liter and 5.7 liter. 1993 and earlier vehicles are equipped with Throttle-body fuel injection (TBI) while 1994 and later vehicles are equipped with Multi-port fuel injection (MPFI).

The engine drives the rear wheels through a four-speed automatic transmission via a driveshaft, differential and axles.

The steering gearbox is mounted to the left of the engine. Power steering is standard equipment on all models.

The front suspension is composed of upper and lower control arms, coil spring/ shock absorber assemblies and a stabilizer

bar. The rear suspension is solid axle with trailing arms, coil springs, shock absorber units and a stabilizer bar.

The brakes are either four wheel disc or disc on the front and drum on the rear wheels, depending on model, with an Anti-lock Brake System (ABS) standard.

Vehicle identification numbers

Modifications are a continuing and unpublicized part of vehicle manufacturing. Since spare parts manuals and lists are compiled on a numerical basis, the individual vehicle numbers are essential to correctly identify the component required.

Vehicle Identification Number (VIN)

This very important identification number is stamped on a plate attached to the left side of the dashboard and is visible through the driver's side of the windshield **(see illustration)**. The VIN also appears on the Vehicle Certificate of Title and Registration. It contains valuable information such as where and when the vehicle was manufactured, the model year and the body style.

VIN engine and model year codes

Two particularly important pieces of information found in the VIN are the engine code and the model year code. Counting from the left, the engine code letter designation is the 8th digit and the model year code letter designation is the 10th digit.

On the models covered by this manual the engine codes are:

E	5.0L V8 TBI
7	5.7L V8 TBI
P	5.7L V8 MPFI
W	4.3L V8 MPFI

On the models covered by this manual the model year codes are:

M	1991
N	1992
P	1993
R	1994
S	1995
T	1996

Engine identification numbers

The engine identification number(s) are found on a pad at the front right side of the block, just above the water pump and on the lower left side of the block by the transmission bellhousing **(see illustration)**. An optional location for the engine identification number is just above the oil pan rail and on the left side of the block.

Automatic transmission number

The automatic transmission ID number is stamped into the casting on the right rear corner of the housing **(see illustration)**.

Vehicle Emissions Control Information label

This label is found in the engine compartment. See Chapter 6 for more information on this label.

The Vehicle Identification Number (VIN) is visible through the driver's side of the windshield

The engine ID number is stamped on a pad at the front of the cylinder block just above the water pump and at the left rear (arrow) of the cylinder block by the transmission bellhousing

The transmission ID number is stamped into the transmission housing on the right side just above the oil pan

Buying parts

Replacement parts are available from many sources, which generally fall into one of two categories - authorized dealer parts departments and independent retail auto parts stores. Our advice concerning these parts is as follows:

Retail auto parts stores: Good auto parts stores will stock frequently needed components which wear out relatively fast, such as clutch components, exhaust systems, brake parts, tune-up parts, etc. These stores often supply new or reconditioned parts on an exchange basis, which can save a considerable amount of money. Discount auto parts stores are often very good places to buy materials and parts needed for general vehicle maintenance such as oil, grease, filters, spark plugs, belts, touch-up paint, bulbs, etc. They also usually sell tools and general accessories, have convenient hours, charge lower prices and can often be found not far from home.

Authorized dealer parts department: This is the best source for parts which are unique to the vehicle and not generally available elsewhere (such as major engine parts, transmission parts, trim pieces, etc.).

Warranty information: If the vehicle is still covered under warranty, be sure that any replacement parts purchased - regardless of the source - do not invalidate the warranty!

To be sure of obtaining the correct parts, have engine and chassis numbers available and, if possible, take the old parts along for positive identification.

Maintenance techniques, tools and working facilities

Maintenance techniques

There are a number of techniques involved in maintenance and repair that will be referred to throughout this manual. Application of these techniques will enable the home mechanic to be more efficient, better organized and capable of performing the various tasks properly, which will ensure that the repair job is thorough and complete.

Fasteners

Fasteners are nuts, bolts, studs and screws used to hold two or more parts together. There are a few things to keep in mind when working with fasteners. Almost all of them use a locking device of some type, either a lockwasher, locknut, locking tab or thread adhesive. All threaded fasteners should be clean and straight, with undamaged threads and undamaged corners on the hex head where the wrench fits. Develop the habit of replacing all damaged nuts and bolts with new ones. Special locknuts with nylon or fiber inserts can only be used once. If they are removed, they lose their locking ability and must be replaced with new ones.

Rusted nuts and bolts should be treated with a penetrating fluid to ease removal and prevent breakage. Some mechanics use turpentine in a spout-type oil can, which works quite well. After applying the rust penetrant, let it work for a few minutes before trying to loosen the nut or bolt. Badly rusted fasteners may have to be chiseled or sawed off or removed with a special nut breaker, available at tool stores.

If a bolt or stud breaks off in an assembly, it can be drilled and removed with a special tool commonly available for this purpose. Most automotive machine shops can perform this task, as well as other repair procedures, such as the repair of threaded holes that have been stripped out.

Flat washers and lockwashers, when removed from an assembly, should always be replaced exactly as removed. Replace any damaged washers with new ones. Never use a lockwasher on any soft metal surface (such as aluminum), thin sheet metal or plastic.

Grade 1 or 2   Grade 5   Grade 8

Bolt strength marking (standard/SAE/USS; bottom - metric)

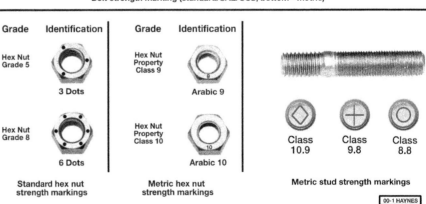

Grade	Identification	Grade	Identification
Hex Nut Grade 5	3 Dots	Hex Nut Property Class 9	Arabic 9
Hex Nut Grade 8	6 Dots	Hex Nut Property Class 10	Arabic 10
Standard hex nut strength markings		**Metric hex nut strength markings**	

Class 10.9   Class 9.8   Class 8.8

Metric stud strength markings

00-1 HAYNES

Fastener sizes

For a number of reasons, automobile manufacturers are making wider and wider use of metric fasteners. Therefore, it is important to be able to tell the difference between standard (sometimes called U.S. or SAE) and metric hardware, since they cannot be interchanged.

All bolts, whether standard or metric, are sized according to diameter, thread pitch and length. For example, a standard 1/2 - 13 x 1 bolt is 1/2 inch in diameter, has 13 threads per inch and is 1 inch long. An M12 - 1.75 x 25 metric bolt is 12 mm in diameter, has a thread pitch of 1.75 mm (the distance between threads) and is 25 mm long. The two bolts are nearly identical, and easily confused, but they are not interchangeable.

In addition to the differences in diameter, thread pitch and length, metric and standard bolts can also be distinguished by examining the bolt heads. To begin with, the distance across the flats on a standard bolt head is measured in inches, while the same dimension on a metric bolt is sized in millimeters (the same is true for nuts). As a result, a standard wrench should not be used on a metric bolt and a metric wrench should not be used on a standard bolt. Also, most standard bolts have slashes radiating out from the center of the head to denote the grade or strength of the bolt, which is an indication of the amount of torque that can be applied to it. The greater the number of slashes, the greater the strength of the bolt. Grades 0 through 5 are commonly used on automobiles. Metric bolts have a property class (grade) number, rather than a slash, molded into their heads to indicate bolt strength. In this case, the higher the number, the stronger the bolt. Property class numbers 8.8, 9.8 and 10.9 are commonly used on automobiles.

Strength markings can also be used to distinguish standard hex nuts from metric hex nuts. Many standard nuts have dots stamped into one side, while metric nuts are marked with a number. The greater the number of dots, or the higher the number, the greater the strength of the nut.

Metric studs are also marked on their ends according to property class (grade). Larger studs are numbered (the same as metric bolts), while smaller studs carry a geometric code to denote grade.

It should be noted that many fasteners, especially Grades 0 through 2, have no distinguishing marks on them. When such is the case, the only way to determine whether it is standard or metric is to measure the thread pitch or compare it to a known fastener of the same size.

Standard fasteners are often referred to as SAE, as opposed to metric. However, it should be noted that SAE technically refers to a non-metric fine thread fastener only. Coarse thread non-metric fasteners are referred to as USS sizes.

Since fasteners of the same size (both standard and metric) may have different

Metric thread sizes

	Ft-lbs	Nm
M-6	6 to 9	9 to 12
M-8	14 to 21	19 to 28
M-10	28 to 40	38 to 54
M-12	50 to 71	68 to 96
M-14	80 to 140	109 to 154

Pipe thread sizes

1/8	5 to 8	7 to 10
1/4	12 to 18	17 to 24
3/8	22 to 33	30 to 44
1/2	25 to 35	34 to 47

U.S. thread sizes

1/4 - 20	6 to 9	9 to 12
5/16 - 18	12 to 18	17 to 24
5/16 - 24	14 to 20	19 to 27
3/8 - 16	22 to 32	30 to 43
3/8 - 24	27 to 38	37 to 51
7/16 - 14	40 to 55	55 to 74
7/16 - 20	40 to 60	55 to 81
1/2 - 13	55 to 80	75 to 108

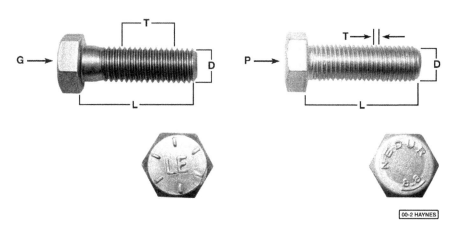

Standard (SAE and USS) bolt dimensions/ grade marks

G   Grade marks (bolt strength)
L   Length (in inches)
T   Thread pitch (number of threads per inch)
D   Nominal diameter (in inches)

Metric bolt dimensions/grade marks

P   Property class (bolt strength)
L   Length (in millimeters)
T   Thread pitch (distance between threads in millimeters)
D   Diameter

strength ratings, be sure to reinstall any bolts, studs or nuts removed from your vehicle in their original locations. Also, when replacing a fastener with a new one, make sure that the new one has a strength rating equal to or greater than the original.

Tightening sequences and procedures

Most threaded fasteners should be tightened to a specific torque value (torque is the twisting force applied to a threaded component such as a nut or bolt). Overtightening the fastener can weaken it and cause it to break, while undertightening can cause it to eventually come loose. Bolts, screws and studs, depending on the material they are made of and their thread diameters, have specific torque values, many of which are noted in the Specifications at the beginning of each Chapter. Be sure to follow the torque recommendations closely. For fasteners not assigned a specific torque, a general torque value chart is presented here as a guide. These torque values are for dry (unlubricated) fasteners threaded into steel or cast iron (not aluminum). As was previously mentioned, the size and grade of a fastener determine the amount of torque that can safely be applied to it. The figures listed here are approximate for Grade 2 and Grade 3 fasteners. Higher grades can tolerate higher torque values.

Fasteners laid out in a pattern, such as cylinder head bolts, oil pan bolts, differential cover bolts, etc., must be loosened or tightened in sequence to avoid warping the com-

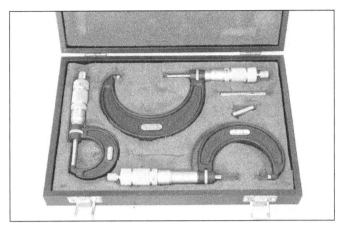

Micrometer set

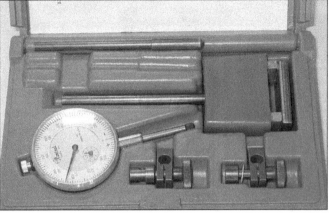

Dial indicator set

ponent. This sequence will normally be shown in the appropriate Chapter. If a specific pattern is not given, the following procedures can be used to prevent warping.

Initially, the bolts or nuts should be assembled finger-tight only. Next, they should be tightened one full turn each, in a crisscross or diagonal pattern. After each one has been tightened one full turn, return to the first one and tighten them all one-half turn, following the same pattern. Finally, tighten each of them one-quarter turn at a time until each fastener has been tightened to the proper torque. To loosen and remove the fasteners, the procedure would be reversed.

Component disassembly

Component disassembly should be done with care and purpose to help ensure that the parts go back together properly. Always keep track of the sequence in which parts are removed. Make note of special characteristics or marks on parts that can be installed more than one way, such as a grooved thrust washer on a shaft. It is a good idea to lay the disassembled parts out on a clean surface in the order that they were removed. It may also be helpful to make sketches or take instant photos of components before removal.

When removing fasteners from a component, keep track of their locations. Sometimes threading a bolt back in a part, or putting the washers and nut back on a stud, can prevent mix-ups later. If nuts and bolts cannot be returned to their original locations, they should be kept in a compartmented box or a series of small boxes. A cupcake or muffin tin is ideal for this purpose, since each cavity can hold the bolts and nuts from a particular area (i.e. oil pan bolts, valve cover bolts, engine mount bolts, etc.). A pan of this type is especially helpful when working on assemblies with very small parts, such as the carburetor, alternator, valve train or interior dash and trim pieces. The cavities can be marked with paint or tape to identify the contents.

Whenever wiring looms, harnesses or connectors are separated, it is a good idea to identify the two halves with numbered pieces of masking tape so they can be easily reconnected.

Gasket sealing surfaces

Throughout any vehicle, gaskets are used to seal the mating surfaces between two parts and keep lubricants, fluids, vacuum or pressure contained in an assembly.

Many times these gaskets are coated with a liquid or paste-type gasket sealing compound before assembly. Age, heat and pressure can sometimes cause the two parts to stick together so tightly that they are very difficult to separate. Often, the assembly can be loosened by striking it with a soft-face hammer near the mating surfaces. A regular hammer can be used if a block of wood is placed between the hammer and the part. Do not hammer on cast parts or parts that could be easily damaged. With any particularly stubborn part, always recheck to make sure that every fastener has been removed.

Avoid using a screwdriver or bar to pry apart an assembly, as they can easily mar the gasket sealing surfaces of the parts, which must remain smooth. If prying is absolutely necessary, use an old broom handle, but keep in mind that extra clean up will be necessary if the wood splinters.

After the parts are separated, the old gasket must be carefully scraped off and the gasket surfaces cleaned. Stubborn gasket material can be soaked with rust penetrant or treated with a special chemical to soften it so it can be easily scraped off. A scraper can be fashioned from a piece of copper tubing by flattening and sharpening one end. Copper is recommended because it is usually softer than the surfaces to be scraped, which reduces the chance of gouging the part. Some gaskets can be removed with a wire brush, but regardless of the method used, the mating surfaces must be left clean and smooth. If for some reason the gasket surface is gouged, then a gasket sealer thick enough to fill scratches will have to be used during reassembly of the components. For most applications, a non-drying (or semi-drying) gasket sealer should be used.

Hose removal tips

Warning: *If the vehicle is equipped with air conditioning, do not disconnect any of the A/C hoses without first having the system depressurized by a dealer service department or a service station.*

Hose removal precautions closely parallel gasket removal precautions. Avoid scratching or gouging the surface that the hose mates against or the connection may leak. This is especially true for radiator hoses. Because of various chemical reactions, the rubber in hoses can bond itself to the metal spigot that the hose fits over. To remove a hose, first loosen the hose clamps that secure it to the spigot. Then, with slip-joint pliers, grab the hose at the clamp and rotate it around the spigot. Work it back and forth until it is completely free, then pull it off. Silicone or other lubricants will ease removal if they can be applied between the hose and the outside of the spigot. Apply the same lubricant to the inside of the hose and the outside of the spigot to simplify installation.

As a last resort (and if the hose is to be replaced with a new one anyway), the rubber can be slit with a knife and the hose peeled from the spigot. If this must be done, be careful that the metal connection is not damaged.

If a hose clamp is broken or damaged, do not reuse it. Wire-type clamps usually weaken with age, so it is a good idea to replace them with screw-type clamps whenever a hose is removed.

Tools

A selection of good tools is a basic requirement for anyone who plans to maintain and repair his or her own vehicle. For the owner who has few tools, the initial investment might seem high, but when compared to the spiraling costs of professional auto maintenance and repair, it is a wise one.

To help the owner decide which tools are needed to perform the tasks detailed in this manual, the following tool lists are offered: *Maintenance and minor repair, Repair/overhaul* and *Special*.

The newcomer to practical mechanics

Dial caliper

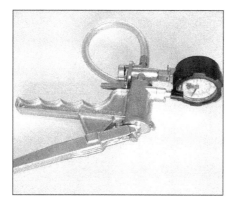

Hand-operated vacuum pump

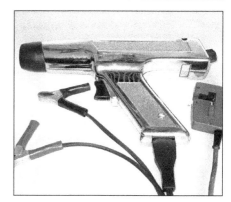

Timing light

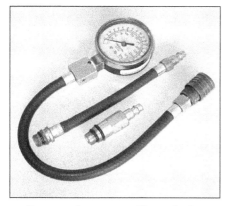

Compression gauge with spark plug
hole adapter

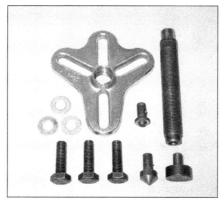

Damper/steering wheel puller

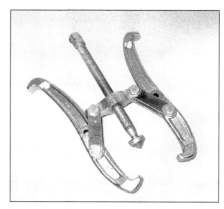

General purpose puller

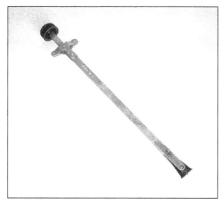

Hydraulic lifter removal tool

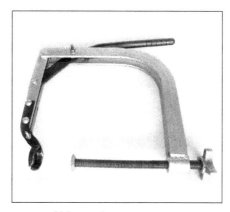

Valve spring compressor

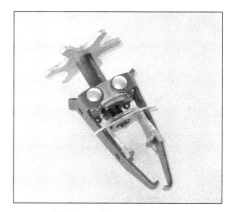

Valve spring compressor

Ridge reamer

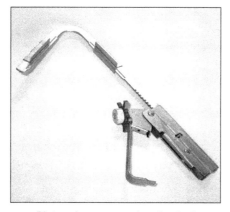

Piston ring groove cleaning tool

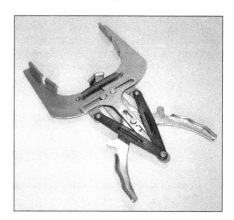

Ring removal/installation tool

Ring compressor

Cylinder hone

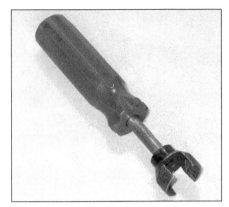

Brake hold-down spring tool

Torque angle gauge

Clutch plate alignment tool

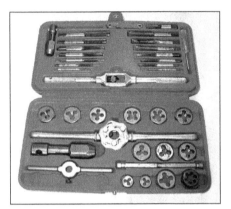

Tap and die set

should start off with the *maintenance and minor repair* tool kit, which is adequate for the simpler jobs performed on a vehicle. Then, as confidence and experience grow, the owner can tackle more difficult tasks, buying additional tools as they are needed. Eventually the basic kit will be expanded into the *repair and overhaul* tool set. Over a period of time, the experienced do-it-yourselfer will assemble a tool set complete enough for most repair and overhaul procedures and will add tools from the special category when it is felt that the expense is justified by the frequency of use.

Maintenance and minor repair tool kit

The tools in this list should be considered the minimum required for performance of routine maintenance, servicing and minor repair work. We recommend the purchase of combination wrenches (box-end and open-end combined in one wrench). While more expensive than open end wrenches, they offer the advantages of both types of wrench.

> *Combination wrench set (1/4-inch to 1 inch or 6 mm to 19 mm)*
> *Adjustable wrench, 8 inch*
> *Spark plug wrench with rubber insert*
> *Spark plug gap adjusting tool*
> *Feeler gauge set*
> *Brake bleeder wrench*

> *Standard screwdriver (5/16-inch x 6 inch)*
> *Phillips screwdriver (No. 2 x 6 inch)*
> *Combination pliers - 6 inch*
> *Hacksaw and assortment of blades*
> *Tire pressure gauge*
> *Grease gun*
> *Oil can*
> *Fine emery cloth*
> *Wire brush*
> *Battery post and cable cleaning tool*
> *Oil filter wrench*
> *Funnel (medium size)*
> *Safety goggles*
> *Jackstands (2)*
> *Drain pan*

Note: *If basic tune-ups are going to be part of routine maintenance, it will be necessary to purchase a good quality stroboscopic timing light and combination tachometer/dwell meter. Although they are included in the list of special tools, it is mentioned here because they are absolutely necessary for tuning most vehicles properly.*

Repair and overhaul tool set

These tools are essential for anyone who plans to perform major repairs and are in addition to those in the maintenance and minor repair tool kit. Included is a comprehensive set of sockets which, though expensive, are invaluable because of their versatility,

especially when various extensions and drives are available. We recommend the 1/2-inch drive over the 3/8-inch drive. Although the larger drive is bulky and more expensive, it has the capacity of accepting a very wide range of large sockets. Ideally, however, the mechanic should have a 3/8-inch drive set and a 1/2-inch drive set.

> *Socket set(s)*
> *Reversible ratchet*
> *Extension - 10 inch*
> *Universal joint*
> *Torque wrench (same size drive as sockets)*
> *Ball peen hammer - 8 ounce*
> *Soft-face hammer (plastic/rubber)*
> *Standard screwdriver (1/4-inch x 6 inch)*
> *Standard screwdriver (stubby - 5/16-inch)*
> *Phillips screwdriver (No. 3 x 8 inch)*
> *Phillips screwdriver (stubby - No. 2)*
> *Pliers - vise grip*
> *Pliers - lineman's*
> *Pliers - needle nose*
> *Pliers - snap-ring (internal and external)*
> *Cold chisel - 1/2-inch*
> *Scribe*
> *Scraper (made from flattened copper tubing)*
> *Centerpunch*
> *Pin punches (1/16, 1/8, 3/16-inch)*
> *Steel rule/straightedge - 12 inch*

*Allen wrench set (1/8 to 3/8-inch or
    4 mm to 10 mm)
A selection of files
Wire brush (large)
Jackstands (second set)
Jack (scissor or hydraulic type)*

Note: *Another tool which is often useful is an electric drill with a chuck capacity of 3/8-inch and a set of good quality drill bits.*

Special tools

The tools in this list include those which are not used regularly, are expensive to buy, or which need to be used in accordance with their manufacturer's instructions. Unless these tools will be used frequently, it is not very economical to purchase many of them. A consideration would be to split the cost and use between yourself and a friend or friends. In addition, most of these tools can be obtained from a tool rental shop on a temporary basis.

This list primarily contains only those tools and instruments widely available to the public, and not those special tools produced by the vehicle manufacturer for distribution to dealer service departments. Occasionally, references to the manufacturer's special tools are included in the text of this manual. Generally, an alternative method of doing the job without the special tool is offered. However, sometimes there is no alternative to their use. Where this is the case, and the tool cannot be purchased or borrowed, the work should be turned over to the dealer service department or an automotive repair shop.

*Valve spring compressor
Piston ring groove cleaning tool
Piston ring compressor
Piston ring installation tool
Cylinder compression gauge
Cylinder ridge reamer
Cylinder surfacing hone
Cylinder bore gauge
Micrometers and/or dial calipers
Hydraulic lifter removal tool
Balljoint separator
Universal-type puller
Impact screwdriver
Dial indicator set
Stroboscopic timing light (inductive
    pick-up)
Hand operated vacuum/pressure pump
Tachometer/dwell meter
Universal electrical multimeter
Cable hoist
Brake spring removal and installation
    tools
Floor jack*

Buying tools

For the do-it-yourselfer who is just starting to get involved in vehicle maintenance and repair, there are a number of options available when purchasing tools. If maintenance and minor repair is the extent of the work to be done, the purchase of individual tools is satisfactory. If, on the other hand, extensive work is planned, it would be a good idea to purchase a modest tool set from one

of the large retail chain stores. A set can usually be bought at a substantial savings over the individual tool prices, and they often come with a tool box. As additional tools are needed, add-on sets, individual tools and a larger tool box can be purchased to expand the tool selection. Building a tool set gradually allows the cost of the tools to be spread over a longer period of time and gives the mechanic the freedom to choose only those tools that will actually be used.

Tool stores will often be the only source of some of the special tools that are needed, but regardless of where tools are bought, try to avoid cheap ones, especially when buying screwdrivers and sockets, because they won't last very long. The expense involved in replacing cheap tools will eventually be greater than the initial cost of quality tools.

Care and maintenance of tools

Good tools are expensive, so it makes sense to treat them with respect. Keep them clean and in usable condition and store them properly when not in use. Always wipe off any dirt, grease or metal chips before putting them away. Never leave tools lying around in the work area. Upon completion of a job, always check closely under the hood for tools that may have been left there so they won't get lost during a test drive.

Some tools, such as screwdrivers, pliers, wrenches and sockets, can be hung on a panel mounted on the garage or workshop wall, while others should be kept in a tool box or tray. Measuring instruments, gauges, meters, etc. must be carefully stored where they cannot be damaged by weather or impact from other tools.

When tools are used with care and stored properly, they will last a very long time. Even with the best of care, though, tools will wear out if used frequently. When a tool is damaged or worn out, replace it. Subsequent jobs will be safer and more enjoyable if you do.

How to repair damaged threads

Sometimes, the internal threads of a nut or bolt hole can become stripped, usually from overtightening. Stripping threads is an all-too-common occurrence, especially when working with aluminum parts, because aluminum is so soft that it easily strips out.

Usually, external or internal threads are only partially stripped. After they've been cleaned up with a tap or die, they'll still work. Sometimes, however, threads are badly damaged. When this happens, you've got three choices:

1) *Drill and tap the hole to the next suitable oversize and install a larger diameter bolt, screw or stud.*
2) *Drill and tap the hole to accept a threaded plug, then drill and tap the plug to the original screw size. You can also buy a plug already threaded to the original size. Then you simply drill a hole to the specified size, then run the threaded*

plug into the hole with a bolt and jam nut. Once the plug is fully seated, remove the jam nut and bolt.
3) *The third method uses a patented thread repair kit like Heli-Coil or Slimsert. These easy-to-use kits are designed to repair damaged threads in straight-through holes and blind holes. Both are available as kits which can handle a variety of sizes and thread patterns. Drill the hole, then tap it with the special included tap. Install the Heli-Coil and the hole is back to its original diameter and thread pitch.*

Regardless of which method you use, be sure to proceed calmly and carefully. A little impatience or carelessness during one of these relatively simple procedures can ruin your whole day's work and cost you a bundle if you wreck an expensive part.

Working facilities

Not to be overlooked when discussing tools is the workshop. If anything more than routine maintenance is to be carried out, some sort of suitable work area is essential.

It is understood, and appreciated, that many home mechanics do not have a good workshop or garage available, and end up removing an engine or doing major repairs outside. It is recommended, however, that the overhaul or repair be completed under the cover of a roof.

A clean, flat workbench or table of comfortable working height is an absolute necessity. The workbench should be equipped with a vise that has a jaw opening of at least four inches.

As mentioned previously, some clean, dry storage space is also required for tools, as well as the lubricants, fluids, cleaning solvents, etc. which soon become necessary.

Sometimes waste oil and fluids, drained from the engine or cooling system during normal maintenance or repairs, present a disposal problem. To avoid pouring them on the ground or into a sewage system, pour the used fluids into large containers, seal them with caps and take them to an authorized disposal site or recycling center. Plastic jugs, such as old antifreeze containers, are ideal for this purpose.

Always keep a supply of old newspapers and clean rags available. Old towels are excellent for mopping up spills. Many mechanics use rolls of paper towels for most work because they are readily available and disposable. To help keep the area under the vehicle clean, a large cardboard box can be cut open and flattened to protect the garage or shop floor.

Whenever working over a painted surface, such as when leaning over a fender to service something under the hood, always cover it with an old blanket or bedspread to protect the finish. Vinyl covered pads, made especially for this purpose, are available at auto parts stores.

Jacking and towing

Jacking

Warning: *The jack supplied with the vehicle should only be used for raising the vehicle when changing a tire or placing jackstands under the frame. Never work under the vehicle or start the engine while the jack is being used as the only means of support.*

The vehicle must be on a level surface with the wheels blocked and the transmission in Park. Apply the parking brake if the front of the vehicle must be raised. Make sure no one is in the vehicle as it's being raised with the jack.

Remove the jack, lug nut wrench and spare tire from the trunk compartment.

To replace the tire, use the tapered end of the lug wrench to pry loose the wheel cover. **Note:** *If the vehicle is equipped with aluminum wheels, it may be necessary to pry out the special lug nut covers. Also, aluminum wheels normally have anti-theft lug nuts (one per wheel) which require using a special "key" between the lug wrench and lug nut. The key is usually in the glove compartment.* Loosen the lug nuts one-half turn, but leave them in place until the tire is raised off the ground.

Position the jack under the side of the vehicle at the indicated jacking points. There's a front and rear jacking point on each side of the vehicle **(see illustration)**.

Turn the jack handle clockwise (the lug wrench also serves as the jack handle) until the tire clears the ground. Remove the lug nuts and pull the tire off. Clean the mating surfaces of the hub and wheel, then install the spare. Replace the lug nuts with the beveled edges facing in and tighten them snugly. Don't attempt to tighten them completely until the vehicle is lowered or it could

slip off the jack.

Turn the jack handle counterclockwise to lower the vehicle. Remove the jack and tighten the lug nuts in a criss-cross pattern. If possible, tighten the nuts with a torque wrench (see Chapter 1 for the torque values). If you don't have access to a torque wrench, have the nuts checked by a service station or repair shop as soon as possible. **Caution:** *The compact spare included with these vehicles is intended for temporary use only. Have the tire repaired and reinstall it on the vehicle at the earliest opportunity and don't exceed 50 mph with the spare tire on the car.*

Install the wheel cover, then stow the tire, jack and wrench and unblock the wheels.

Towing

We recommend these vehicles be towed from the rear, with the rear wheels off the

ground. If it's absolutely necessary, these vehicles can be towed from the front with the front wheels off the ground, provided that speeds don't exceed 35 mph and the distance is less than 50 miles; the transmission can be damaged if these mileage/speed limitations are exceeded.

Equipment specifically designed for towing should be used. It must be attached to the main structural members of the vehicle, not the bumpers or brackets.

Safety is a major consideration when towing and all applicable state and local laws must be obeyed. A safety chain must be used at all times.

The parking brake must be released and the transmission must be in Neutral. The steering must be unlocked (ignition switch in the Off position). Remember that power steering and power brakes won't work with the engine off.

The head of the jack should be placed on the frame rail at either the front or the rear of the vehicle

Anti-theft audio system

General information

Some of these models are equipped with the Delco Loc II (early models) or THEFT-LOCK (later models) audio systems, which include an anti-theft feature that will render the stereo inoperative if stolen. If the power source to the stereo is cut with the anti-theft feature activated, the stereo will be inoperative. Even if the power source is immediately re-connected, the stereo will not function.

If your vehicle is equipped with this anti-theft system, do not disconnect the battery, remove the stereo or disconnect related components unless you have either turned off the feature or have the individual ID (code) number for the stereo.

Delco-Loc II

Disabling the anti-theft feature

1   Press the stereo's 1 and 4 preset buttons at the same time for five seconds with the ignition on and the radio power off. The display will show SEC, indicating the unit is in the secure mode (anti-theft feature enabled).
2   Press the SET button. The display will show "000".
3   Press the SCAN button to make the first number appear.
4   Press SEEK button right or left until the second and third digit of your code appear. The numbers will be displayed as entered.
5   Press the AM-FM button. "000" will be displayed.
6   Press the SCAN button to make the fourth number appear.

7   Press SEEK button right or left until the fifth and sixth digits of the code are displayed.
8   Press the AM-FM button. If the display shows "_ _," you have successfully disabled the anti-theft feature. If SEC is displayed, the code you entered was incorrect and the anti-theft feature is still enabled.

Unlocking the stereo after a power loss

9   When power is restored to the stereo, the stereo won't turn on and LOC will appear on the display. Enter your ID code as follows; pause no more than 15 seconds between Steps.
10   Turn the ignition switch to ON, but leave the stereo off.
11   Press the SET button. "000" should display.
12   Press the SCAN button to make the first number appear, then release it.
13   Press the SEEK button right or left to make sure the second and third numbers agree with your code.
14   Repeat Steps 7, 8 and 9 for the last three digits of your code.
15   Press the AM-FM button. If SEC appears, the numbers you entered were correct and the stereo will work. If LOC appears, the numbers you entered were not correct and the stereo is still inoperative

THEFTLOCK

Disabling the anti-theft feature

16   Press the stereo's 1 and 4 buttons at the

same time for five seconds with the ignition on and the radio power off. The display will show SEC, indicating the unit is in the secure mode (anti-theft feature enabled).
17   Press the MIN button. The display will show "000."
18   Press the MIN button to make the last two numbers appear.
19   Press HR to display the first one or two numbers of your code. The numbers will be displayed as entered.
20   Press AM/FM. If the display shows "_ _," you have successfully disabled the anti-theft feature. If SEC is displayed, the code you entered was incorrect and the anti-theft feature is still enabled.

Unlocking the stereo after a power loss

21   When power is restored to the stereo, the stereo won't turn on and LOC will appear on the display. Enter your ID code as follows; pause no more than 15 seconds between Steps.
22   Turn the ignition switch to ON, but leave the stereo off.
23   Press the MIN button. "000" should display.
24   Press the HR button to make the last two numbers appear, then release the button.
25   Press the HR button until the first one or two numbers appear.
26   Press AM/FM. SEC should appear, indicating the stereo is unlocked. If LOC appears, the numbers you entered were not correct and the stereo is still inoperative.

Booster battery (jump) starting

Observe these precautions when using a booster battery to start a vehicle:

a) *Before connecting the booster battery, make sure the ignition switch is in the Off position.*
b) *Turn off the lights, heater and other electrical loads.*
c) *Your eyes should be shielded. Safety goggles are a good idea.*
d) *Make sure the booster battery is the same voltage as the dead one in the vehicle.*
e) *The two vehicles MUST NOT TOUCH each other!*
f) *Make sure the transaxle is in Neutral (manual) or Park (automatic).*
g) *If the booster battery is not a maintenance-free type, remove the vent caps and lay a cloth over the vent holes.*

Connect the red jumper cable to the positive (+) terminals of each battery **(see illustration)**.

Connect one end of the black jumper cable to the negative (-) terminal of the booster battery. The other end of this cable should be connected to a good ground on the vehicle to be started, such as a bolt or bracket on the body.

Start the engine using the booster battery, then, with the engine running at idle speed, disconnect the jumper cables in the reverse order of connection.

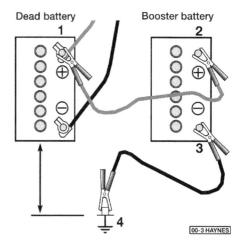

Make the booster battery cable connections in the numerical order shown (note that the negative cable of the booster battery is NOT attached to the negative terminal of the dead battery)

Automotive chemicals and lubricants

A number of automotive chemicals and lubricants are available for use during vehicle maintenance and repair. They include a wide variety of products ranging from cleaning solvents and degreasers to lubricants and protective sprays for rubber, plastic and vinyl.

Cleaners

Carburetor cleaner and choke cleaner is a strong solvent for gum, varnish and carbon. Most carburetor cleaners leave a dry-type lubricant film which will not harden or gum up. Because of this film it is not recommended for use on electrical components.

Brake system cleaner is used to remove brake dust, grease and brake fluid from the brake system, where clean surfaces are absolutely necessary. It leaves no residue and often eliminates brake squeal caused by contaminants.

Electrical cleaner removes oxidation, corrosion and carbon deposits from electrical contacts, restoring full current flow. It can also be used to clean spark plugs, carburetor jets, voltage regulators and other parts where an oil-free surface is desired.

Demoisturants remove water and moisture from electrical components such as alternators, voltage regulators, electrical connectors and fuse blocks. They are non-conductive and non-corrosive.

Degreasers are heavy-duty solvents used to remove grease from the outside of the engine and from chassis components. They can be sprayed or brushed on and, depending on the type, are rinsed off either with water or solvent.

Lubricants

Motor oil is the lubricant formulated for use in engines. It normally contains a wide variety of additives to prevent corrosion and reduce foaming and wear. Motor oil comes in various weights (viscosity ratings) from 0 to 50. The recommended weight of the oil depends on the season, temperature and the demands on the engine. Light oil is used in cold climates and under light load conditions. Heavy oil is used in hot climates and where high loads are encountered. Multi-viscosity oils are designed to have characteristics of both light and heavy oils and are available in a number of weights from 0W-20 to 20W-50.

Gear oil is designed to be used in differentials, manual transmissions and other areas where high-temperature lubrication is required.

Chassis and wheel bearing grease is a heavy grease used where increased loads and friction are encountered, such as for wheel bearings, balljoints, tie-rod ends and universal joints.

High-temperature wheel bearing grease is designed to withstand the extreme temperatures encountered by wheel bearings in disc brake equipped vehicles. It usually contains molybdenum disulfide (moly), which is a dry-type lubricant.

White grease is a heavy grease for metal-to-metal applications where water is a problem. White grease stays soft under both low and high temperatures (usually from -100 to +190-degrees F), and will not wash off or dilute in the presence of water.

Assembly lube is a special extreme pressure lubricant, usually containing moly, used to lubricate high-load parts (such as main and rod bearings and cam lobes) for initial start-up of a new engine. The assembly lube lubricates the parts without being squeezed out or washed away until the engine oiling system begins to function.

Silicone lubricants are used to protect rubber, plastic, vinyl and nylon parts.

Graphite lubricants are used where oils cannot be used due to contamination problems, such as in locks. The dry graphite will lubricate metal parts while remaining uncontaminated by dirt, water, oil or acids. It is electrically conductive and will not foul electrical contacts in locks such as the ignition switch.

Moly penetrants loosen and lubricate frozen, rusted and corroded fasteners and prevent future rusting or freezing.

Heat-sink grease is a special electrically non-conductive grease that is used for mounting electronic ignition modules where it is essential that heat is transferred away from the module.

Sealants

RTV sealant is one of the most widely used gasket compounds. Made from silicone, RTV is air curing, it seals, bonds, waterproofs, fills surface irregularities, remains flexible, doesn't shrink, is relatively easy to remove, and is used as a supplementary sealer with almost all low and medium temperature gaskets.

Anaerobic sealant is much like RTV in that it can be used either to seal gaskets or to form gaskets by itself. It remains flexible, is solvent resistant and fills surface imperfections. The difference between an anaerobic sealant and an RTV-type sealant is in the curing. RTV cures when exposed to air, while an anaerobic sealant cures only in the absence of air. This means that an anaerobic sealant cures only after the assembly of parts, sealing them together.

Thread and pipe sealant is used for sealing hydraulic and pneumatic fittings and vacuum lines. It is usually made from a Teflon compound, and comes in a spray, a paint-on liquid and as a wrap-around tape.

Chemicals

Anti-seize compound prevents seizing, galling, cold welding, rust and corrosion in fasteners. High-temperature anti-seize, usually made with copper and graphite lubricants, is used for exhaust system and exhaust manifold bolts.

Anaerobic locking compounds are used to keep fasteners from vibrating or working loose and cure only after installation, in the absence of air. Medium strength locking compound is used for small nuts, bolts and screws that may be removed later. High-strength locking compound is for large nuts, bolts and studs which aren't removed on a regular basis.

Oil additives range from viscosity index improvers to chemical treatments that claim to reduce internal engine friction. It should be noted that most oil manufacturers caution against using additives with their oils.

Gas additives perform several functions, depending on their chemical makeup. They usually contain solvents that help dissolve gum and varnish that build up on carburetor, fuel injection and intake parts. They also serve to break down carbon deposits that form on the inside surfaces of the combustion chambers. Some additives contain upper cylinder lubricants for valves and piston rings, and others contain chemicals to remove condensation from the gas tank.

Miscellaneous

Brake fluid is specially formulated hydraulic fluid that can withstand the heat and pressure encountered in brake systems. Care must be taken so this fluid does not come in contact with painted surfaces or plastics. An opened container should always be resealed to prevent contamination by water or dirt.

Weatherstrip adhesive is used to bond weatherstripping around doors, windows and trunk lids. It is sometimes used to attach trim pieces.

Undercoating is a petroleum-based, tar-like substance that is designed to protect metal surfaces on the underside of the vehicle from corrosion. It also acts as a sound-deadening agent by insulating the bottom of the vehicle.

Waxes and polishes are used to help protect painted and plated surfaces from the weather. Different types of paint may require the use of different types of wax and polish. Some polishes utilize a chemical or abrasive cleaner to help remove the top layer of oxidized (dull) paint on older vehicles. In recent years many non-wax polishes that contain a wide variety of chemicals such as polymers and silicones have been introduced. These non-wax polishes are usually easier to apply and last longer than conventional waxes and polishes.

Conversion factors

Length (distance)

Inches (in)	X	25.4	= Millimeters (mm)	X 0.0394	= Inches (in)
Feet (ft)	X	0.305	= Meters (m)	X 3.281	= Feet (ft)
Miles	X	1.609	= Kilometers (km)	X 0.621	= Miles

Volume (capacity)

Cubic inches (cu in; in³)	X	16.387	= Cubic centimeters (cc; cm³)	X 0.061	= Cubic inches (cu in; in³)
Imperial pints (Imp pt)	X	0.568	= Liters (l)	X 1.76	= Imperial pints (Imp pt)
Imperial quarts (Imp qt)	X	1.137	= Liters (l)	X 0.88	= Imperial quarts (Imp qt)
Imperial quarts (Imp qt)	X	1.201	= US quarts (US qt)	X 0.833	= Imperial quarts (Imp qt)
US quarts (US qt)	X	0.946	= Liters (l)	X 1.057	= US quarts (US qt)
Imperial gallons (Imp gal)	X	4.546	= Liters (l)	X 0.22	= Imperial gallons (Imp gal)
Imperial gallons (Imp gal)	X	1.201	= US gallons (US gal)	X 0.833	= Imperial gallons (Imp gal)
US gallons (US gal)	X	3.785	= Liters (l)	X 0.264	= US gallons (US gal)

Mass (weight)

Ounces (oz)	X	28.35	= Grams (g)	X 0.035	= Ounces (oz)
Pounds (lb)	X	0.454	= Kilograms (kg)	X 2.205	= Pounds (lb)

Force

Ounces-force (ozf; oz)	X	0.278	= Newtons (N)	X 3.6	= Ounces-force (ozf; oz)
Pounds-force (lbf; lb)	X	4.448	= Newtons (N)	X 0.225	= Pounds-force (lbf; lb)
Newtons (N)	X	0.1	= Kilograms-force (kgf; kg)	X 9.81	= Newtons (N)

Pressure

Pounds-force per square inch (psi; lbf/in²; lb/in²)	X	0.070	= Kilograms-force per square centimeter (kgf/cm²; kg/cm²)	X 14.223	= Pounds-force per square inch (psi; lbf/in²; lb/in²)
Pounds-force per square inch (psi; lbf/in²; lb/in²)	X	0.068	= Atmospheres (atm)	X 14.696	= Pounds-force per square inch (psi; lbf/in²; lb/in²)
Pounds-force per square inch (psi; lbf/in²; lb/in²)	X	0.069	= Bars	X 14.5	= Pounds-force per square inch (psi; lbf/in²; lb/in²)
Pounds-force per square inch (psi; lbf/in²; lb/in²)	X	6.895	= Kilopascals (kPa)	X 0.145	= Pounds-force per square inch (psi; lbf/in²; lb/in²)
Kilopascals (kPa)	X	0.01	= Kilograms-force per square centimeter (kgf/cm²; kg/cm²)	X 98.1	= Kilopascals (kPa)

Torque (moment of force)

Pounds-force inches (lbf in; lb in)	X	1.152	= Kilograms-force centimeter (kgf cm; kg cm)	X 0.868	= Pounds-force inches (lbf in; lb in)
Pounds-force inches (lbf in; lb in)	X	0.113	= Newton meters (Nm)	X 8.85	= Pounds-force inches (lbf in; lb in)
Pounds-force inches (lbf in; lb in)	X	0.083	= Pounds-force feet (lbf ft; lb ft)	X 12	= Pounds-force inches (lbf in; lb in)
Pounds-force feet (lbf ft; lb ft)	X	0.138	= Kilograms-force meters (kgf m; kg m)	X 7.233	= Pounds-force feet (lbf ft; lb ft)
Pounds-force feet (lbf ft; lb ft)	X	1.356	= Newton meters (Nm)	X 0.738	= Pounds-force feet (lbf ft; lb ft)
Newton meters (Nm)	X	0.102	= Kilograms-force meters (kgf m; kg m)	X 9.804	= Newton meters (Nm)

Vacuum

Inches mercury (in. Hg)	X	3.377	= Kilopascals (kPa)	X 0.2961	= Inches mercury
Inches mercury (in. Hg)	X	25.4	= Millimeters mercury (mm Hg)	X 0.0394	= Inches mercury

Power

Horsepower (hp)	X	745.7	= Watts (W)	X 0.0013	= Horsepower (hp)

Velocity (speed)

Miles per hour (miles/hr; mph)	X	1.609	= Kilometers per hour (km/hr; kph)	X 0.621	= Miles per hour (miles/hr; mph)

Fuel consumption*

Miles per gallon, Imperial (mpg)	X	0.354	= Kilometers per liter (km/l)	X 2.825	= Miles per gallon, Imperial (mpg)
Miles per gallon, US (mpg)	X	0.425	= Kilometers per liter (km/l)	X 2.352	= Miles per gallon, US (mpg)

Temperature

Degrees Fahrenheit = (°C x 1.8) + 32

Degrees Celsius (Degrees Centigrade; °C) = (°F - 32) x 0.56

*It is common practice to convert from miles per gallon (mpg) to liters/100 kilometers (l/100km), where mpg (Imperial) x l/100 km = 282 and mpg (US) x l/100 km = 235

Safety first!

Regardless of how enthusiastic you may be about getting on with the job at hand, take the time to ensure that your safety is not jeopardized. A moment's lack of attention can result in an accident, as can failure to observe certain simple safety precautions. The possibility of an accident will always exist, and the following points should not be considered a comprehensive list of all dangers. Rather, they are intended to make you aware of the risks and to encourage a safety conscious approach to all work you carry out on your vehicle.

Essential DOs and DON'Ts

DON'T rely on a jack when working under the vehicle. Always use approved jackstands to support the weight of the vehicle and place them under the recommended lift or support points.

DON'T attempt to loosen extremely tight fasteners (i.e. wheel lug nuts) while the vehicle is on a jack - it may fall.

DON'T start the engine without first making sure that the transmission is in Neutral (or Park where applicable) and the parking brake is set.

DON'T remove the radiator cap from a hot cooling system - let it cool or cover it with a cloth and release the pressure gradually.

DON'T attempt to drain the engine oil until you are sure it has cooled to the point that it will not burn you.

DON'T touch any part of the engine or exhaust system until it has cooled sufficiently to avoid burns.

DON'T siphon toxic liquids such as gasoline, antifreeze and brake fluid by mouth, or allow them to remain on your skin.

DON'T inhale brake lining dust - it is potentially hazardous (see *Asbestos* below).

DON'T allow spilled oil or grease to remain on the floor - wipe it up before someone slips on it.

DON'T use loose fitting wrenches or other tools which may slip and cause injury.

DON'T push on wrenches when loosening or tightening nuts or bolts. Always try to pull the wrench toward you. If the situation calls for pushing the wrench away, push with an open hand to avoid scraped knuckles if the wrench should slip.

DON'T attempt to lift a heavy component alone - get someone to help you.

DON'T *rush or take unsafe shortcuts to finish a job.*

DON'T allow children or animals in or around the vehicle while you are working on it.

DO wear eye protection when using power tools such as a drill, sander, bench grinder, etc. and when working under a vehicle.

DO keep loose clothing and long hair well out of the way of moving parts.

DO make sure that any hoist used has a safe working load rating adequate for the job.

DO get someone to check on you periodically when working alone on a vehicle.

DO carry out work in a logical sequence and make sure that everything is correctly assembled and tightened.

DO keep chemicals and fluids tightly capped and out of the reach of children and pets.

DO remember that your vehicle's safety affects that of yourself and others. If in doubt on any point, get professional advice.

Steering, suspension and brakes

These systems are essential to driving safety, so make sure you have a qualified shop or individual check your work. Also, compressed suspension springs can cause injury if released suddenly - be sure to use a spring compressor.

Airbags

Airbags are explosive devices that can **CAUSE** injury if they deploy while you're working on the vehicle. Follow the manufacturer's instructions to disable the airbag whenever you're working in the vicinity of airbag components.

Asbestos

Certain friction, insulating, sealing, and other products - such as brake linings, brake bands, clutch linings, torque converters, gaskets, etc. - may contain asbestos or other hazardous friction material. Extreme care must be taken to avoid inhalation of dust from such products, since it is hazardous to health. If in doubt, assume that they do contain asbestos.

Fire

Remember at all times that gasoline is highly flammable. Never smoke or have any kind of open flame around when working on a vehicle. But the risk does not end there. A spark caused by an electrical short circuit, by two metal surfaces contacting each other, or even by static electricity built up in your body under certain conditions, can ignite gasoline vapors, which in a confined space are highly explosive. Do not, under any circumstances, use gasoline for cleaning parts. Use an approved safety solvent.

Always disconnect the battery ground (-) cable at the battery before working on any part of the fuel system or electrical system. Never risk spilling fuel on a hot engine or exhaust component. It is strongly recommended that a fire extinguisher suitable for use on fuel and electrical fires be kept handy in the garage or workshop at all times. Never try to extinguish a fuel or electrical fire with water.

Fumes

Certain fumes are highly toxic and can quickly cause unconsciousness and even death if inhaled to any extent. Gasoline vapor falls into this category, as do the vapors from some cleaning solvents. Any draining or pouring of such volatile fluids should be done in a well ventilated area.

When using cleaning fluids and solvents, read the instructions on the container carefully. Never use materials from unmarked containers.

Never run the engine in an enclosed space, such as a garage. Exhaust fumes contain carbon monoxide, which is extremely poisonous. If you need to run the engine, always do so in the open air, or at least have the rear of the vehicle outside the work area.

The battery

Never create a spark or allow a bare light bulb near a battery. They normally give off a certain amount of hydrogen gas, which is highly explosive.

Always disconnect the battery ground (-) cable at the battery before working on the fuel or electrical systems.

If possible, loosen the filler caps or cover when charging the battery from an external source (this does not apply to sealed or maintenance-free batteries). Do not charge at an excessive rate or the battery may burst.

Take care when adding water to a non maintenance-free battery and when carrying a battery. The electrolyte, even when diluted, is very corrosive and should not be allowed to contact clothing or skin.

Always wear eye protection when cleaning the battery to prevent the caustic deposits from entering your eyes.

Household current

When using an electric power tool, inspection light, etc., which operates on household current, always make sure that the tool is correctly connected to its plug and that, where necessary, it is properly grounded. Do not use such items in damp conditions and, again, do not create a spark or apply excessive heat in the vicinity of fuel or fuel vapor.

Secondary ignition system voltage

A severe electric shock can result from touching certain parts of the ignition system (such as the spark plug wires) when the engine is running or being cranked, particularly if components are damp or the insulation is defective. In the case of an electronic ignition system, the secondary system voltage is much higher and could prove fatal.

Hydrofluoric acid

This extremely corrosive acid is formed when certain types of synthetic rubber, found in some O-rings, oil seals, fuel hoses, etc. are exposed to temperatures above 750-degrees F (400-degrees C). The rubber changes into a charred or sticky substance containing the acid. *Once formed, the acid remains dangerous for years. If it gets onto the skin, it may be necessary to amputate the limb concerned.*

When dealing with a vehicle which has suffered a fire, or with components salvaged from such a vehicle, wear protective gloves and discard them after use.

Troubleshooting

Contents

Engine and performance

1 Engine will not rotate when attempting to start

1   Battery terminal connections loose or corroded. Check the cable terminals at the battery; tighten cable clamp and/or clean off corrosion as necessary (see Chapter 1).
2   Battery discharged or faulty. If the cable ends are clean and tight on the battery posts, turn the key to the On position and switch on the headlights or windshield wipers. If they won't run, the battery is discharged.
3   Automatic transmission not engaged in park (P) or Neutral (N).
4   Broken, loose or disconnected wires in the starting circuit. Inspect all wires and connectors at the battery, starter solenoid and ignition switch (on steering column).
5   Starter motor pinion jammed in flywheel ring gear. Remove starter (Chapter 5) and inspect pinion and flywheel (Chapter 2) at earliest convenience.
6   Starter solenoid faulty (Chapter 5).
7   Starter motor faulty (Chapter 5).
8   Ignition switch faulty (Chapter 12).
9   Engine seized. Try to turn the crankshaft with a large socket and breaker bar on the pulley bolt.

2 Engine rotates but will not start

1   Fuel tank empty.
2   Battery discharged (engine rotates slowly). Check the operation of electrical components as described in previous Section.
3   Battery terminal connections loose or corroded. See previous Section.
4   Fuel not reaching fuel injectors. Check for clogged fuel filter or lines and defective fuel pump. Also make sure the tank vent lines aren't clogged (Chapter 4).
5   Faulty distributor (if equipped) components. Check the cap and rotor (Chapter 1).
6   Low cylinder compression. Check as described in Chapter 2.
7   Water in fuel. Drain tank and fill with new fuel.
8   Dirty or clogged fuel injectors.
9   Wet or damaged ignition components (Chapters 1 and 5).
10   Worn, faulty or incorrectly gapped spark plugs (Chapter 1).
11   Broken, loose or disconnected wires in the ignition circuit.
12   Broken, loose or disconnected wires at the ignition coil(s) or faulty coil(s) (Chapter 5).
13   Timing chain failure or wear affecting valve timing (Chapter 2).

3 Starter motor operates without turning engine

1   Starter pinion sticking. Remove the starter (Chapter 5) and inspect.

2   Starter pinion or flywheel/driveplate teeth worn or broken. Remove the inspection cover on the left side of the engine and inspect.

4 Engine hard to start when cold

1   Battery discharged or low. Check as described in Chapter 1.
2   Fuel not reaching the fuel injectors. Check the fuel filter and lines (Chapters 1 and 4).
3   Defective spark plugs (Chapter 1).
4   Defective ignition system components (Chapter 5).
5   Defective engine control system (coolant temperature sensor, Throttle Position Sensor, etc.) (Chapter 6).
6   Intake manifold vacuum leaks. Make sure all mounting bolts/nuts are tight and all vacuum hoses connected to the manifold are attached properly and in good condition.

5 Engine hard to start when hot

1   Air filter dirty (Chapter 1).
2   Bad engine ground connection.
3   Fuel not reaching the injectors (Chapter 4).
4   Defective ignition system components (Chapter 5).
5   Defective engine control system (coolant temperature sensor, Throttle Position Sensor, etc.) (Chapter 6).
6   Intake manifold vacuum leaks. Make sure all mounting bolts/nuts are tight and all vacuum hoses connected to the manifold are attached properly and in good condition.

6 Starter motor noisy or engages roughly

1   Pinion or flywheel/driveplate teeth worn or broken. Remove the inspection cover and inspect.
2   Starter motor mounting bolts loose or missing.

7 Engine starts but stops immediately

1   Defective ignition system components (Chapter 5).
2   Intake manifold vacuum leaks. Make sure all mounting bolts/nuts are tight and all vacuum hoses connected to the manifold are attached properly and in good condition.
3   Defective engine control system (coolant temperature sensor, Throttle Position Sensor, etc.) (Chapter 6).

8 Engine lopes while idling or idles erratically

1   Vacuum leaks. Check mounting bolts at the intake manifold or plenum for tightness. Make sure that all vacuum hoses are connected and in good condition. Use a stethoscope or a length of fuel hose held against your ear to listen for vacuum leaks while the engine is running. A hissing sound will be heard. A soapy water solution will also detect leaks. Check the intake manifold or plenum gasket surfaces.
2   Leaking EGR valve or plugged PCV valve (Chapter 6).
3   Air filter clogged (Chapter 1).
4   Leaking head gasket. Perform a cylinder compression check (Chapter 2).
5   Defective timing chain (Chapter 2).
6   Camshaft lobes worn (Chapter 2).
7   Valves burned or otherwise leaking (Chapter 2).
8   Ignition system not operating properly (Chapters 1 and 5).
9   Dirty or clogged injectors (Chapter 4).
10   Idle speed out of adjustment (Chapter 4).
11   Defective engine control system (coolant temperature sensor, Throttle Position Sensor, etc.) (Chapter 6).
12   Defective Idle Air Control valve (Chapter 4).

9 Engine misses at idle speed

1   Spark plugs faulty or not gapped properly (Chapter 1).
2   Faulty spark plug wires (Chapter 1).
3   Wet or damaged distributor components (Chapter 1).
4   Defective ignition coil (Chapter 5).
5   Sticking or faulty emissions system components (Chapter 6).
6   Clogged fuel filter and/or foreign matter in fuel. Replace the fuel filter (Chapter 1).
7   Vacuum leaks at intake manifold or plenum or hose connections. Check as described in Section 8.
8   Worn camshaft lobes or defective valve train (Chapter 2).
9   Low or uneven cylinder compression. Check as described in Chapter 2.
10   Clogged, dirty or defective fuel injectors (Chapter 4).
11   Leaky EGR valve (Chapter 6).

10 Excessively high idle speed

1   Sticking throttle linkage (Chapter 4).
2   Defective engine control system (coolant temperature sensor, Throttle Position Sensor, etc.) (Chapter 6).
3   Defective Idle Air Control valve (Chapter 4).
4   Vacuum leaks at intake manifold or plenum or hose connections. Check as described in Section 8.

11 Battery will not hold a charge

1　Drivebelt defective or not adjusted properly (Chapter 1).
2　Battery cables loose or corroded (Chapter 1).
3　Alternator not charging properly (Chapter 5).
4　Loose, broken or faulty wires in the charging circuit (Chapter 5).
5　Short circuit causing a continuous drain on the battery (Chapter 12).
6　Battery defective internally.

12 Alternator light stays on

1　Fault in alternator or charging circuit (Chapter 5).
2　Drivebelt defective or not properly adjusted (Chapter 1).

13 Alternator light fails to come on when key is turned on

1　Faulty bulb (Chapter 12).
2　Defective alternator (Chapter 5).
3　Fault in the printed circuit, dash wiring or bulb holder (Chapter 12).

14 Engine misses throughout driving speed range

1　Fuel filter clogged and/or impurities in the fuel system. Check fuel filter (Chapter 1) or clean system (Chapter 4).
2　Faulty or incorrectly gapped spark plugs (Chapter 1).
3　Incorrect ignition timing (Chapter 5).
4　Cracked distributor cap (if equipped) or disconnected ignition system wires or damaged ignition system components (Chapter 1).
5　Defective spark plug wires (Chapter 1).
6　Emissions system components faulty (Chapter 6).
7　Low or uneven cylinder compression pressures. Check as described in Chapter 2.
8　Weak or faulty ignition coil (Chapter 5).
9　Camshaft lobes worn or defective valve train (Chapter 2).
10　Vacuum leaks at intake manifold or plenum or vacuum hoses (see Section 8).
11　Dirty or clogged fuel injector (Chapter 4).

15 Hesitation or stumble during acceleration

1　Ignition timing incorrect (Chapter 5).
2　Ignition system not operating properly (Chapter 5).
3　Dirty or clogged fuel injectors (Chapter 4).
4　Low fuel pressure. Check for proper operation of the fuel pump and for restrictions in the fuel filter and lines (Chapter 4).
5　Malfunction in the emissions or engine control systems (Chapter 6).

16 Engine stalls

1　Idle speed incorrect (Chapter 4).
2　Fuel filter clogged and/or water and impurities in the fuel system (Chapter 1).
3　Damaged or wet ignition system wires or components.
4　Emissions system components faulty (Chapter 6).
5　Faulty or incorrectly gapped spark plugs (Chapter 1). Also check the spark plug wires (Chapter 1).
6　Vacuum leak at the intake manifold or plenum or vacuum hoses. Check as described in Section 8.

17 Engine lacks power

1　Incorrect ignition timing (Chapter 5).
2　Check for faulty distributor cap, wires, etc. (Chapter 1).
3　Faulty or incorrectly gapped spark plugs (Chapter 1).
4　Air filter dirty (Chapter 1).
5　Spark timing control system not operating properly (Chapter 5).
6　Faulty ignition coil(s) (Chapter 5).
7　Brakes binding (Chapters 1 and 9).
8　Automatic transmission fluid level incorrect, causing slippage (Chapter 1).
9　Fuel filter clogged and/or impurities in the fuel system (Chapters 1 and 4).
10　EGR system not functioning properly (Chapter 6).
11　Use of sub-standard fuel. Fill tank with proper octane fuel.
12　Low or uneven cylinder compression pressures. Check as described in Chapter 2.
13　Air (vacuum) leak at intake manifold or plenum (check as described in Section 8).

18 Engine backfires

1　EGR system not functioning properly (Chapter 6).
2　Ignition timing incorrect (Chapter 5).
3　Vacuum leak (refer to Section 8).
4　Damaged valve springs, sticking or burnt valves (Chapter 2).
5　Intake air (vacuum) leak (see Section 8).
6　Camshaft lobes worn (Chapter 2).
7　Defective air injection check valve (Chapter 6).

19 Engine surges while holding accelerator steady

1　Intake air (vacuum) leak (see Section 8).

2　Fuel pump not working properly.
3　Malfunction in the emissions or engine control systems (Chapter 6).

20 Pinging or knocking engine sounds when engine is under load

1　Incorrect grade of fuel. Fill tank with fuel of the proper octane rating.
2　Ignition timing incorrect (Chapter 5) or problem in the ignition system (Chapter 5).
3　Carbon build-up in combustion chambers. Remove cylinder head and clean combustion chambers (Chapter 2).
4　Incorrect spark plugs (Chapter 1).
5　Malfunction in the emissions or engine control systems (Chapter 6).
6　Check for causes of overheating (Section 27).

21 Engine continues to run after being turned off

1　Idle speed too high (Chapter 4).
2　Ignition timing incorrect (Chapter 5).
3　Incorrect spark plug heat range (Chapter 1).
4　Intake air (vacuum) leak (see Section 8).
5　Carbon build-up in combustion chambers. Remove the cylinder head and clean the combustion chambers (Chapter 2).
6　Valves sticking (Chapter 2).
7　EGR system not operating properly (Chapter 6).
8　Leaking fuel injector(s) (Chapter 4).
9　Check for causes of overheating (Section 27).

22 Low oil pressure

1　Improper grade of oil.
2　Oil pump regulator valve not operating properly (Chapter 2).
3　Oil pump worn or damaged (Chapter 2).
4　Engine overheating (refer to Section 27).
5　Clogged oil filter (Chapter 1).
6　Clogged oil strainer (Chapter 2).
7　Oil pressure gauge not working properly (Chapter 2).

23 Excessive oil consumption

1　Loose oil drain plug.
2　Loose bolts or damaged oil pan gasket (Chapter 2).
3　Loose bolts or damaged front cover gasket (Chapter 2).
4　Front or rear crankshaft oil seal leaking (Chapter 2).
5　Loose bolts or damaged valve cover gasket (Chapter 2).
6　Loose oil filter (Chapter 1).

7　Loose or damaged oil pressure switch (Chapter 2).
8　Pistons and cylinders excessively worn (Chapter 2).
9　Piston rings not installed correctly on pistons (Chapter 2).
10　Worn or damaged piston rings (Chapter 2).
11　Intake and/or exhaust valve oil seals worn or damaged (Chapter 2).
12　Worn valve stems.
13　Worn or damaged valves/guides (Chapter 2).

24　Excessive fuel consumption

1　Dirty or clogged air filter element (Chapter 1).
2　Incorrect ignition timing (Chapter 5).
3　Incorrect idle speed (Chapter 4).
4　Low tire pressure or incorrect tire size (Chapter 10).
5　Fuel leakage. Check all connections, lines and components in the fuel system (Chapter 4).
6　Dirty or clogged fuel injectors (Chapter 4).
7　Malfunction in the emissions or engine control systems (Chapter 6).

25　Fuel odor

1　Fuel leakage. Check all connections, lines and components in the fuel system (Chapter 4).
2　Fuel tank overfilled. Fill only to automatic shut-off.
3　Charcoal canister filter in Evaporative Emissions Control system clogged (Chapter 6).
4　Vapor leaks from Evaporative Emissions Control system lines (Chapter 6).

26　Miscellaneous engine noises

1　A strong dull noise that becomes more rapid as the engine accelerates indicates worn or damaged crankshaft bearings or an unevenly worn crankshaft. To pinpoint the trouble spot, remove the spark plug wire from one plug at a time and crank the engine over. If the noise stops, the cylinder with the removed plug wire indicates the problem area. Replace the bearing and/or service or replace the crankshaft (Chapter 2).
2　A similar (yet slightly higher pitched) noise to the crankshaft knocking described in the previous paragraph, that becomes more rapid as the engine accelerates, indicates worn or damaged connecting rod bearings (Chapter 2). The procedure for locating the problem cylinder is the same as described in Paragraph 1.
3　An overlapping metallic noise that increases in intensity as the engine speed

increases, yet diminishes as the engine warms up indicates abnormal piston and cylinder wear (Chapter 2).To locate the problem cylinder, use the procedure described in Paragraph 1.
4　A rapid clicking noise that becomes faster as the engine accelerates indicates a worn piston pin or piston pin hole. This sound will happen each time the piston hits the highest and lowest points in the stroke (Chapter 2). The procedure for locating the problem piston is described in Paragraph 1.
5　A metallic clicking noise coming from the water pump indicates worn or damaged water pump bearings or pump. Replace the water pump with a new one (Chapter 3).
6　A rapid tapping sound or clicking sound that becomes faster as the engine speed increases indicates "valve tapping" or stuck valve lifters. This can be identified by holding one end of a section of hose to your ear and placing the other end at different spots along the rocker arm cover. The point where the sound is loudest indicates the problem valve. Adjust the valve clearance (Chapter 1).
7　A steady metallic rattling or rapping sound coming from the area of the timing chain cover indicates a worn, damaged or out-of-adjustment timing chain. Service or replace the chain and related components (Chapter 2).

Cooling system

27　Overheating

1　Insufficient coolant in system (Chapter 1).
2　Drivebelt defective or not adjusted properly (Chapter 1).
3　Radiator core blocked or radiator grille dirty and restricted (Chapter 3).
4　Thermostat faulty (Chapter 3).
5　Fan not functioning properly (Chapter 3).
6　Radiator cap not maintaining proper pressure. Have cap pressure tested by gas station or repair shop.
7　Ignition timing incorrect (Chapter 5).
8　Defective water pump (Chapter 3).
9　Improper grade of engine oil.
10　Inaccurate temperature gauge (Chapter 3).

28　Overcooling

1　Thermostat faulty (Chapter 3).
2　Inaccurate temperature gauge (Chapter 3).

29　External coolant leakage

1　Deteriorated or damaged hoses. Loose clamps at hose connections (Chapter 1).
2　Water pump seals defective. If this is the

case, water will drip from the weep hole in the water pump body (Chapter 3).
3　Leakage from radiator core or header tank. This will require the radiator to be professionally repaired (see Chapter 3 for removal procedures).
4　Engine drain plugs or water jacket freeze plugs leaking (see Chapters 1 and 2).
5　Leak from coolant temperature switch (Chapter 3).
6　Leak from damaged gaskets or small cracks (Chapter 2).
7　Damaged head gasket. This can be verified by checking the condition of the engine oil as noted in Section 30.

30　Internal coolant leakage

Note: *Internal coolant leaks can usually be detected by examining the oil. Check the dipstick and inside the valve cover for water deposits and an oil consistency like that of a milkshake.*

1　Leaking intake manifold or cylinder head gaskets. Have the system pressure tested or remove the intake manifold and cylinder heads (Chapter 2) and inspect.
2　Cracked cylinder bore or cylinder head. Dismantle engine and inspect (Chapter 2).
3　Loose cylinder head bolts (tighten as described in Chapter 2).

31　Abnormal coolant loss

1　Overfilling system (Chapter 1).
2　Coolant boiling away due to overheating (see causes in Section 27).
3　Internal or external leakage (see Sections 29 and 30).
4　Faulty radiator cap. Have the cap pressure tested.
5　Cooling system being pressurized by engine compression. This could be due to a cracked head or block or leaking head gasket(s).

32　Poor coolant circulation

1　Inoperative water pump. A quick test is to pinch the top radiator hose closed with your hand while the engine is idling, then release it. You should feel a surge of coolant if the pump is working properly (Chapter 3).
2　Restriction in cooling system. Drain, flush and refill the system (Chapter 1). If necessary, remove the radiator (Chapter 3) and have it reverse flushed or professionally cleaned.
3　Thermostat sticking (Chapter 3).
4　Insufficient coolant (Chapter 1).

33　Corrosion

1　Excessive impurities in the water. Soft,

clean water is recommended. Distilled or rain-water is satisfactory.

2 Insufficient antifreeze solution (refer to Chapter 1 for the proper type and ratio of water to antifreeze).

3 Infrequent flushing and draining of system. Regular flushing of the cooling system should be carried out at the specified intervals as described in Chapter 1.

Automatic transmission

Note: *Due to the complexity of the automatic transmission, it's difficult for the home mechanic to properly diagnose and service. For problems other than the following, the vehicle should be taken to a reputable mechanic.*

34 Fluid leakage

1 Automatic transmission fluid is a deep red color, and fluid leaks should not be confused with engine oil which can easily be blown by air flow to the transmission.

2 To pinpoint a leak, first remove all built-up dirt and grime from the transmission. Degreasing agents and/or steam cleaning will achieve this. With the underside clean, drive the vehicle at low speeds so the air flow will not blow the leak far from its source. Raise the vehicle and determine where the leak is located. Common areas of leakage are:

a) *Fluid pan: tighten mounting bolts and/or replace pan gasket as necessary (Chapter 1).*

b) *Rear extension: tighten bolts and/or replace oil seal as necessary.*

c) *Filler pipe: replace the rubber oil seal where pipe enters transmission case.*

d) *Transmission oil lines: tighten fittings where lines enter transmission case and/or replace lines.*

e) *Vent pipe: transmission overfilled and/or water in fluid (see checking procedures, Chapter 1).*

f) *Speedometer connector: replace the O-ring where speedometer sensor enters transmission case.*

35 General shift mechanism problems

Chapter 7 deals with checking and adjusting the shift linkage on automatic transmissions. Common problems which may be caused by out of adjustment linkage are:

a) *Engine starting in gears other than P (Park) or N (Neutral).*

b) *Indicator pointing to a gear other than the one actually engaged.*

c) *Vehicle moves with transmission in P (Park) position.*

36 Transmission will not downshift with the accelerator pedal pressed to the floor

Chapter 7 deals with adjusting the throttle valve cable installed on automatic transmissions.

37 Engine will start in gears other than Park or Neutral

Chapter 7 deals with adjusting the Neutral start switch installed on automatic transmissions.

38 Transmission slips, shifts rough, is noisy or has no drive in forward or Reverse gears

1 There are many probable causes for the above problems, but the home mechanic should concern himself only with one possibility; fluid level.

2 Before taking the vehicle to a shop, check the fluid level and condition as described in Chapter 1. Add fluid, if necessary, or change the fluid and filter if needed. If problems persist, have a professional diagnose the transmission.

Driveshaft

39 Leaks at front of driveshaft

Defective transmission rear seal. See Chapter 7 for replacement procedure. As this is done, check the splined yoke for burrs or roughness that could damage the new seal. Remove burrs with a fine file or whetstone.

40 Knock or clunk when transmission is under initial load (just after transmission is put into gear)

1 Loose or disconnected rear suspension components. Check all mounting bolts and bushings (Chapters 1 and 11).

2 Loose driveshaft bolts. Inspect all bolts and nuts and tighten them securely.

3 Worn or damaged universal joint bearings. Replace driveshaft universal joints (Chapter 8).

4 Worn sleeve yoke and mainshaft spline.

41 Metallic grating sound consistent with vehicle speed

Pronounced wear in the universal joint

bearings. Replace U-joints or driveshaft, as necessary.

42 Vibration

1 Install a tachometer inside the vehicle to monitor engine speed as the vehicle is driven. Drive the vehicle and note the engine speed at which the vibration (roughness) is most pronounced. Now shift the transmission to a different gear and bring the engine speed to the same point.

2 If the vibration occurs at the same engine speed (rpm) regardless of which gear the transmission is in, the driveshaft is NOT at fault since the driveshaft speed varies.

3 If the vibration decreases or is eliminated when the transmission is in a different gear at the same engine speed, refer to the following probable causes.

4 Bent or dented driveshaft. Inspect and replace as necessary.

5 Undercoating or built-up dirt, etc. on the driveshaft. Clean the shaft thoroughly.

6 Worn universal joint bearings. Replace the U-joints or driveshaft as necessary.

7 Driveshaft and/or companion flange out of balance. Check for missing weights on the shaft. Remove driveshaft and reinstall 180-degrees from original position, then recheck. Have the driveshaft balanced if problem persists.

8 Loose driveshaft mounting bolts/nuts.

9 Worn transmission rear bushing.

43 Scraping noise

Make sure the dust cover on the sleeve yoke isn't rubbing on the transmission extension housing.

Rear axle and differential

Note: *For differential servicing information, refer to Chapter 8, unless otherwise specified.*

44 Noise - same when in drive as when vehicle is coasting

1 Road noise. No corrective action available.

2 Tire noise. Inspect tires and check tire pressures (Chapter 1).

3 Front wheel bearings worn or damaged (Chapter 10).

4 Insufficient differential oil (Chapter 1).

5 Defective differential.

45 Knocking sound when starting or shifting gears

Defective or incorrectly adjusted differential.

46   Noise when turning

Defective differential.

47   Vibration

See probable causes under Driveshaft. Proceed under the guidelines listed for the driveshaft. If the problem persists, check the rear wheel bearings by raising the rear of the vehicle and spinning the wheels by hand. Listen for evidence of rough (noisy) bearings. Remove and inspect (Chapter 8).

48   Oil leaks

1   Pinion oil seal damaged (Chapter 8).
2   Axleshaft or driveaxle oil seals damaged (Chapter 8).
3   Loose cover bolts or filler plug on differential (Chapter 1).
4   Clogged or damaged breather on differential.

Brakes

Note: *Before assuming a brake problem exists, make sure the tires are in good condition and inflated properly, the front end alignment is correct and the vehicle is not loaded with weight in an unequal manner. All service procedures for the brakes are included in Chapter 9, unless otherwise noted.*

49   Vehicle pulls to one side during braking

1   Defective, damaged or oil contaminated brake pad on one side. Inspect as described in Chapter 1. Refer to Chapter 9 if replacement is required.
2   Excessive wear of brake pad material or disc on one side. Inspect and repair as necessary.
3   Loose or disconnected front suspension components. Inspect and tighten all bolts securely (Chapters 1 and 10).
4   Defective caliper assembly. Remove caliper and inspect for stuck piston or damage.
5   Brake pad-to-disc adjustment needed. Inspect automatic adjusting mechanism for proper operation.
6   Scored or out of round disc.
7   Loose caliper mounting bolts.

50   Noise (high-pitched squeal)

1   Front brake pads worn out. This noise comes from the wear sensor rubbing against the disc. Replace pads with new ones immediately!

2   Glazed or contaminated pads.
3   Dirty or scored disc.
4   Bent support plate.

51   Excessive brake pedal travel

1   Partial brake system failure. Inspect entire system and correct as required.
2   Insufficient fluid in master cylinder. Check (Chapter 1) and add fluid - bleed system if necessary.
3   Air in system. Bleed system.
4   Excessive lateral disc play.
5   Brakes out of adjustment. Check the operation of the automatic adjusters.
6   Defective check valve. Replace valve and bleed system.

52   Brake pedal feels spongy when depressed

1   Air in brake lines. Bleed the brake system.
2   Deteriorated rubber brake hoses. Inspect all system hoses and lines. Replace parts as necessary.
3   Master cylinder mounting nuts loose. Inspect master cylinder bolts (nuts) and tighten them securely.
4   Master cylinder faulty.
5   Incorrect shoe or pad clearance.
6   Defective check valve. Replace valve and bleed system.
7   Clogged reservoir cap vent hole.
8   Deformed rubber brake lines.
9   Soft or swollen caliper seals.
10   Poor quality brake fluid. Bleed entire system and fill with new approved fluid.

53   Excessive effort required to stop vehicle

1   Power brake booster not operating properly.
2   Excessively worn pads. Check and replace if necessary.
3   One or more caliper pistons seized or sticking. Inspect and rebuild as required.
4   Brake pads contaminated with oil or grease. Inspect and replace as required.
5   New pads installed and not yet seated. It'll take a while for the new material to seat against the disc.
6   Worn or damaged master cylinder or caliper assemblies. Check particularly for frozen pistons.
7   Also see causes listed under Section 52.

54   Pedal travels to the floor with little resistance

Little or no fluid in the master cylinder reservoir caused by leaking caliper piston(s)

or loose, damaged or disconnected brake lines. Inspect entire system and repair as necessary.

55   Brake pedal pulsates during brake application

Note: *Brake pedal pulsation during operation of the Anti-Lock Brake System (ABS) is normal.*
1   Wheel bearings damaged or worn (Chapter 10).
2   Caliper not sliding properly due to improper installation or obstructions. Remove and inspect.
3   Disc not within specifications. Remove the disc and check for excessive lateral runout and parallelism. Have the discs resurfaced or replace them with new ones. Also make sure that all discs are the same thickness.

56   Brakes drag (indicated by sluggish engine performance or wheels being very hot after driving)

1   Output rod adjustment incorrect at the brake pedal.
2   Obstructed master cylinder compensator. Disassemble master cylinder and clean.
3   Master cylinder piston seized in bore. Overhaul master cylinder.
4   Caliper assembly in need of overhaul.
5   Brake pads or shoes worn out.
6   Piston cups in master cylinder or caliper assembly deformed. Overhaul master cylinder.
7   Parking brake assembly will not release.
8   Clogged brake lines.
9   Brake pedal height improperly adjusted.

57   Rear brakes lock up under light brake application

1   Tire pressures too high.
2   Tires excessively worn (Chapter 1).
3   Defective ABS system.

58   Rear brakes lock up under heavy brake application

1   Tire pressures too high.
2   Tires excessively worn (Chapter 1).
3   Front brake pads contaminated with oil, mud or water. Clean or replace the pads.
4   Front brake pads excessively worn.
5   Defective master cylinder or caliper assembly.
6   Defective ABS system.

Suspension and steering

Note: *All service procedures for the suspension and steering systems are included in Chapter 10, unless otherwise noted.*

59   Vehicle pulls to one side

1   Tire pressures uneven (Chapter 1).
2   Defective tire (Chapter 1).
3   Excessive wear in suspension or steering components (Chapter 10).
4   Front end alignment incorrect.
5   Front brakes dragging. Inspect as described in Section 56.
6   Wheel lug nuts loose.
7   Worn upper or lower link or strut rod bushings.

60   Shimmy, shake or vibration

1   Tire or wheel out of balance or out of round. Have them balanced on the vehicle.
2   Worn or damaged wheel bearings (Chapter 10).
3   Shock absorbers and/or suspension components worn or damaged. Check for worn bushings in the upper and lower links.
4   Wheel lug nuts loose.
5   Incorrect tire pressures.
6   Excessively worn or damaged tire.
7   Loosely mounted steering gear housing.
8   Steering gear improperly adjusted.
9   Loose, worn or damaged steering components.
10   Worn balljoint.

61   Excessive pitching and/or rolling around corners or during braking

1   Defective shock absorbers. Replace as a set.
2   Broken or weak springs and/or suspension components.
3   Worn or damaged stabilizer bar or bushings.
4   Worn or damaged upper or lower links or bushings.

62   Wandering or general instability

1   Improper tire pressures.
2   Worn or damaged bushings.
3   Incorrect front end alignment.
4   Worn or damaged steering linkage.
5   Improperly adjusted steering gear.
6   Out of balance wheels.
7   Loose wheel lug nuts.
8   Worn rear shock absorbers.
9   Fatigued or damaged rear springs.

63   Excessively stiff steering

1   Lack of lubricant in power steering fluid reservoir, where appropriate (Chapter 1).
2   Incorrect tire pressures (Chapter 1).
3   Front end out of alignment.
4   Steering gear out of adjustment or lacking lubrication.
5   Worn or damaged wheel bearings.
6   Worn or damaged steering gear.
7   Interference of steering column with turn signal switch.
8   Low tire pressures.
9   Worn or damaged balljoints.
10   Worn or damaged steering linkage.

64   Excessive play in steering

1   Worn wheel bearings (Chapter 10).
3   Steering gear improperly adjusted.
2   Incorrect front end alignment.
3   Steering gear mounting bolts loose.
4   Worn steering linkage.

65   Lack of power assistance

1   Drivebelt faulty or not adjusted properly (Chapter 1).
2   Fluid level low (Chapter 1).
3   Hoses or pipes restricting the flow. Inspect and replace parts as necessary.
4   Air in power steering system. Bleed system.
5   Defective power steering pump.

66   Steering wheel fails to return to straight-ahead position

1   Incorrect front end alignment.
2   Tire pressures low.
3   Steering gears improperly engaged.
4   Steering column out of alignment.
5   Worn or damaged balljoint.
6   Worn or damaged steering linkage.
7   Insufficient oil in steering gear.
8   Lack of fluid in power steering pump.

67   Steering effort not the same in both directions (power system)

1   Leaks in steering gear.
2   Clogged fluid passage in steering gear.

68   Noisy power steering pump

1   Insufficient oil in pump.

2   Clogged hoses or oil filter in pump.
3   Loose pulley.
4   Worn or improperly adjusted drivebelt (Chapter 1).
5   Defective pump.

69   Miscellaneous noises

1   Improper tire pressures.
2   Insufficiently lubricated balljoint or steering linkage.
3   Loose or worn steering gear, steering linkage or suspension components.
4   Defective shock absorber.
5   Defective wheel bearing.
6   Damaged spring.
7   Loose wheel lug nuts.
8   Worn or damaged rear axleshaft spline.
9   Worn or damaged rear shock absorber mounting bushing.
10   Worn rear axle bearing.
11   See also causes of noises at the rear axle and driveshaft.

70   Excessive tire wear (not specific to one area)

1   Incorrect tire pressures.
2   Tires out of balance. Have them balanced on the vehicle.
3   Wheels damaged. Inspect and replace as necessary.
4   Suspension or steering components worn (Chapter 1).

71   Excessive tire wear on outside edge

1   Incorrect tire pressure
2   Excessive speed in turns.
3   Wheel alignment incorrect.

72   Excessive tire wear on inside edge

1   Incorrect tire pressure.
2   Wheel alignment incorrect.
3   Loose or damaged steering components (Chapter 1).

73   Tire tread worn in one place

1   Tires out of balance. Have them balanced on the vehicle.
2   Damaged or buckled wheel. Inspect and replace if necessary.
3   Defective tire.

Notes

Chapter 1
Tune-up and routine maintenance

Contents

Specifications

Recommended lubricants and fluids

Note: *Listed here are manufacturer recommendations at the time this manual was written. Manufacturers occasionally upgrade their fluid and lubricant specifications, so check with your local auto parts store for current recommendations.*

Engine oil
  Type .... API grade SG, SG/CC or SG/CD multigrade and fuel-efficient oil
  Viscosity .... See accompanying chart

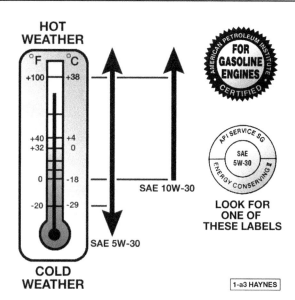

Engine oil viscosity chart - For best fuel economy and cold starting, select the lowest SAE viscosity grade for the expected temperature range

Recommended lubricants and fluids (continued)

Automatic transmission fluid..	Dexron IIE or III Automatic Transmission Fluid (ATF)
Engine coolant...	50/50 mixture of water and the specified ethylene glycol-based (green color) antifreeze or "DEX-COOL, silicate-free (orange-color) coolant - DO NOT mix the two types (refer to Sections 4, 12 and 31)
Brake fluid	DOT 3 fluid
Power steering fluid...	GM power steering fluid or equivalent
Chassis grease	SAE NLGI no. 2 chassis grease
Differential lubricant	SAE 80W-90 GL-5 gear lubricant*

Limited slip axles add 4 oz. of friction modifier when oil is changed.

Capacities*

Engine oil (with filter change)...	4.5 to 5.0 qts
Fuel tank	
Sedan	23 gallons
Station wagon	21 gallons
Cooling system	
1991 thru 1993	
5.0L engine	
Standard duty..	16.7 qts
Heavy duty	17.3 qts
5.7L engine	
Standard duty..	14.3 qts
Heavy duty	14.6 qts
1994 thru 1995	
Standard duty..	14.3 qts
Heavy duty..	14.6 qts
1996	
4.3L engine	
Standard duty..	17.9 qts
Heavy duty	18.1 qts
5.7L engine	
Standard duty..	16.4 qts
Heavy duty	16.9 qts
Automatic transmission (fluid and filter replacement)	10 pts

* All capacities approximate. Add as necessary to bring to appropriate level.

Ignition system

Spark plug type and gap	
1991 thru 1993	
5.7L..	AC type CR43TS or equivalent @ 0.035 in.
5.0L..	AC type CR45TS or equivalent @ 0.035 in.
1994 and 1995..	AC type 41-906 or equivalent @ 0.050 in.
1996	AC type 41-943 or equivalent @ 0.050 in.
Firing order	1-8-4-3-6-5-7-2

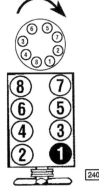

Cylinder location and distributor rotation -
1991 thru 1993

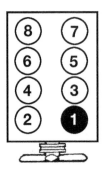
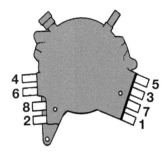

Cylinder and spark plug wire terminal locations -
1994 thru 1996

General

Radiator cap pressure rating...	15 psi
Brake pad wear limit...	1/8 inch

Torque specifications

	Ft-lbs (unless otherwise indicated)
Automatic transmission pan bolts..	96 to 120 in-lbs
Differential filler plug..	26
Engine oil drain plug..	15 to 20
Spark plugs	11
Throttle body nuts/bolts	
1993 and earlier	16
1994 and later	18
Wheel lug nuts...	100

Typical engine compartment layout

1	Engine oil dipstick		8	Lower radiator hose
2	Engine oil filler cap		9	Drivebelt
3	Automatic transmission dipstick		10	Battery
4	Brake master cylinder reservoir		11	Pressurized engine coolant reservoir
5	Windshield washer fluid reservoir		12	Power steering fluid reservoir
6	Air filter housing		13	Fuse block
7	Upper radiator hose			

Typical front underside components

1	Drivebelt	4	Exhaust system	7	Automatic transmission fluid pan
2	Sway bar bushing	5	Engine oil filter	8	Spring and shock absorber assembly
3	Lower balljoint grease fitting	6	Engine oil drain plug	9	Front brake caliper

Typical rear underside components

1	Fuel tank	3	Rear brake assembly	5	Muffler	7	Lower control arm bushing
2	Tail pipe	4	Shock absorber	6	Driveshaft	8	Coil spring

1 Chevrolet Impala SS, Caprice and Buick Roadmaster Maintenance schedule

The following maintenance intervals are based on the assumption that the vehicle owner will be doing the maintenance or service work, as opposed to having a dealer service department do the work. Although the time/mileage intervals are loosely based on factory recommendations, most have been shortened to ensure, for example, that such items as lubricants and fluids are checked/changed at intervals that promote maximum engine/driveline service life. Also, subject to the preference of the individual owner interested in keeping his or her vehicle in peak condition at all times, and with the vehicle's ultimate resale in mind, many of the maintenance procedures may be performed more often than recommended in the following schedule. We encourage such owner initiative.

When the vehicle is new it should be serviced initially by a factory authorized dealer service department to protect the factory warranty. In many cases the initial maintenance check is done at no cost to the owner (check with your dealer service department for more information).

Every 250 miles or weekly, whichever comes first

Check the engine oil level (Section 4)
Check the engine coolant level (Section 4)
Check the windshield washer fluid level (Section 4)
Check the brake fluid level (Section 4)
Check the tires and tire pressures (Section 5)

Every 3000 miles or 3 months, whichever comes first

All items listed above plus:
Check the power steering fluid level (Section 6)
Check the automatic transmission fluid level (Section 7)
Change the engine oil and filter (Section 8)
Lubricate the chassis (Section 9)
Check the differential lubricant level (Section 10)

Every 7500 miles or 6 months, whichever comes first

All items listed above plus:
Check and service the battery (Section 11)
Check the cooling system (Section 12)
Inspect and replace, if necessary, all underhood hoses (Section 13)
Check the engine drivebelt (Section 14)
Inspect the suspension and steering components (Section 15)
Check the brakes (Section 16)*
Rotate the tires (Section 17)
Inspect the exhaust system (Section 18)

Every 15,000 miles or 12 months, whichever comes first

Check the throttle body nut torque (Section 19)
Inspect and replace, if necessary, the windshield wiper blades (Section 20)
Inspect the seat belts (Section 21)

Every 30,000 miles or 24 months, whichever comes first

All items listed above plus:
Replace the air filter (Section 22)
Inspect and replace, if necessary, the PCV valve (Section 23)
Check the EGR valve (Section 24)
Replace the spark plugs (conventional [non-platinum] spark plugs) (Section 25)
Inspect the spark plug wires, distributor cap and wires (Section 26)
Replace the fuel filter (Section 27)
Inspect the fuel system (Section 28)
Check and repack the front wheel bearings (Section 29)
Change the automatic transmission fluid and filter (Section 30)**
Service the cooling system (drain, flush and refill) (green-colored ethylene glycol anti-freeze only) (Section 31)
Change the differential lubricant (Section 32)

Every 100,000 miles or 5 years, whichever comes first

Replace the spark plugs (platinum-tipped spark plugs) (Section 25)
Service the cooling system (drain, flush and refill) (orange-colored "DEX-COOL" silicate-free coolant only) (Section 31)

* *If the vehicle frequently tows a trailer, is operated primarily in stop-and-go conditions or its brakes receive severe usage for any other reason, check the brakes every 3000 miles or three months.*

** *If operated under one or more of the following conditions, change the automatic transmission fluid every 15,000 miles:*

In heavy city traffic where the outside temperature regularly reaches 90-degrees F (32-degrees C) or higher
In hilly or mountainous terrain
Frequent trailer pulling

2   Introduction

This Chapter is designed to help the home mechanic maintain the Chevrolet Caprice, Impala SS and Buick Roadmaster models with the goals of maximum performance, economy, safety and reliability in mind.

Included is a master maintenance schedule (page 1-6), followed by procedures dealing specifically with each item on the schedule. Visual checks, adjustments, component replacement and other helpful items are included. Refer to the **accompanying illustrations** of the engine compartment and the underside of the vehicle for the locations of various components.

Servicing your vehicle in accordance with the mileage/time maintenance schedule and the step-by-step procedures will result in a planned maintenance program that should produce a long and reliable service life. Keep in mind that it's a comprehensive plan, so maintaining some items but not others at the specified intervals will not produce the same results.

As you service your vehicle, you'll discover that many of the procedures can - and should - be grouped together because of the nature of the particular procedure you're performing or because of the close proximity of two otherwise unrelated components to one another.

For example, if the vehicle is raised, you should inspect the exhaust, suspension, steering and fuel systems while you're under the vehicle. When you're rotating the tires, it makes good sense to check the brakes since the wheels are already removed. Finally, let's suppose you have to borrow or rent a torque wrench. Even if you only need it to tighten the spark plugs, you might as well check the torque of as many critical fasteners as time allows.

The first step in this maintenance program is to prepare yourself before the actual work begins. Read through all the procedures you're planning to do, then gather up all the parts and tools needed. If it looks like you might run into problems during a particular job, seek advice from a mechanic or an experienced do-it-yourselfer.
Caution: *The stereo in your vehicle may be equipped with an anti-theft system. Refer to the information at the front of this manual before performing any procedure which requires disconnecting the battery cable.*

3   Tune-up general information

The term tune-up is used in this manual to represent a combination of individual operations rather than one specific procedure.
If, from the time the vehicle is new, the routine maintenance schedule is followed closely and frequent checks are made of fluid levels and high wear items, as suggested through-

4.2 The engine oil dipstick (arrow) is located on the passenger's side of the engine

out this manual, the engine will be kept in relatively good running condition and the need for additional work will be minimized due to lack of regular maintenance. This is even more likely if a used vehicle, which has not received regular and frequent maintenance checks, is purchased. In such cases, an engine tune-up will be needed outside of the regular routine maintenance intervals.

The first step in any tune-up or diagnostic procedure to help correct a poor running engine is a cylinder compression check. A compression check (see Chapter 2, Part B) will help determine the condition of internal engine components and should be used as a guide for tune-up and repair procedures. If, for instance, a compression check indicates serious internal engine wear, a conventional tune-up won't improve the performance of the engine and would be a waste of time and money. Because of its importance, the compression check should be done by someone with the right equipment and the knowledge to use it properly.

The following procedures are those most often needed to bring a generally poor running engine back into a proper state of tune.

Minor tune-up

Check all engine related fluids (Section 4)
Clean, inspect and test the battery (Section 11)
Check the cooling system (Section 12)
Check all underhood hoses (Section 13)
Check and adjust the drivebelts (Section 14)
Check the air filter (Section 22)
Check the PCV valve (Section 23)
Replace the spark plugs (Section 25)
Inspect the spark plug wires (Section 26)

Major tune-up

All items listed under Minor tune-up plus . . .
Replace the air filter (Section 22)
Replace the spark plug wires (Section 26)
Replace the fuel filter (Section 27)

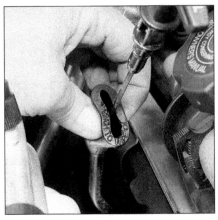

4.4a On some engines the dipstick tube incorporates a built-in wiper

Check the fuel system (Section 28)
Check the ignition timing (Chapter 5)
Check the charging system (Chapter 5)
Check the EGR system (Chapter 6)

4   Fluid level checks (every 250 miles or weekly)

Note: *The following are fluid level checks to be done on a 250 mile or weekly basis. Additional fluid level checks can be found in specific maintenance procedures which follow. Regardless of intervals, be alert to fluid leaks under the vehicle which would indicate a problem to be corrected immediately.*
1   Fluids are an essential part of the lubrication, cooling, brake and windshield washer systems. Because the fluids gradually become depleted and/or contaminated during normal operation of the vehicle, they must be periodically replenished. See *Recommended lubricants and fluids* at the beginning of this Chapter before adding fluid to any of the following components. **Note:** *The vehicle must be on level ground when fluid levels are checked.*

Engine oil

Refer to illustrations 4.2, 4.4a, 4.4b and 4.6
2   The engine oil level is checked with a dipstick **(see illustration)**. The dipstick extends through a metal tube down into the oil pan.
3   The oil level should be checked before the vehicle has been driven, or about 15 minutes after the engine has been shut off. If the oil is checked immediately after driving the vehicle, some of the oil will remain in the upper part of the engine, resulting in an inaccurate reading on the dipstick.
4   Pull the dipstick from the tube and wipe all the oil from the end with a clean rag or paper towel. On some V8 engines the dipstick tube incorporates a wiper **(see illustration)**. Insert the clean dipstick all the way back into the tube and pull it out again. Note the oil at the end of the dipstick. Add oil as

necessary to keep the level above the ADD mark in the cross hatched area of the dipstick **(see illustration)**.

5   Do not overfill the engine by adding too much oil since this may result in oil fouled spark plugs, oil leaks or oil seal failures.

6   Oil is added to the engine after removing a twist-off cap located on the engine **(see illustration)**. A funnel may help to reduce spills.

7   Checking the oil level is an important preventive maintenance step. A consistently low oil level indicates oil leakage through damaged seals, defective gaskets or past worn rings or valve guides. If the oil looks milky in color or has water droplets in it, the cylinder head gasket may be blown or the head or block may be cracked. The engine should be checked immediately. The condition of the oil should also be checked. Whenever you check the oil level, slide your thumb and index finger up the dipstick before wiping off the oil. If you see small dirt or metal particles clinging to the dipstick, the oil should be changed (see Section 8).

Engine coolant

Refer to illustration 4.9

Warning: *Do not allow antifreeze to come in contact with your skin or painted surfaces of the vehicle. Flush contaminated areas immediately with plenty of water. Do not store new coolant or leave old coolant lying around where it's accessible to children or pets - they're attracted by its sweet smell. Ingestion of even a small amount of coolant can be fatal! Wipe up garage floor and drip pan coolant spills immediately. Keep antifreeze containers covered and repair leaks in the cooling system immediately.*

Caution: *Never mix green-colored ethylene glycol anti-freeze and orange-colored "DEX-COOL" silicate-free coolant because doing so will destroy the efficiency of the "DEX-COOL" coolant which is designed to last for 100,000 miles or five years.*

8   Two types of cooling systems are used on the vehicles covered by this manual:

1991 through 1993 models are equipped with a pressurized coolant recovery system. A plastic coolant reservoir located at the front of the engine compartment is connected by a hose to the radiator assembly. As the engine warms up and the coolant expands, it escapes through a valve in the radiator cap and travels through the hose into the reservoir. As the engine cools, the coolant is automatically drawn back into the cooling system to maintain the correct level

On later models (1994 thru 1996), the plastic coolant reservoir is a pressurized part of the cooling system - it is NOT a recovery reservoir. Do not remove the cap at the top of this reservoir until the engine has cooled completely.

9   The coolant level in the reservoir should be checked regularly. **Warning:** *Do not remove the radiator cap (1991 through 1993 models) or the reservoir cap (1994 and later models) to check the coolant level when the engine is warm. The level in the reservoir varies with the temperature of the engine. When the engine is cold, the coolant level should be at or slightly above the FULL COLD mark on the reservoir* **(see illustration)**. *Once the engine has warmed up, the level should be at or near the FULL HOT*

mark. If it isn't, add coolant to the reservoir. On 1991 through 1993 models, simply flip up the cap and add coolant. On 1994 and later models, allow the engine to cool, then unscrew the cap from the reservoir and add a 50/50 mixture of ethylene glycol based green-colored antifreeze or orange-colored "DEX-COOL" silicate-free coolant and water (see **Caution** above).

10   Drive the vehicle and recheck the coolant level. If only a small amount of coolant is required to bring the system up to the proper level, water can be used. However, repeated additions of water will dilute the antifreeze and water solution. In order to maintain the proper ratio of antifreeze and water, always top up the coolant level with the correct mixture. An empty plastic milk jug or bleach bottle makes an excellent container for mixing coolant. Do not use rust inhibitors or additives.

11   If the coolant level drops consistently, there may be a leak in the system. Inspect the radiator, hoses, filler cap, drain plugs and water pump (see Section 12). If no leaks are noted, have the radiator cap or coolant reservoir cap pressure tested by a service station.

12   If you have to remove the radiator cap, wait until the engine has cooled completely, then wrap a thick cloth around the cap and turn it to the first stop. If coolant or steam escapes, let the engine cool down longer, then remove the cap.

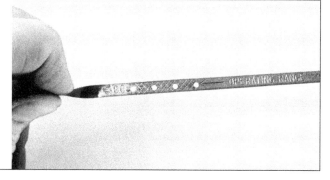

4.4b The oil level should be at or near the upper hole or cross-hatched area on the dipstick - if it's below the ADD line, add enough oil to bring the level into the upper hole or cross-hatched area

4.6 The engine oil filler cap (arrow) is clearly marked and threads into the tube on the valve cover - turn it counter clockwise to remove it

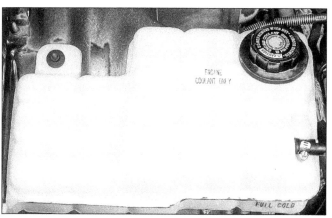

4.9 The coolant reservoir is located in the right (passenger's side) front corner of the engine compartment. Shown is a pressurized reservoir used on 1994 and later models. On this later-type reservoir, do not remove the reservoir cap until the engine cools completely. The coolant level on both early and later reservoirs can be checked by observing it through the translucent plastic

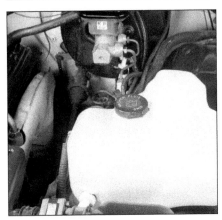

4.14 Flip the windshield washer fluid cap up to add fluid

4.16 If equipped with a maintenance type battery, remove the cell caps to check the battery electrolyte level

4.18 The fluid level inside the brake fluid reservoir can easily be checked by observing the level from the outside - fluid can be added to the reservoir after the cover is removed by prying up on the tabs

13   Check the condition of the coolant as well. It should be relatively clear. If it is brown or rust colored, the system should be drained, flushed and refilled. Even if the coolant appears to be normal, the corrosion inhibitors wear out, so it must be replaced at the specified intervals.

Windshield washer fluid

Refer to illustration 4.14

14   Fluid for the windshield washer system is located in a plastic reservoir on the left side of the engine compartment **(see illustration)**. In milder climates, plain water can be used in the reservoir, but it should be kept no more than two-thirds full to allow for expansion if the water freezes. In colder climates, use windshield washer system antifreeze, available at any auto parts store, to lower the freezing point of the fluid. Mix the antifreeze with water in accordance with the manufacturer's directions on the container. **Caution:** *Do not use cooling system antifreeze - it will damage the vehicle's paint.*

15   To help prevent icing in cold weather, warm the windshield with the defroster before using the washer.

Battery electrolyte

Refer to illustration 4.16

16   All vehicles covered by this manual are equipped with a battery which is permanently sealed (except for vent holes) and has no filler caps. Water does not have to be added to these batteries at any time; however, if a maintenance-type battery has been installed on the vehicle since it was new, remove all the cell caps on top of the battery **(see illustration)** (usually there are two caps that cover three cells each). If the electrolyte level is low, add distilled water until the level is above the plates . There is usually a split-ring indicator in each cell to help you judge when enough water has been added. Add water until the electrolyte level is just up to the bottom of the split ring indicator. Do not overfill the battery or it will spew out electrolyte when it is charging.

Brake fluid

Refer to illustration 4.18

17   The brake fluid level is checked by looking through the plastic reservoir mounted on the master cylinder. The master cylinder is mounted on the front of the power booster unit in the left (driver's side) rear corner of the engine compartment.

18   The fluid level should be between the MAX and MIN lines on the side of the reservoir **(see illustration)**. If the fluid level is low, wipe the top of the reservoir and the lid with a clean rag to prevent contamination of the system as the lid is pried off.

19   When adding fluid, pour it carefully into the reservoir to avoid spilling it on surrounding painted surfaces. Be sure the specified fluid is used, since mixing different types of brake fluid can cause damage to the system. See *Recommended lubricants and fluids* at the front of this Chapter or your owner's manual. **Warning:** *Brake fluid can harm your eyes and damage painted surfaces, so use extreme caution when handling or pouring it. Do not use brake fluid that has been standing open or is more than one year old. Brake fluid absorbs moisture from the air. Excess moisture can cause a dangerous loss of braking effectiveness.*

20   At this time the fluid and master cylinder can be inspected for contamination. The system should be drained and refilled if deposits, dirt particles or water droplets are seen in the fluid.

21   After filling the reservoir to the proper level, make sure the lid completely snaps in place to prevent fluid leakage.

22   The brake fluid level in the master cylinder will drop slightly as the pads at each wheel wear down during normal operation. If the master cylinder requires repeated replenishing to keep it at the proper level, this is an indication of leakage in the brake system, which should be corrected immediately. Check all brake lines and connections (see Section 16 for more information).

23   If, when checking the master cylinder fluid level, you discover one or both reservoirs empty or nearly empty, the brake system should be bled (see Chapter 9).

5   Tire and tire pressure checks (every 250 miles or weekly)

Refer to illustrations 5.2, 5.3, 5.4a, 5.4b and 5.8

1   Periodic inspection of the tires may spare you the inconvenience of being stranded with a flat tire. It can also provide you with vital information regarding possible problems in the steering and suspension systems before major damage occurs.

2   The original tires on this vehicle are equipped with 1/2-inch wide bands that appear when tread depth reaches 1/16-inch, indicating the tires are worn out. Tread wear can be monitored with a simple, inexpensive device known as a tread depth indicator **(see illustration)**.

3   Note any abnormal tread wear **(see illustration)**. Tread pattern irregularities such as cupping, flat spots and more wear on one side than the other are indications of front end alignment and/or balance problems. If

5.2 Use a tire tread depth gauge to monitor tire wear - they are available at auto parts stores and service stations and cost very little

UNDERINFLATION

OVERINFLATION

CUPPING

Cupping may be caused by:
- Underinflation and/or mechanical irregularities such as out-of-balance condition of wheel and/or tire, and bent or damaged wheel.
- Loose or worn steering tie-rod or steering idler arm.
- Loose, damaged or worn front suspension parts.

INCORRECT TOE-IN OR EXTREME CAMBER

FEATHERING DUE TO MISALIGNMENT

5.3 This chart will help you determine the condition of the tires, the probable cause(s) of abnormal wear and the corrective action necessary

any of these conditions are noted, take the vehicle to a tire shop or service station to correct the problem.

4   Look closely for cuts, punctures and embedded nails or tacks. Sometimes a tire will hold air pressure for short time or leak down very slowly after a nail has embedded itself in the tread. If a slow leak persists, check the valve stem core to make sure it's tight **(see illustration)**. Examine the tread for an object that may have embedded itself in the tire or for a "plug" that may have begun to leak (radial tire punctures are repaired with a plug that's installed in a puncture). If a punc-

ture is suspected, it can be easily verified by spraying a solution of soapy water onto the suspected area **(see illustration)**. The soapy solution will bubble if there's a leak. Unless the puncture is unusually large, a tire shop or service station can usually repair the tire.

5   Carefully inspect the inner sidewall of each tire for evidence of brake fluid. If you see any, inspect the brakes immediately.

6   Correct air pressure adds miles to the lifespan of the tires, improves mileage and enhances overall ride quality. Tire pressure cannot be accurately estimated by looking at a tire, especially if it's a radial. A tire pressure

gauge is essential. Keep an accurate gauge in the vehicle. The pressure gauges attached to the nozzles of air hoses at gas stations are often inaccurate.

7   Always check tire pressure when the tires are cold. Cold, in this case, means the vehicle has not been driven over a mile in the three hours preceding a tire pressure check. A pressure rise of four to eight pounds is not uncommon once the tires are warm.

8   Unscrew the valve cap protruding from the wheel or hubcap and push the gauge firmly onto the valve stem **(see illustration)**. Note the reading on the gauge and compare

5.4a If a tire looses air on a steady basis, check the valve core first to make sure it's snug (special inexpensive wrenches are commonly available at auto parts stores)

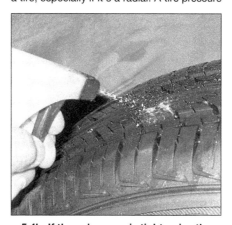

5.4b If the valve core is tight, raise the corner of the vehicle with the low tire and spray a soapy water solution onto the tread as the tire is turned slowly - leaks will cause small bubbles to appear

5.8 To extend the life of the tires, check the air pressure at least once a week with an accurate gauge (don't forget the spare)

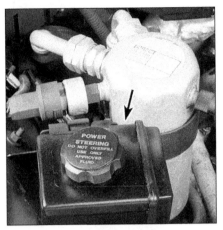

6.2 The power steering fluid reservoir (arrow) is located on the right (passenger's side) of the engine compartment - turn the cap counterclockwise for removal

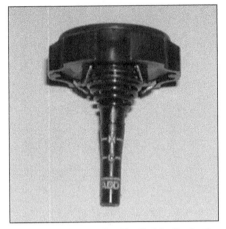

6.6 The marks on the dipstick indicate the safe fluid range

7.3 The automatic transmission fluid dipstick (arrow) is located at the rear of the engine compartment

7.6 With the automatic transmission at normal operating temperature, the fluid level must be maintained within the cross-hatched area on the dipstick, between the upper and lower holes

the figure to the recommended tire pressure shown on the label attached to the inside of the glove compartment door. Be sure to reinstall the valve cap to keep dirt and moisture out of the valve stem mechanism. Check all four tires and, if necessary, add enough air to bring them up to the recommended pressure.

9   Don't forget to keep the spare tire inflated to the specified pressure (refer to your owner's manual or the tire sidewall).

6   Power steering fluid level check (every 3000 miles or 3 months)

Refer to illustrations 6.2 and 6.6

1   Unlike manual steering, the power steering system relies on fluid which may, over a period of time, require replenishing.

2   The fluid reservoir for the power steering pump is located on the passenger side inner fenderwell of the engine compartment **(see illustration)**.

3   For the check, the front wheels should be pointed straight ahead and the engine should be off.

4   Use a clean rag to wipe off the reservoir cap and the area around the cap. This will help prevent any foreign matter from entering the reservoir during the check.

5   Twist off the cap and check the temperature of the fluid at the end of the dipstick with your finger.

6   Wipe off the fluid with a clean rag, reinsert it, then withdraw it and read the fluid level. The level should be at the HOT mark if the fluid was hot to the touch **(see illustration)**. It should be at the COLD mark if the fluid was cool to the touch.

7   If additional fluid is required, pour the specified type directly into the reservoir, using a funnel to prevent spills.

8   If the reservoir requires frequent fluid additions, all power steering hoses, hose connections, the power steering pump and the rack and pinion assembly should be carefully checked for leaks.

7   Automatic transmission fluid level check (every 3000 miles or 3 months)

Refer to illustrations 7.3 and 7.6

1   The automatic transmission fluid level should be carefully maintained. Low fluid level can lead to slipping or loss of drive, while overfilling can cause foaming and loss of fluid.

2   With the parking brake set, start the engine, then move the shift lever through all the gear ranges, ending in Park. The fluid level must be checked with the vehicle level and the engine running at idle. **Note:** *Incorrect fluid level readings will result if the vehicle has just been driven at high speeds for an extended period, in hot weather in city traffic, or if it has been pulling a trailer. If any of these conditions apply, wait until the fluid has cooled (about 30 minutes).*

3   With the transmission at normal operating temperature, remove the dipstick from the filler tube. The dipstick is located at the rear of the engine compartment **(see illustration)**.

4   Carefully touch the fluid at the end of the

dipstick to determine if the fluid is cool, warm or hot. Wipe the fluid from the dipstick with a clean rag and push it back into the filler tube until the cap seats.

5   Pull the dipstick out again and note the fluid level.

6   If the fluid felt cool, the level should be within the lower marks on the dipstick **(see illustration)**. If it felt warm or hot, the level should be within the cross-hatched upper areas on the dipstick. If additional fluid is required, pour it directly into the tube using a funnel. It takes about one pint to raise the level from the lower mark to the upper edge of the cross-hatched area with a hot transmission, so add the fluid a little at a time and keep checking the level until it's correct.

7   The condition of the fluid should also be checked along with the level. If the fluid at the end of the dipstick is a dark reddish-brown color, or if the fluid has a burned smell, the fluid should be changed. If you're in doubt about the condition of the fluid, purchase some new fluid and compare the two for color and smell.

8   Engine oil and filter change (every 3000 miles or 3 months)

Refer to illustrations 8.2, 8.7, 8.12 and 8.14

1   Frequent oil changes are the best preventive maintenance the home mechanic can give the engine, because aging oil becomes

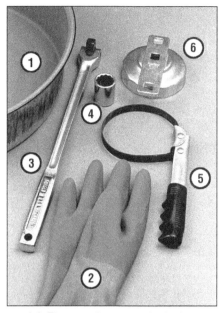

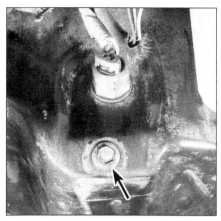

8.7 The engine oil drain plug (arrow) is located at the rear of the oil pan - it is usually very tight, so use a box-end wrench to avoid rounding off the hex

8.12 The oil filter is usually on very tight as well and will require a special wrench for removal - DO NOT use the wrench to tighten the new filter!

8.2 These tools are required when changing the engine oil and filter

1   **Drain pan** - *It should be fairly shallow in depth, but wide to prevent spills*
2   **Rubber gloves** - *When removing the drain plug and filter, you will get oil on your hands (the gloves will prevent burns)*
3   **Breaker bar** - *Sometimes the oil drain plug is tight, and a long breaker bar is needed to loosen it*
4   **Socket** - *To be used with the breaker bar or a ratchet (must be the correct size to fit the drain plug)*
5   **Filter wrench** - *This is a metal band-type wrench, which requires clearance around the filter to be effective*
6   **Filter wrench** - *This type fits on the bottom of the filter and can be turned with a ratchet or breaker bar (different-size wrenches are available for different types of filters)*

the engine and allow it to reach its normal operating temperature. Warm oil and sludge will flow out more easily. Turn off the engine when it's warmed up. Remove the filler cap from the valve cover.
6   Raise the vehicle and support it securely on jackstands. **Warning:** *Never get beneath the vehicle when it is supported only by a jack. The jack provided with your vehicle is designed solely for raising the vehicle to remove and replace the wheels. Always use jackstands to support the vehicle when it becomes necessary to place your body underneath the vehicle.*
7   Being careful not to touch the hot exhaust components, place the drain pan under the drain plug in the bottom of the pan and remove the plug **(see illustration)**. You may want to wear gloves while unscrewing the plug the final few turns if the engine is hot.
8   Allow the old oil to drain into the pan. It may be necessary to move the pan farther under the engine as the oil flow slows to a trickle. Inspect the old oil for the presence of metal shavings and chips.
9   After all the oil has drained, wipe off the drain plug with a clean rag. Even minute metal particles clinging to the plug would immediately contaminate the new oil.
10   Clean the area around the drain plug opening, reinstall the plug and tighten it securely, but do not strip the threads.
11   Move the drain pan into position under the oil filter.
12   Loosen the oil filter **(see illustration)** by turning it counterclockwise with the filter wrench. Use a quality filter wrench of the correct size and be careful not to collapse the canister as you apply pressure. Once the filter is loose, use your hands to unscrew it from the block. Just as the filter is detached from the block, immediately tilt the open end up to prevent the oil inside the filter from spilling out. **Warning:** *The exhaust system may still be hot, so be careful.*
13   With a clean rag, wipe off the mounting surface on the block. If a residue of old oil is

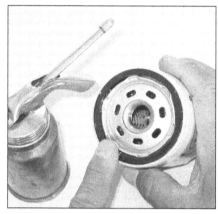

8.14 Lubricate the oil filter gasket with clean engine oil before installing the filter on the engine

allowed to remain, it will smoke when the block is heated up. Also make sure that none of the old gasket remains stuck to the mounting surface. It can be removed with a scraper if necessary.
14   Compare the old filter with the new one to make sure they are the same type. Smear some clean engine oil on the rubber gasket of the new filter and screw it into place **(see illustration)**. Because overtightening the filter will damage the gasket, do not use a filter wrench to tighten the filter. Tighten it by hand until the gasket contacts the seating surface. Then seat the filter by giving it an additional 3/4-turn.
15   Remove all tools, rags, etc. from under the vehicle, being careful not to spill the oil in the drain pan, then lower the vehicle.
16   Add new oil to the engine through the oil filler cap in the valve cover. Use a funnel, if necessary, to prevent oil from spilling onto the top of the engine. Pour three quarts of fresh oil into the engine. Wait a few minutes to allow the oil to drain into the pan, then check the level on the oil dipstick (see Section 4 if necessary). If the oil level is at or near the upper hole on the dipstick, install the filler

diluted and contaminated, which leads to premature engine wear.
2   Make sure you have all the necessary tools before you begin this procedure **(see illustration)**. You should also have plenty of rags or newspapers handy for mopping up any spills.
3   Access to the underside of the vehicle is greatly improved if the vehicle can be lifted on a hoist, driven onto ramps or supported by jackstands. **Warning:** *Do not work under a vehicle which is supported only by a hydraulic or scissors-type jack.*
4   If this is your first oil change, get under the vehicle and familiarize yourself with the locations of the oil drain plug and the oil filter. The engine and exhaust components will be warm during the actual work, so try to anticipate any potential problems before the engine and accessories are hot.
5   Park the vehicle on a level spot. Start

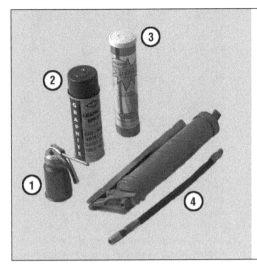

9.1 Materials required for chassis and body lubrication

1   **Engine oil** - Light engine oil in a can like this can be used for door and hood hinges
2   **Graphite spray** - Used to lubricate lock cylinders
3   **Grease** - Grease, in a variety of types and weights, is available for use in a grease gun. Check the Specification for your requirements
4   **Grease gun** - A common grease gun, shown here with a detachable hose and nozzle, is needed for chassis lubrication. After use, clean it thoroughly!

container (capped plastic jugs, topped bottles, milk cartons, etc.) for transport to one of these disposal sites.

9   Chassis lubrication (every 3000 miles or 3 months)

Refer to illustrations 9.1 and 9.2

1   Refer to *Recommended lubricants and fluids* at the front of this Chapter to obtain the necessary grease, etc. You'll also need a grease gun **(see illustration)**. Occasionally plugs will be installed rather than grease fittings. If so, grease fittings will have to be purchased and installed.

2   Look under the vehicle and see if grease fittings or plugs are installed **(see illustration)**. If there are plugs, remove them and buy grease fittings, which will thread into the component. A dealer or auto parts store will be able to supply the correct fittings. Straight, as well as angled, fittings are available.

3   For easier access under the vehicle, raise it with a jack and place jackstands under the frame. Make sure it's safely supported by the stands. If the wheels are to be removed at this interval for tire rotation or brake inspection, loosen the lug nuts slightly while the vehicle is still on the ground.

cap hand tight, start the engine and allow the new oil to circulate.

17   Allow the engine to run for about a minute. While the engine is running, look under the vehicle and check for leaks at the oil pan drain plug and around the oil filter. If either is leaking, stop the engine and tighten the plug or filter.

18   Wait a few minutes to allow the oil to trickle down into the pan, then recheck the level on the dipstick and, if necessary, add enough oil to bring the level to the upper hole.

19   During the first few trips after an oil change, make it a point to check frequently for leaks and proper oil level.

20   The old oil drained from the engine cannot be re-used in its present state and should be discarded. Check with your local refuse disposal company, disposal facility or environmental agency to see whether they will accept the oil for recycling. Don't pour used oil into drains or onto the ground. After the oil has cooled, it can be drained into a suitable

9.2 Front suspension grease fitting locations - Upper balljoint grease fittings (A) not visible in this photograph

10.2a Use a 3/8 drive ratchet and extension to remove the differential check/fill plug

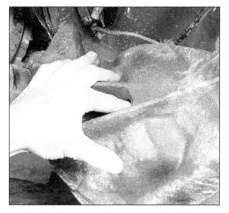

10.2b Use your finger as a dipstick to check the differential lubricant level

4   Before beginning, force a little grease out of the nozzle to remove any dirt from the end of the gun. Wipe the nozzle clean with a rag.

5   With the grease gun and plenty of clean rags, crawl under the vehicle and begin lubricating the components.

6   Wipe one of the grease fittings clean and push the nozzle firmly over it. Pump the gun until the balljoint rubber seal is firm to the touch. Do not pump too much grease into the fitting as it could rupture the seal.

7   Wipe the excess grease from the components and the grease fitting. Repeat the procedure for the remaining fitting.

8   Clean and lubricate the parking brake cable, along with the cable guides and levers. This can be done by smearing some of the chassis grease onto the cable and its related parts with your fingers.

9   Open the hood and smear a little chassis grease on the hood latch mechanism. Have an assistant pull the hood release lever

from inside the vehicle as you lubricate the cable at the latch.

10   Lubricate all the hinges (door, hood, etc.) with engine oil to keep them in proper working order.

11   The key lock cylinders can be lubricated with spray graphite or silicone lubricant, which is available at auto parts stores.

12   Lubricate the door weatherstripping with silicone spray. This will reduce chafing and retard wear.

10   Differential lubricant level check (every 3000 miles or 3 months)

Refer to illustrations 10.2a and 10.2b

1   The differential has a filler plug which must be removed to check the lubricant level. If the vehicle is raised to gain access to the plug, be sure to support it safely on jackstands - DO NOT crawl under the vehicle when it's supported only by the jack.

2   Remove the filler plug from the front of the differential **(see illustrations)**.

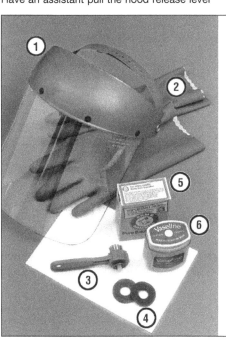

11.1 Tools and materials required for battery maintenance

1   *Face shield/safety goggles - When removing corrosion with a brush, the acidic particles can easily fly up into your eyes*

2   *Rubber gloves - Another safety item to consider when servicing the battery - remember that's acid inside the battery!*

3   *Battery terminal/cable cleaner - This wire brush cleaning tool will remove all traces of corrosion from the battery and cable*

4   *Treated felt washers - Placing one of these on each terminal, directly under the cable end, will help prevent corrosion (be sure to get the correct type for side-terminal batteries)*

5   *Baking soda - A solution of baking soda and water can be used to neutralize corrosion*

6   *Petroleum jelly - A layer of this on the battery terminal bolts will help prevent corrosion*

3   The lubricant level should be at the bottom of the plug opening. If not, use a syringe to add the recommended lubricant until it just starts to run out of the opening. On some models a tag is located in the area of the plug which gives information regarding lubricant type, particularly on models equipped with a limited slip differential.

4   Install the plug and tighten it securely.

11   Battery check, maintenance and charging (every 7500 miles or 6 months)

Refer to illustrations 11.1, 11.4, 11.5a, 11.5b and 11.5c

Warning: *Hydrogen gas is produced by the battery, so keep open flames and lighted tobacco away from it at all times. Always wear eye protection when working around the battery. Rinse off spilled electrolyte immediately with large amounts of water. When removing the battery cables, always detach the negative cable first and hook it up last!*

Caution: *If the radio in your vehicle is equipped with an anti-theft system, make sure you have the correct activation code before disconnecting the battery.*

1   Battery maintenance is an important procedure which will help ensure you aren't stranded because of a dead battery. Several tools are required for this procedure **(see illustration)**.

2   A sealed battery is standard equipment on all vehicles covered by this manual. Although this type of battery has many advantages over the older, capped cell type, and never requires the addition of water, it should still be routinely maintained according to the procedures which follow.

Check

3   The battery is located in the right front corner of the engine compartment. The exterior of the battery should be inspected periodically for damage such as a cracked case or cover.

4   Check the tightness of the battery cable terminals and connections to ensure good

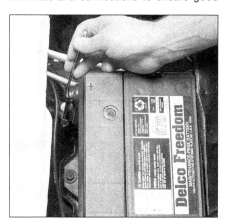

11.4 Check the tightness of the battery cable terminal bolts

11.5a A tool like this one (available at auto parts stores) is used to clean the side terminal type battery contact area

11.5b Use the brush to finish the cleaning job

11.5c The result should be a clean, shiny terminal area

electrical connections and check the entire length of each cable for cracks and frayed conductors **(see illustration)**.

5   If corrosion (visible as white, fluffy deposits) is evident, remove the cables from the terminals, clean them with a battery brush and reinstall the cables **(see illustrations)**. Corrosion can be kept to a minimum by using special treated fiber washers available at auto parts stores or by applying a layer of petroleum jelly to the terminals and cables after they are assembled.

6   Make sure that the battery tray is in good condition and the hold-down clamp bolt is tight. If the battery is removed from the tray, make sure no parts remain in the bottom of the tray when the battery is reinstalled. When reinstalling the hold-down clamp bolt, do not overtighten it.

7   Information on removing and installing the battery can be found in Chapter 5. Information on jump starting can be found at the front of this manual. For more detailed battery checking procedures, refer to the *Haynes Automotive Electrical Manual*.

Cleaning

8   Corrosion on the hold-down components, battery case and surrounding areas can be removed with a solution of water and baking soda. Thoroughly rinse all cleaned areas with plain water.

9   Any metal parts of the vehicle damaged by corrosion should be covered with a zinc-based primer, then painted.

Charging

Warning: *When batteries are being charged, hydrogen gas, which is very explosive and flammable, is produced. Do not smoke or allow open flames near a charging or a recently charged battery. Wear eye protection when near the battery during charging. Also, make sure the charger is unplugged before connecting or disconnecting the battery from the charger.*

10   Slow-rate charging is the best way to restore a battery that's discharged to the

point where it will not start the engine. It's also a good way to maintain the battery charge in a vehicle that's only driven a few miles between starts. Maintaining the battery charge is particularly important in the winter when the battery must work harder to start the engine and electrical accessories that drain the battery are in greater use.

11   It's best to use a one or two-amp battery charger (sometimes called a "trickle" charger). They are the safest and put the least strain on the battery. They are also the least expensive. For a faster charge, you can use a higher amperage charger, but don't use one rated more than 1/10th the amp/hour rating of the battery. Rapid boost charges that claim to restore the power of the battery in one to two hours are hardest on the battery and can damage batteries that aren't in good condition. This type of charging should only be used in emergency situations.

12   The average time necessary to charge a battery should be listed in the instructions that come with the charger. As a general rule, a trickle charger will charge a battery in 12 to 16 hours.

13   Remove all of the cell caps (if equipped) and cover the holes with a clean cloth to prevent spattering electrolyte. Disconnect the negative battery cable and hook the battery charger leads to the battery posts (positive to positive, negative to negative), then plug in the charger. Make sure it is set at 12-volts if it has a selector switch.

14   If you're using a charger with a rate higher than two amps, check the battery regularly during charging to make sure it doesn't overheat. If you're using a trickle charger, you can safely let the battery charge overnight after you've checked it regularly for the first couple of hours.

15   If the battery has removable cell caps, measure the specific gravity with a hydrometer every hour during the last few hours of the charging cycle. Hydrometers are available inexpensively from auto parts stores - follow the instructions that come with the hydrometer. Consider the battery charged when there's no change in the specific gravity read-

ing for two hours and the electrolyte in the cells is gassing (bubbling) freely. The specific gravity reading from each cell should be very close to the others. If not, the battery probably has a bad cell(s).

16   Some batteries with sealed tops have built-in hydrometers on the top that indicate the state of charge by the color displayed in the hydrometer window. Normally, a bright-colored hydrometer indicates a full charge and a dark hydrometer indicates the battery still needs charging. Check the battery manufacturer's instructions to be sure you know what the colors mean.

17   If the battery has a sealed top and no built-in hydrometer, you can hook up a digital voltmeter across the battery terminals to check the charge. A fully charged battery should read 12.5-volts or higher.

12   Cooling system check (every 7500 miles or 6 months)

Refer to illustration 12.4

Caution: *Never mix green-colored ethylene glycol anti-freeze and orange-colored "DEX-COOL" silicate-free coolant because doing so will destroy the efficiency of the "DEX-COOL" coolant which is designed to last for 100,000 miles or five years.*

1   Many major engine failures can be attributed to a faulty cooling system. If the vehicle is equipped with an automatic transmission, the cooling system also cools the transmission fluid and plays an important role in prolonging transmission life.

2   The cooling system should be checked with the engine cold. Do this before the vehicle is driven for the day or after the engine has been shut off for at least three hours.

3   Remove the radiator cap (if equipped) by turning it to the left until it reaches a stop. If you hear any hissing sounds (indicating there is still pressure in the system), wait until it stops. Now press down on the cap with the palm of your hand and continue turning to the left until the cap can be removed. Thoroughly clean the cap, inside and out, with clean

Check for a chafed area that could fail prematurely.

Check for a soft area indicating the hose has deteriorated inside.

Overtightening the clamp on a hardened hose will damage the hose and cause a leak.

Check each hose for swelling and oil-soaked ends. Cracks and breaks can be located by squeezing the hose.

12.4 Hoses, like drivebelts, have a habit of failing at the worst possible time - to prevent the inconvenience of a blown radiator or heater hose, inspect them carefully as shown here

water. Also clean the filler neck on the radiator. All traces of corrosion should be removed. The coolant inside the radiator should be relatively transparent. If it is rust colored, the system should be drained and refilled (see Section 31). If the coolant level is not up to the top, add additional anti-freeze/coolant mixture (see Section 4).

4   Carefully check the large upper and lower radiator hoses along with any smaller diameter heater hoses which run from the engine to the firewall. Inspect each hose along its entire length, replacing any hose which is cracked, swollen or shows signs of deterioration. Cracks may become more apparent if the hose is squeezed **(see illustration)**.

5   Make sure all hose connections are tight. A leak in the cooling system will usually show up as white or rust colored deposits on the areas adjoining the leak. If wire-type clamps are used at the ends of the hoses, it

may be wise to replace them with more secure screw-type clamps.

6   Use compressed air or a soft brush to remove bugs, leaves, etc. from the front of the radiator or air conditioning condenser. Be careful not to damage the delicate cooling fins or cut yourself on them.

7   Every other inspection, or at the first indication of cooling system problems, have the cap and system pressure tested. If you don't have a pressure tester, most gas stations and repair shops will do this for a minimal charge.

13 Underhood hose check and replacement (every 7500 miles or 6 months)

General

1   **Warning:** *Replacement of air conditioning hoses must be left to a dealer service department or air conditioning shop that has the equipment to depressurize the system safely. Never remove air conditioning components or hoses until the system has been depressurized.*

2   High temperatures under the hood can cause the deterioration of the rubber and plastic hoses used for engine, accessory and emission systems operation. Periodic inspection should be made for cracks, loose clamps, material hardening and leaks. Information specific to the cooling system hoses can be found in Section 12.

3   Some, but not all, hoses are secured to the fittings with clamps. Where clamps are used, check to be sure they haven't lost their tension, allowing the hose to leak. If clamps aren't used, make sure the hose hasn't expanded and/or hardened where it slips over the fitting, allowing it to leak.

Vacuum hoses

4   It's quite common for vacuum hoses, especially those in the emissions system, to be color coded or identified by colored stripes molded into each hose. Various systems require hoses with different wall thicknesses, collapse resistance and temperature resistance. When replacing hoses, be sure the new ones are made of the same material.

5   Often the only effective way to check a hose is to remove it completely from the vehicle. If more than one hose is removed, be sure to label the hoses and fittings to ensure correct installation.

6   When checking vacuum hoses, be sure to include any plastic T-fittings in the check. Inspect the fittings for cracks and the hose where it fits over the fitting for distortion, which could cause leakage.

7   A small piece of vacuum hose (1/4-inch inside diameter) can be used as a stethoscope to detect vacuum leaks. Hold one end of the hose to your ear and probe around vacuum hoses and fittings, listening for the "hissing" sound characteristic of a vacuum leak. **Warning:** *When probing with the vac-*

uum hose stethoscope, be careful not to allow your body or the hose to come into contact with moving engine components such as the drivebelt, cooling fan, etc.

Fuel hose

Warning: *Gasoline is extremely flammable, so take extra precautions when you work on any part of the fuel system. Don't smoke or allow open flames or bare light bulbs near the work area, and don't work in a garage where a natural gas-type appliance (such as a water heater or clothes dryer) with a pilot light is present. If you spill any fuel on your skin, rinse it off immediately with soap and water. When you perform any kind of work on the fuel system, wear safety glasses and have a Class B type fire extinguisher on hand. The fuel system is under pressure, so if any lines must be disconnected, the pressure in the system must be relieved first (see Chapter 4 for more information).*

8   Check all rubber fuel lines for deterioration and chafing. Check especially for cracks in areas where the hose bends and just before fittings, such as where a hose attaches to the fuel filter and fuel injection unit.

9   High quality fuel line, usually identified by the word *Fluoroelastomer* printed on the hose, should be used for fuel line replacement. Never, under any circumstances, use unreinforced vacuum line, clear plastic tubing or water hose for fuel lines.

10   Spring-type clamps are commonly used on fuel lines. These clamps often lose their tension over a period of time, and can be "sprung" during the removal process. As a result spring-type clamps should be replaced with screw-type clamps whenever a hose is replaced.

Metal lines

11   Sections of steel tubing often used for fuel line between the fuel pump and fuel injection unit. Check carefully for cracks, kinks and flat spots in the line.

12   If a section of metal fuel line must be replaced, only seamless steel tubing should be used, since copper and aluminum tubing do not have the strength necessary to withstand normal engine vibration.

13   Check the metal brake lines where they enter the master cylinder and brake proportioning unit (if used) for cracks in the lines and loose fittings. Any sign of brake fluid leakage calls for an immediate thorough inspection of the brake system.

14 Drivebelt check and replacement (every 7500 miles or 6 months)

Refer to illustrations 14.2, 14.5 and 14.7

1   A single serpentine drivebelt is located at the front of the engine and plays an important role in the overall operation of the engine and its components. Due to its function and material make up, the belt is prone to wear and should be periodically inspected. The serpentine belt drives the alternator, power

ACCEPTABLE

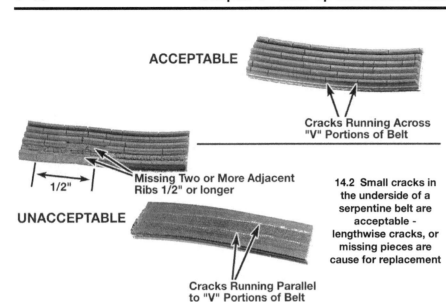

Cracks Running Across
"V" Portions of Belt

1/2"

Missing Two or More Adjacent
Ribs 1/2" or longer

UNACCEPTABLE

Cracks Running Parallel
to "V" Portions of Belt

14.2 Small cracks in
the underside of a
serpentine belt are
acceptable -
lengthwise cracks, or
missing pieces are
cause for replacement

14.5 Rotate the tensioner clockwise to
remove or install the belt

ACCESSORY DRIVE BELT ROUTING

PRINTED IN U.S.A.   PT.NO. 1010536

14.7 The serpentine drivebelt routing
diagram is located on the radiator shroud
(later model diagram shown)

steering pump, water pump and air conditioning compressor.

2   With the engine off, open the hood and use your fingers (and a flashlight, if necessary), to move along the belt checking for cracks and separation of the belt plies. Also check for fraying and glazing, which gives the belt a shiny appearance **(see illustration)**. Both sides of the belt should be inspected, which means you will have to twist the belt to check the underside.

3   Check the ribs on the underside of the belt. They should all be the same depth, with none of the surface uneven.

4   The tension of the belt is maintained by the tensioner assembly and isn't adjustable. The belt should be checked at the mileage specified in the maintenance schedule at the front of this chapter, if the belt shows noticeable damage or wear during these checks it should be replaced.

5   To replace the belt, rotate the tensioner clockwise to release belt tension **(see illustration)**.

6   Remove the belt from the auxiliary components and slowly release the tensioner.

7   Route the new belt over the various pulleys, again rotating the tensioner to allow the belt to be installed, then release the belt tensioner. **Note:** *These models have a drivebelt routing decal on the radiator shroud to help during drivebelt installation* **(see illustration)**.

15   Suspension and steering check (every 7500 miles or 6 months)

Refer to illustrations 15.3 and 15.4

1   Indications of a fault in these systems are excessive play in the steering wheel before the front wheels react, excessive sway around corners, body movement over rough roads or binding at some point as the steering wheel is turned.

2   Raise the front of the vehicle periodically and visually check the suspension and steering components for wear. Because of the

work to be done, make sure the vehicle cannot fall from the stands.

3   Check the wheel bearings. Do this by spinning the front wheels. Listen for any abnormal noises and watch to make sure the wheel spins true (doesn't wobble). Grab the top and bottom of the tire and pull in-and-out on it. Notice any movement which would indicate a loose wheel bearing assembly **(see illustration)**. If the bearings are suspect, refer to Section 29 for more information.

15.3 Grab the top and bottom of the tire and pull in-and-out on it -
Check for any noticeable movement which would indicate a loose
wheel bearing assembly

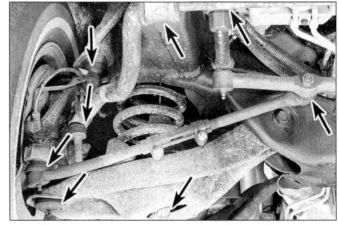

15.4 Check for deteriorated rubber bushings and damaged
grease seals in each of the arrowed positions

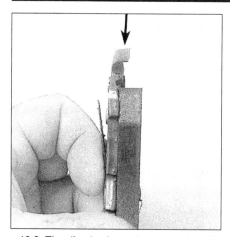

16.3 The disc brake pads are equipped with wear indicators that contact the disc and make a squealing sound when the pad has worn to its limit

16.5 With the wheels removed, the brake pad lining can be inspected through the caliper window (arrow) and at each end of the caliper

16.7 Check for any sign of brake fluid leakage at the line fittings and the brake hoses

4   From under the vehicle check for loose bolts, broken or disconnected parts and deteriorated rubber bushings on all suspension and steering components **(see illustration)**. Check the power steering hoses and connections for leaks. Check the shock absorbers or leaking fluid or damage.

5   Have an assistant turn the steering wheel from side-to-side and check the steering components for free movement, chafing and binding. If the steering doesn't react with the movement of the steering wheel, try to determine where the slack is located.

16   Brake check (every 7500 miles or 6 months)

Note: *For detailed information of the brake system,* refer to Chapter 9.

Warning: *Brake system dust may contain asbestos, which is hazardous to your health. DO NOT blow it out with compressed air, inhale it or use gasoline or solvents to remove it. Use brake system cleaner only.*

1   In addition to the specified intervals, the brakes should be inspected every time the wheels are removed or whenever a defect is suspected. Raise the vehicle and place it securely on jackstands. Remove the wheels (see *Jacking and towing* at the front of the manual, if necessary).

Disc brakes

Refer to illustrations 16.3, 16.5 and 16.7

2   Disc brakes can be checked without removing any parts except the wheels. Extensive disc damage can occur if the pads are not replaced when needed.

3   The disc brake pads have built-in wear indicators **(see illustration)** which make a high-pitched squealing sound when the pads are worn. **Caution:** *Expensive damage to the disc can result if the pads are not replaced soon after the wear indicators start squealing.*

4   The disc brake calipers, which contain the brake pads, have an inner pad and outer pad in each caliper. All pads should be inspected.

5   Each caliper has a "window" to inspect the pads **(see illustration)**. If the pad material has worn to about 1/8-inch thick or less, the pads should be replaced.

6   If you're unsure about the exact thickness of the remaining lining material, remove the pads for further inspection or replacement (see Chapter 9).

7   Before installing the wheels, check for leakage and/or damage at the brake hoses and connections **(see illustration)**. Replace the hose or fittings as necessary, (see Chapter 9).

8   Check the condition of the brake disc. Look for score marks, deep scratches and overheated areas (they will appear blue or discolored). If damage or wear is noted, the disc can be removed and resurfaced by an automotive machine shop or replaced with a new one. See Chapter 9 for more detailed inspection and repair procedures.

Drum brakes

Refer to illustrations 16.14 and 16.16

9   Raise the vehicle and support it securely on jackstands. Block the front tires to prevent the vehicle from rolling; however, don't apply the parking brake or it will lock the drums in place.

10   Remove the wheels, referring to *Jacking and towing* at the front of this manual if necessary.

11   Mark the hub so it can be reinstalled in the same position. Use a scribe, chalk, etc. on the drum, hub and backing plate.

12   Remove the brake drum.

13   With the drum removed, carefully clean the brake assembly with brake system cleaner. **Warning:** *Don't blow the dust out with compressed air and don't inhale any of it (it may contain asbestos, which is harmful to your health).*

14   Note the thickness of the lining material on both front and rear brake shoes. If the material has worn away to within 1/8-inch of the recessed rivets or metal backing, the shoes should be replaced **(see illustration)**. The shoes should also be replaced if they're cracked, glazed (shiny areas), or covered with brake fluid.

15   Make sure all the brake assembly springs are connected and in good condition.

16   Check the brake components for signs of fluid leakage. With your finger or a small screwdriver, carefully pry back the rubber cups on the wheel cylinder located at the top

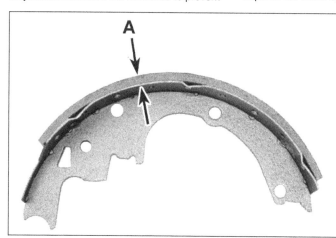

16.14 If the lining is bonded to the brake shoe, measure the lining thickness from the outer surface to the metal shoe, as shown here; if the lining is riveted to the shoe, measure from the lining outer surface to the rivet head

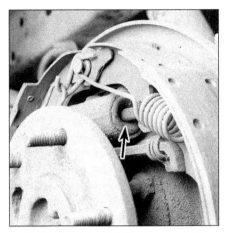

16.16 Check for fluid leakage at both ends of the wheel cylinder dust covers (arrow)

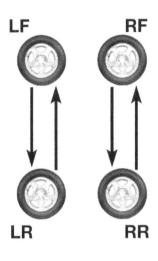

1-AJ HAYNES

17.2 Tire rotation diagram

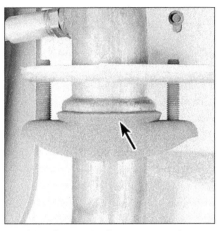

18.2a Check the flange connections (arrows) for exhaust leaks - also check that the retaining nuts are securely tightened

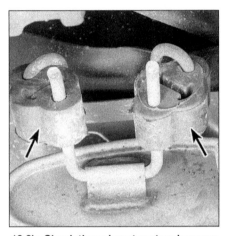

18.2b Check the exhaust system hangers (arrows) for damage and cracks

of the brake shoes **(see illustration)**. Any leakage here is an indication that the wheel cylinders should be overhauled immediately (see Chapter 9). Also, check all hoses and connections for signs of leakage.

17   Wipe the inside of the drum with a clean rag and denatured alcohol or brake cleaner. Again, be careful not to breathe the dangerous asbestos dust.

18   Check the inside of the drum for cracks, score marks, deep scratches and "hard spots" which will appear as small discolored areas. If imperfections cannot be removed with fine emery cloth, the drum must be taken to an automotive machine shop for resurfacing.

19   Repeat the procedure for the remaining wheel. If the inspection reveals that all parts are in good condition, reinstall the brake drums, install the wheels and lower the vehicle to the ground.

Parking brake

20   The parking brake is operated by a hand lever and locks the rear brake system. The easiest, and perhaps most obvious, method of periodically checking the operation of the parking brake assembly is to park the vehicle on a steep hill with the parking brake set and the transmission in Neutral (be sure to stay in the vehicle during this check!). If the parking brake cannot prevent the vehicle from rolling, it needs service (see Chapter 9).

17   Tire rotation (every 7500 miles or 6 months)

Refer to illustration 17.2

1   The tires should be rotated at the specified intervals and whenever uneven wear is noticed.

2   Refer to the **accompanying illustration** for the preferred tire rotation pattern.

3   Refer to the information in *Jacking and towing* at the front of this manual for the proper procedures to follow when raising the

vehicle and changing a tire. If the brakes are to be checked, don't apply the parking brake as stated. Make sure the tires are blocked to prevent the vehicle from rolling as it's raised.

4   Preferably, the entire vehicle should be raised at the same time. This can be done on a hoist or by jacking up each corner and then lowering the vehicle onto jackstands placed under the frame rails. Always use four jackstands and make sure the vehicle is safely supported.

5   After rotation, check and adjust the tire pressures as necessary and be sure to properly tighten the lug nuts.

18   Exhaust system check (every 7500 miles or 6 months)

Refer to illustrations 18.2a and 18.2b

1   With the engine cold (at least three hours after the vehicle has been driven), check the complete exhaust system from the engine to the end of the tailpipe. Ideally, the inspection should be done with the vehicle on a hoist to permit unrestricted access. If a hoist is not available, raise the vehicle and support it securely on jackstands.

2   Check the exhaust pipes and connections for evidence of leaks, severe corrosion and damage. Make sure that all brackets and hangers are in good condition and tight **(see illustrations)**.

3   At the same time, inspect the underside of the body for holes, corrosion, open seams, etc. which may allow exhaust gases to enter the interior. Seal all body openings with silicone or body putty.

4   Rattles and other noises can often be traced to the exhaust system, especially the mounts and hangers. Try to move the pipes, muffler and catalytic converter. If the components can come in contact with the body or

suspension parts, secure the exhaust system with new mounts.

19   Throttle body mounting nut torque check (every 15,000 miles or 12 months)

Refer to illustration 19.4

1   The throttle body unit is attached to the top of the intake manifold by several bolts or nuts. These fasteners can sometimes work loose from vibration and temperature changes during normal engine operation and cause a vacuum leak.

2   If you suspect that a vacuum leak exists at the bottom of the throttle body, obtain a length of hose about the diameter of fuel hose. Start the engine and place one end of the hose next to your ear as you probe around the base with the other end. You will hear a hissing sound if a leak exists (be careful of hot or moving engine components).

3   Remove the air cleaner assembly, tagging each hose to be disconnected with a

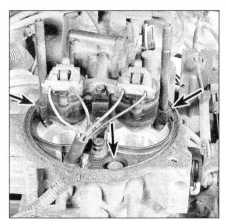

19.4 Tighten the throttle body mounting bolts (arrows) to the torque listed in this Chapter's Specifications

20.3 Gently pry off the trim cap and check the tightness of the wiper arm retaining nut

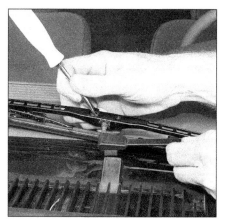

20.5 Lift the release lever with a flat-bladed screwdriver, then slide the blade assembly off the pin on the end of the wiper arm

piece of numbered tape to make reassembly easier.

4   Locate the mounting nuts or bolts at the base of the throttle body **(see illustration)**. Decide what special tools or adapters will be necessary, if any, to tighten the fasteners.

5   Tighten the nuts or bolts to the torque listed in this Chapter's Specifications. Do not overtighten them, as the threads could strip.

6   If, after the nuts or bolts are properly tightened, a vacuum leak still exists, the throttle body must be removed and a new gasket installed. See Chapter 4 for more information.

7   After tightening the fasteners, reinstall the air cleaner and return all hoses to their original positions.

20   Wiper blade inspection and replacement (every 15,000 miles or 12 months)

Refer to illustrations 20.3, 20.5 and 20.7

1   The windshield wiper and blade assemblies should be inspected periodically for damage, loose components and cracked or worn blade elements.

2   Road film can build up on the wiper blades and affect their efficiency, so they

should be washed regularly with a mild detergent solution.

3   The action of the wiping mechanism can loosen the bolts, nuts and fasteners, so they should be checked and tightened, as necessary, at the same time the wiper blades are checked **(see illustration)**.

4   If the wiper blade elements (sometimes called inserts) are cracked, worn or warped, they should be replaced with new ones.

5   Remove the wiper blade assembly from the wiper arm by inserting a screwdriver and lifting the release lever while pulling on the blade to release it **(see illustration)**.

6   With the blade removed from the vehicle, you can remove the rubber element from the blade.

7   Grasp the end of the wiper bridge securely with one hand and the element with the other. Detach the end of the element from the bridge claw and slide to free it, then slide the element out **(see illustration)**.

8   Compare the new element with the old for length, design, etc.

9   Slide the new element into the claw into place, notched end last and secure the claw into the notches.

10   Reinstall the blade assembly on the arm, wet the windshield and test for proper operation.

21   Seat belt check (every 15,000 miles or 12 months)

1   Check the seat belts, buckles, latch plates and guide loops for obvious damage and signs of wear.

2   See if the seat belt reminder light comes on when the key is turned to the Run or Start position. A chime should also sound.

3   The seat belts are designed to lock up during a sudden stop or impact, yet allow free movement during normal driving. Make sure the retractors return the belt against your chest while driving and rewind the belt fully when the buckle is unlatched.

4   If any of the above checks reveal problems with the seat belt system, replace parts as necessary.

22   Air filter replacement (every 30,000 miles or 24 months)

Refer to illustrations 22.2, 22.3a and 22.3b

1   At the specified intervals, the air filter should be replaced with a new one. The filter should be inspected between changes.

2   On early models equipped with throttle

20.7 Squeeze the end of the wiper element to free it from the bridge claw, then slide the element out

22.2 On early models, detach the clips and separate the cover halves to access the air filter

22.3a On later models, detach the wing nuts . . .

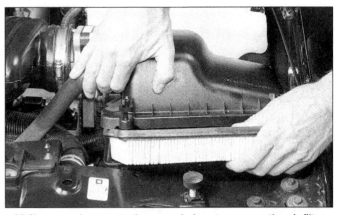

22.3b . . . and separate the cover halves to remove the air filter

23.1a On early models the PCV valve is located in the left valve cover

23.1b On later models the PCV valve is located in the left side of the intake plenum. On all models, remove the valve and place your finger over the end of the valve to check for vacuum

24.1 On later models the EGR valve is located at the back of the intake manifold next to the firewall

body injection the air filter is located inside the air cleaner housing which is mounted directly on top of the engine. Grasp the air cleaner housing and pull it up for access **(see illustration)**. It may be necessary to remove the clamp and detach the air cleaner duct from the housing.

3   On later models equipped with multi-port injection the air filter is located inside the air cleaner housing which is mounted in the left front corner of the engine compartment. Remove the wing nuts, separate the housing halves and lift the filter out **(see illustrations)**.

4   While the filter housing cover is off, be careful not to drop anything down into the air cleaner assembly.

5   Wipe out the inside of the air cleaner housing with a clean rag.

6   Place the new filter in the air cleaner housing. Make sure it seats properly.

7   The remainder of installation is the reverse of removal.

23   Positive Crankcase Ventilation (PCV) valve check and replacement (every 30,000 miles or 24 months)

Refer to illustrations 23.1a and 23.1b

1   On early models equipped with throttle body injection the PCV valve is located in the

left valve cover. On later models equipped with multi-port injection the PCV valve is located on the left side of the intake plenum **(see illustrations)**.

2   With the engine idling at normal operating temperature, pull the valve (with hose attached) out of the rubber grommet in the intake plenum or valve cover.

3   Place your finger over the end of the valve. If there is no vacuum at the valve, check for a plugged hose, manifold port, or the valve itself. Replace any plugged or deteriorated hoses.

4   Turn off the engine and shake the PCV valve, listening for a rattle. If the valve doesn't rattle, replace it with a new one.

5   To replace the valve, pull it out of the end of the hose, noting its installed position and direction.

6   When purchasing a replacement PCV valve, make sure it's for your particular vehicle, model year and engine size. Compare the old valve with the new one to make sure they are the same.

7   Push the valve into the end of the hose until it's seated.

8   Inspect the rubber grommet for damage and replace it with a new one if necessary.

9   Push the PCV valve and hose securely into position.

24   Exhaust Gas Recirculation (EGR) system check (every 30,000 miles or 24 months)

Refer to illustration 24.1

1   The EGR valve is usually located on the back half of the intake manifold. Most of the time when a problem develops in this emissions system, it's due to a stuck or corroded EGR valve **(see illustration)**.

2   With the engine cold, to prevent burns, push on the EGR valve diaphragm. Using moderate pressure, you should be able to press the diaphragm in and out within the housing.

3   If the diaphragm doesn't move or moves only with much effort, replace the EGR valve with a new one. If in doubt about the condition of the valve, compare the free movement of your EGR valve with a new valve.

4   Refer to Chapter 6 for more information on the EGR system.

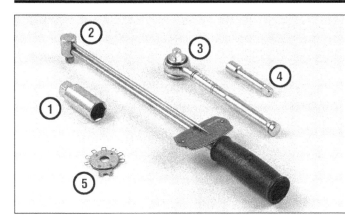

25.2 Tools required for changing spark plugs

1 **Spark plug socket** - This will have special padding inside to protect the spark plug's porcelain insulator
2 **Torque wrench** - Although not mandatory, using this tool is the best way to ensure the plugs are tightened properly
3 **Ratchet** - Standard hand tool to fit the spark plug socket
4 **Extension** - Depending on model and accessories, you may need special extensions and universal joints to reach one or more of the plugs
5 **Spark plug gap gauge** - This gauge for checking the gap comes in a variety of styles. Make sure the gap for your engine is included

25.5a Spark plug manufacturers recommend using a wire type gauge when checking the gap - if the wire does not slide between the electrodes with a slight drag, adjustment is required

25.5b To change the gap, bend the *side* electrode only, as indicated by the arrows, and be very careful not to crack or chip the porcelain insulator surrounding the center electrode

The plug type can be found in the Specifications at the front of this Chapter and on the Emission Control Information label located under the hood. If these two sources list different plug types, consider the emission control label correct.

4   Allow the engine to cool completely before attempting to remove any of the plugs. While you are waiting for the engine to cool, check the new plugs for defects and adjust the gaps.

5   Check the gap by inserting the proper thickness gauge between the electrodes at the tip of the plug **(see illustration)**. The gap between the electrodes should be the same as the one specified on the Emissions Control Information label or in Chapter 5. The wire should slide between the electrodes with a slight amount of drag. If the gap is incorrect, use the adjuster on the gauge body to bend the curved side electrode slightly until the proper gap is obtained **(see illustration)**. If the side electrode is not exactly over the center electrode, bend it with the adjuster until it is. Check for cracks in the porcelain insulator (if any are found, the plug should not be used).

6   With the engine cool, remove the spark plug wire from one spark plug. Pull only on the boot at the end of the wire - do not pull on the wire. A plug wire removal tool should be used if available **(see illustration)**. Some

25 Spark plug replacement (see maintenance schedule for service intervals)

Refer to illustrations 25.2, 25.5a, 25.5b, 25.6a, 25.6b, 25.9 and 25.10

1   The spark plugs are located at the sides of the engine. The front spark plugs can be reached from the engine compartment. **Note:** *Replace the rear spark plugs from underneath. Raise the vehicle and support it securely on jackstands.*

2   In most cases, the tools necessary for spark plug replacement include a spark plug socket which fits onto a ratchet (spark plug sockets are padded inside to prevent dam-

age to the porcelain insulators on the new plugs), various extensions and a gap gauge to check and adjust the gaps on the new plugs **(see illustration)**. A special plug wire removal tool is available for separating the wire boots from the spark plugs, and is a good idea on these models because the boots fit very tightly. A torque wrench should be used to tighten the new plugs. It is a good idea to allow the engine to cool before removing or installing the spark plugs.

3   The best approach when replacing the spark plugs is to purchase the new ones in advance, adjust them to the proper gap and replace the plugs one at a time. When buying the new spark plugs, be sure to obtain the correct plug type for your particular engine.

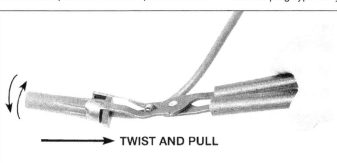

25.6a When removing the spark plug wires, pull only on the boot and use a twisting, pulling motion

TWIST AND PULL

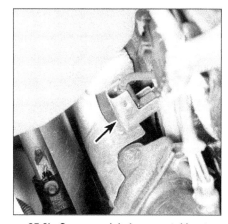

25.6b Some models have metal heat shields to protect the boots (arrow) - these simply pull off (be sure to reinstall them after spark plug replacement)

25.9 Apply a thin coat of anti-seize compound to the spark plug threads

25.10 A length of 3/8-inch ID rubber hose will save time and prevent damaged threads when installing the spark plugs

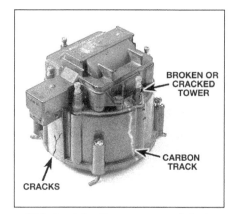

26.6a Look for these common defects when inspecting the outside of the distributor cap

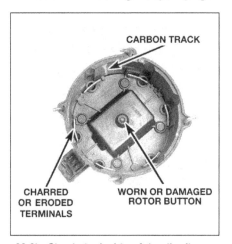

26.6b Check the inside of the distributor cap for these conditions

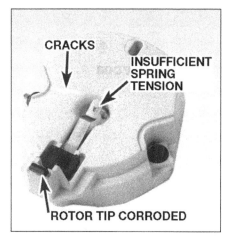

26.7 The ignition rotor should be checked for wear and corrosion as indicated above (if in doubt about its condition, buy a new one)

boots have a metal heat shield that must be removed before pulling the boot off **(see illustration)**. Be sure to replace the shield after reinstalling the boot.

7   If compressed air is available, use it to blow any dirt or foreign material away from the spark plug hole. A common bicycle pump will also work. The idea here is to eliminate the possibility of debris falling into the cylinder as the spark plug is removed.

8   The spark plugs on these models are, for the most part, difficult to reach so a spark plug socket incorporating a universal joint will be necessary. Place the spark plug socket over the plug and remove it from the engine by turning it in a counterclockwise direction. On some models where the plugs are unusually hard to get to, it may be easier to raise the vehicle, support it securely on jackstands and access the plugs from underneath.

9   Compare the spark plug with the chart shown on the inside back cover of this manual to get an indication of the general running condition of the engine. Before installing the new plugs, it is a good idea to apply a thin coat of anti-seize compound to the threads **(see illustration)**.

10   Thread one of the new plugs into the hole until you can no longer turn it with your fingers, then tighten it with a torque wrench (if available) or the ratchet. It's a good idea to slip a short length of rubber hose over the end of the plug to use as a tool to thread it into place **(see illustration)**. The hose will grip the plug well enough to turn it, but will start to slip if the plug begins to cross-thread in the hole - this will prevent damaged threads and the accompanying repair costs.

11   Before pushing the spark plug wire onto the end of the plug, inspect it following the procedures outlined in Section 26

12   Attach the plug wire to the new spark plug, again using a twisting motion on the boot until it's seated on the spark plug.

13   Repeat the procedure for the remaining spark plugs, replacing them one at a time to prevent mixing up the spark plug wires.

26 Spark plug wire and distributor cap and rotor check and replacement (every 30,000 miles or 24 months)

Refer to illustrations 26.6a, 26.6b and 26.7
Note: *Only early model vehicles equipped with 5.0L and 5.7L engines use a standard type distributor cap and rotor assembly which will be covered in this Section. Later model vehicles equipped with 4.3L and 5.7L engines use a distributor assembly mounted at the front of the engine which requires disassembly of major engine components to access. For further information on these models see Chapter 5. However the spark plug wires on later models can still be inspected and replaced as described below.*

1   The spark plug wires should be checked at the recommended intervals and whenever new spark plugs are installed in the engine.

2   The wires should be inspected one at a time to prevent mixing up the order, which is essential for proper engine operation.

3   Disconnect the plug wire from the spark plug. To do this, grab the rubber boot, twist

slightly and pull the wire off. Do not pull on the wire itself, only on the rubber boot.

4   Check inside the boot for corrosion, which will look like a white crusty powder. Push the wire and boot back onto the end of the spark plug. It should be a tight fit on the plug. If it isn't, remove the wire and use pliers to carefully crimp the metal connector inside the boot until it fits securely on the end of the spark plug.

5   Using a clean rag, wipe the entire length of the wire to remove any built-up dirt and grease. Once the wire is clean, check for burns, cracks and other damage. Do not bend the wire excessively or pull the wire lengthwise - the conductor inside might break.

6   Detach the clips or loosen the screws and remove the distributor cap. Check the cap for cracks, carbon tracks and worn, burned or loose terminal **(see illustrations)** Check for corrosion and a tight fit in the same manner as the spark plug end. On the later model engines disconnect the wire from the distributor at the front of the engine. Check for corrosion and a tight fit in the same manner as the spark plug end.

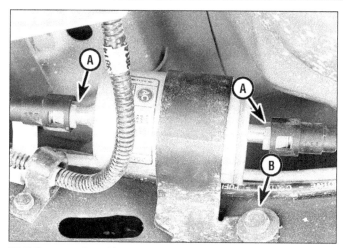

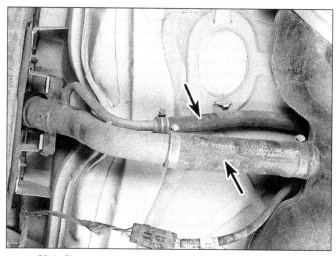

27.5 Squeeze the white plastic quick-disconnect tabs (A) together and pull the fuel lines away from the filter, then detach the fuel filter bracket mounting bolt (B) to remove the fuel filter

28.4 Check the fuel filler lines (arrows) for cracks and deterioration and the hose clamps for tightness

7   Check the distributor rotor for cracks and carbon tracks **(see illustration)**. Make sure the center terminal tension is adequate and look for corrosion and wear on the rotor tip.

8   Check the remaining spark plug wires one at a time, making sure they are securely fastened at the distributor and the spark plug when the check is complete.

9   If new spark plug wires are required, purchase a set for your specific engine model. Wire sets are available pre-cut, with the rubber boots already installed. Remove and replace the wires one at a time to avoid mix-ups in the firing order. **Note:** *If an accidental mix-up occurs, refer to the firing order diagrams at the beginning of this Chapter.*

---

27   Fuel filter replacement (every 30,000 miles or 24 months)

Refer to illustration 27.5
Warning: *Gasoline is extremely flammable, so take extra precautions when you work on any part of the fuel system. Don't smoke or allow open flames or bare light bulbs near the work area, and don't work in a garage where a natural gas-type appliance (such as a water heater or clothes dryer) with a pilot light is present. Since gasoline is carcinogenic, wear latex gloves when there's a possibility of being exposed to fuel, and, if you spill any fuel on your skin, rinse it off immediately with soap and water. Mop up any spills immediately and do not store fuel-soaked rags where they could ignite. The fuel system is under constant pressure, so, if any fuel lines are to be disconnected, the fuel pressure in the system must be relieved first (see Chapter 4 for more information). When you perform any kind of work on the fuel system, wear safety glasses and have a Class B type fire extinguisher on hand.*

1   Relieve the fuel system pressure (see Chapter 4).

2   Raise the vehicle and support it securely on jackstands.

3   The fuel filter is located on the inside of right frame rail in front of the fuel tank.

4   Use compressed air or carburetor cleaner to clean any dirt surrounding the fuel inlet and outlet line fittings.

5   Once all the dirt has been removed, depress the white plastic quick-disconnect tabs and detach the inlet and outlet lines from the fuel filter **(see illustration)**. **Note:** *Have spare rags or a small container to catch or wipe up extra gasoline which will spill from the filter assembly.*

6   Detach the fuel filter mounting bracket bolt and remove the fuel filter.

7   Installation is the reverse of removal. **Note:** *Make sure the directional arrow on the fuel filter points in the direction of fuel flow.*

---

28   Fuel system check (every 30,000 miles or 24 months)

Refer to illustration 28.4
Warning: *Gasoline is extremely flammable, so take extra precautions when you work on any part of the fuel system. Don't smoke or allow open flames or bare light bulbs near the work area, and don't work in a garage where a natural gas-type appliance (such as a water heater or clothes dryer) with a pilot light is present. Since gasoline is carcinogenic, wear latex gloves when there's a possibility of being exposed to fuel, and, if you spill any fuel on your skin, rinse it off immediately with soap and water. Mop up any spills immediately and do not store fuel-soaked rags where they could ignite. The fuel system is under constant pressure, so, if any fuel lines are to be disconnected, the fuel pressure in the system must be relieved first (see Chapter 4 for more information). When you perform any kind of work on the fuel system, wear safety glasses and have a Class B type fire extinguisher on hand.*

1   The fuel system is most easily checked with the vehicle raised on a hoist so the com-

ponents underneath the vehicle are readily visible and accessible.

2   If the smell of gasoline is noticed while driving or after the vehicle has been in the sun, the system should be thoroughly inspected immediately.

3   Remove the gas tank cap and check for damage, corrosion and an unbroken sealing imprint on the gasket. Replace the cap with a new one if necessary.

4   With the vehicle raised, inspect the gas tank and filler neck for cracks and other damage **(see illustration)**. The connection between the filler neck and tank is especially critical. Sometimes a filler neck will leak due to cracks, problems a home mechanic can't repair. **Warning:** *Do not, under any circumstances, try to repair a fuel tank yourself (except rubber components). A welding torch or any open flame can easily cause the fuel vapors to explode if the proper precautions are not taken.*

5   Carefully check all rubber hoses and metal lines leading away from the fuel tank. Check for loose connections, deteriorated hoses, crimped lines and other damage. Follow the lines to the front of the vehicle, carefully inspecting them all the way. Repair or replace damaged sections as necessary.

---

29   Front wheel bearing check, repack and adjustment (every 30,000 miles or 24 months)

Check and repack
Refer to illustrations 29.1, 29.6, 29.7, 29.9, 29.11 and 29.15

1   In most cases the front wheel bearings will not need servicing until the brake pads are changed. However, the bearings should be checked whenever the front of the vehicle is raised for any reason. Several items, includ-

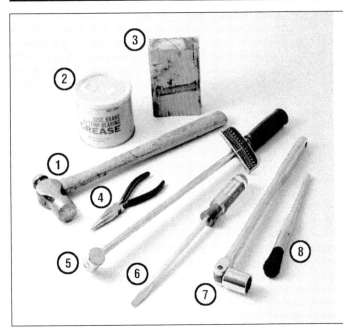

29.1 Tools and materials needed for front wheel bearing maintenance

1   **Hammer** - A common hammer will do just fine
2   **Grease** - High-temperature grease that is formulated specially for front wheel bearings should be used
3   **Wood block** - If you have a+ scrap piece of 2x4, it can be used to drive the new seal into the hub
4   **Needle-nose pliers** - Used to straighten and remove the cotter pin in the spindle
5   **Torque wrench** - This is very important in this procedure; if the bearing is too tight, the wheel won't turn freely - if it's too loose, the wheel will "wobble" on the spindle. Either way, it could mean extensive damage
6   **Screwdriver** - Used to remove the seal from the hub (a long screwdriver is preferred)
7   **Socket/breaker bar** - Needed to loosen the nut on the spindle if it's extremely tight
8   **Brush** - Together with some clean solvent, this will be used to remove old grease from the hub and spindle

ing a torque wrench and special grease, are required for this procedure **(see illustration)**.
2   With the vehicle securely supported on jackstands, spin each wheel and check for noise, rolling resistance and freeplay.
3   Grasp the top of each tire with one hand and the bottom with the other. Move the wheel in-and-out on the spindle. If there's any noticeable movement, the bearings should be checked and then repacked with grease or replaced if necessary.
4   Remove the wheel.
5   Remove the brake caliper rotor (see Chapter 9) and hang it out of the way on a piece of wire. A wood block of the appropriate width can be slid between the brake pads to keep them separated, if necessary.
6   Dislodge the dust cap from the hub/disc assembly using a screwdriver or hammer and chisel **(see illustration)**.
7   Straighten the bent ends of the cotter

pin, then pull the cotter pin out of the nut lock **(see illustration)**. Discard the cotter pin and use a new one during reassembly.
8   Remove the nut lock, nut and washer from the end of the spindle .
9   Pull the hub/disc assembly out slightly, then push it back into its original position. This should force the outer bearing off the spindle enough so it can be removed **(see illustration)**.
10   Pull the disc assembly off the spindle.
11   Use a screwdriver or a seal puller tool to pry the seal out of the rear of the rotor **(see illustration)**. Note how the seal is installed.
12   Remove the inner wheel bearing from the rotor.
13   Use solvent to remove all traces of the old grease from the bearings, hub and spindle. A small brush may prove helpful; however make sure no bristles from the brush embed themselves inside the bearing rollers.

Allow the parts to air dry.
14   Carefully inspect the bearings for cracks, heat discoloration, worn rollers, etc. Check the bearing races inside the hub for wear and damage. If the bearing races are defective, the hubs should be taken to a machine shop with the facilities to remove the old races and press new ones in. Note that the bearings and races come as matched sets and old bearings should never be installed on new races.
15   Use high-temperature front wheel bearing grease to pack the bearings. Work the grease completely into the bearings, forcing it between the rollers, cone and cage from the back side **(see illustration)**.
16   Apply a thin coat of grease to the spindle at the outer bearing seat, inner bearing seat, shoulder and seal seat.
17   Put a small quantity of grease inboard of each bearing race inside the hub. Using your

29.6 Dislodge the dust cap by working around the outer circumference with a hammer and chisel

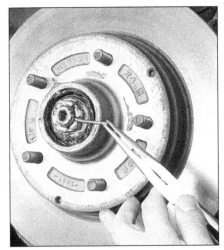

29.7 Remove the cotter pin and discard it - use a new one when the disc assembly is reinstalled

29.9 Pull the disc assembly outward slightly, then push it back into position to dislodge the washer and outer wheel bearing

29.11 Use a seal puller or large screwdriver to remove the inner grease seal - note the seal installed position

29.15 Work the grease completely into the bearing rollers

29.23 Tighten the spindle nut to 12 ft-lbs while rotating the hub to seat the bearings - then back off the nut 1/4 turn. Using your hand, tighten the nut until it's snug. Align the spindle nut so that it lines up with the cotter pin hole

finger, form a dam at these points to provide extra grease availability and to keep thinned grease from flowing out of the bearing.

18   Place the grease-packed inner bearing into the rear of the hub and put a little more grease outboard of the bearing.

19   Place a new seal over the inner bearing and tap the seal evenly into place until it's flush with the hub.

20   Carefully place the hub assembly onto the spindle and push the grease-packed outer bearing into position.

Adjustment

Refer to illustration 29.23

21   Install the washer and spindle nut. Tighten the nut only slightly.

22   Spin the hub in a forward direction while tightening the spindle nut to approximately 12 ft-lbs to seat the bearings and remove any grease or burrs which could cause excessive bearing play later.

23   Loosen the spindle nut 1/4-turn, then using your hand (not a wrench of any kind), tighten the nut until it's snug. Install a new cotter pin through the hole in the spindle and the slots in the nut. If the slots in the nut don't line up with the hole in the spindle, turn the nut slightly until they do **(see illustration)**.

24   Check that the hub/disc assembly spins freely with no noticeable free play. If freeplay exists repeat Steps 22 and 23 until proper adjustment is obtained.

25   Bend the ends of the cotter pin until they're flat against the nut. Cut off any extra length which could interfere with the dust cap.

26   Install the dust cap, lightly tapping it into place with a hammer.

27   Install brake caliper in the reverse order of removal (see Chapter 9).

28   Install the wheel on the hub and tighten the lug nuts.

29   Lower the vehicle.

30   Automatic transmission fluid and filter change (every 30,000 miles or 24 months)

Refer to illustrations 30.6, 30.9 and 30.10

1   At the specified intervals, the transmission fluid should be drained and replaced. Since the fluid will remain hot long after driving, perform this procedure only after the engine has cooled down completely.

2   Before beginning work, purchase the specified transmission fluid (see *Recom-*

mended lubricants and fluids at the front of this Chapter) and a new filter.

3   Other tools necessary for this job include a floor jack, jackstands to support the vehicle in a raised position, a drain pan capable of holding at least eight pints, newspapers and clean rags.

4   Raise the vehicle and support it securely on jackstands.

5   Place the drain pan underneath the transmission pan. Remove the front and side pan mounting bolts, but only loosen the rear pan bolts approximately four turns.

6   Carefully pry the transmission pan loose with a screwdriver, allowing the fluid to drain **(see illustration)**.

7   Remove the remaining bolts, pan and gasket. Carefully clean the gasket surface of the transmission to remove all traces of the old gasket and sealant.

8   Drain the fluid from the transmission pan, clean it with solvent and dry it with compressed air.

9   Remove the filter from the mount inside the transmission **(see illustration)**.

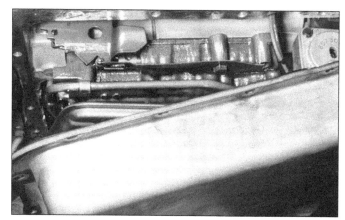

30.6 With the rear bolts in place but loose, pull the front of the pan down to drain the transmission fluid

30.9 Rotate the filter out of the retaining clip, then lower it from the transmission

30.10 If necessary, use a screwdriver to remove the seal from the transmission - be careful not to gouge the aluminum housing

31.6a The drain plug (arrow) is located at the lower left corner of the radiator

31.6b Use a screwdriver to open the bleeder screw (arrow)

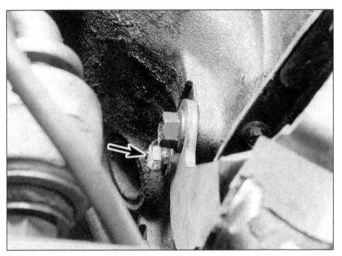

31.7a Early models are equipped with block drain plugs (arrow) - block drain plugs are located on each side of the block about one to two inches above the oil pan rail

10   If the seal did not come out with the filter, remove it from the transmission (**see illustration**). Install a new filter and seal.

11   Make sure the gasket surface on the transmission pan is clean, then install a new gasket on the pan. Put the pan in place against the transmission and, working around the pan, tighten each bolt a little at a time until the final torque figure is reached.

12   Lower the vehicle and add approximately 3-1/2 quarts of the specified type of automatic transmission fluid through the filler tube (Section 7).

13   With the transmission in Park and the parking brake set, run the engine at a fast idle, but don't race it.

14   Move the gear selector through each range and back to Park. Check the fluid level. It will probably be low. Add enough fluid to bring the level up to the COLD FULL range on the dipstick.

15   Check under the vehicle for leaks during the first few trips.

31   Cooling system servicing (draining, flushing and refilling) (see maintenance schedule for service intervals)

Refer to illustrations 31.6a, 31.6b, 31.7a and 31.7b

Warning: *Make sure the engine is completely cool before performing this procedure.*

Caution: *Never mix green-colored ethylene glycol anti-freeze and orange-colored "DEX-COOL" silicate-free coolant because doing so will destroy the efficiency of the "DEX-COOL" coolant which is designed to last for 100,000 miles or five years.*

1   Periodically, the cooling system should be drained, flushed and refilled to replenish the antifreeze mixture and prevent formation of rust and corrosion, which can impair the performance of the cooling system and cause engine damage.

2   At the same time the cooling system is serviced, all hoses and the radiator cap should be inspected and replaced if defective (see Section 12).

3   Since antifreeze is a corrosive and poisonous solution, be careful not to spill any of the coolant mixture on the vehicle's paint or your skin. If this happens, rinse it off immediately with plenty of clean water. Consult local authorities about the dumping of antifreeze before draining the cooling system. In many areas, reclamation centers have been set up to collect automobile oil and drained antifreeze/water mixtures, rather than allowing them to be added to the sewage system.

4   With the engine cold, remove the radiator cap and reservoir cap (1993 and earlier models) or the reservoir pressure cap (1994 and later models).

5   Move a large container under the radiator to catch the coolant as it's drained.

6   Drain the radiator by opening the drain plug at the bottom on the left side (see illus-

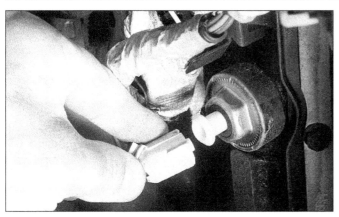

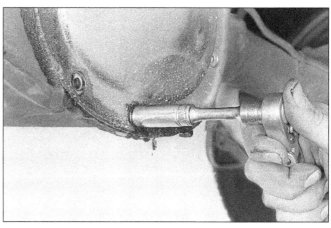

31.7b Later models have knock sensors which screw into the block drain holes - unplug the connector and unscrew the sensors to drain the coolant from the block

32.8a Remove the bolts from the lower edge of the cover . . .

tration). If the drain plug is corroded and can't be turned easily, or if the radiator isn't equipped with a plug, disconnect the lower radiator hose to allow the coolant to drain. Be careful not to get antifreeze on your skin or in your eyes. Some cooling systems are equipped with a bleeder screw, use a screwdriver to open the screw two or three turns **(see illustration)**.

7   After the coolant stops flowing out of the radiator, move the container under the engine block drain plug(s), then remove the plugs, or knock sensors (which double as plugs on some models), on both sides of the block **(see illustrations)**.

8   Disconnect the hose from the coolant reservoir and remove the reservoir (see Chapter 3). Flush it out with clean water.

9   Place a garden hose in the radiator filler neck (if equipped) and flush the system until the water runs clear at all drain points. On models not equipped with a radiator fill neck, remove the upper radiator hose from the radiator, then place a garden hose in the upper radiator opening and flush the system until the water runs clear at all drain points.

10   In severe cases of contamination or clogging of the radiator, remove it (see Chapter 3) and reverse flush it. This involves inserting the hose in the bottom radiator outlet to allow the water to run against the normal flow, draining through the top. A radiator repair shop should be consulted if further cleaning or repair is necessary.

11   When the coolant is regularly drained and the system refilled with the correct antifreeze/water mixture, there should be no need to use chemical cleaners or descalers.

12   To refill the system, install the block plugs or knock sensors, reconnect any radiator hoses and install the reservoir and the overflow hose.

13   On later models, make sure to use the proper coolant (see **Caution** above). The manufacturer recommends adding cooling system sealer, available at most auto parts stores, any time the coolant is changed. Slowly fill the radiator with the recommended mixture of antifreeze and water to the base of the filler neck. On models without a radiator cap, add coolant to the coolant reservoir.

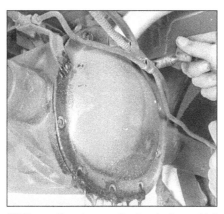

32.8b . . . then loosen the top bolts and let the lubricant drain out

Wait two minutes and recheck the coolant level, adding if necessary. Close the bleed screws (if equipped) when the coolant issuing from them is free of bubbles. **Note:** *The low coolant light may illuminate after the draining and flushing procedure has been completed. Start the engine and let it run until it reaches normal operating temperature, then let it completely cool down. Repeat this procedure two more times until the light goes out.*

14   Keep a close watch on the coolant level and the cooling system hoses during the first few miles of driving. Tighten the hose clamps and/or add more coolant as necessary. The coolant level should be a little above the HOT mark on the reservoir with the engine at normal operating temperature.

---

32   Differential lubricant change (every 30,000 miles or 24 months)

Refer to illustrations 32.8a, 32.8b, 32.8c and 32.10

1   This procedure should be performed after the vehicle has been driven so the lubricant will be warm and therefore will flow out of the differential more easily.

2   Raise the vehicle and support it securely on jackstands.

32.8c After the lubricant has drained, remove the bolts and the cover

3   The easiest way to drain the differential is to remove the lubricant through the filler plug hole with a suction pump. If the differential's bolt-on cover gasket is leaking, it will be necessary to remove the cover to drain the lubricant (which will also allow you to inspect the differential.

Changing the lubricant with a suction pump

4   Remove the fill plug from the differential (see Section 10).

5   Insert the flexible hose. Work the hose down to the bottom of the differential housing and pump the lubricant out.

6   Use a hand pump, syringe or funnel to fill the differential housing with the specified lubricant until it's level with the bottom of the plug hole. Install the fill plug.

Changing lubricant by removing the cover

7   Move a drain pan, rags, newspapers and wrenches under the vehicle.

8   Remove the bolts on the lower half of the plate **(see illustration)**. Loosen the bolts on the upper half and use them to keep the cover loosely attached **(see illustration)**. Allow the oil to drain into the pan, then completely remove the cover **(see illustration)**.

9   Using a lint-free rag, clean the inside of the cover and the accessible areas of the differential housing. As this is done, check for chipped gears and metal particles in the lubricant, indicating that the differential should be more thoroughly inspected and/or repaired.

10   Thoroughly clean the gasket mating surfaces of the differential housing and the cover plate. Use a gasket scraper or putty knife to remove all traces of the old gasket **(see illustration)**.

11   Apply a thin layer of RTV sealant to the cover flange, then press a new gasket into position on the cover. Make sure the bolt holes align properly.

12   Use a hand pump, syringe or funnel to fill the differential housing with the specified lubricant until it's level with the bottom of the plug hole. Install the fill plug.

32.10 Carefully scrape the old gasket material off to ensure a leak-free seal

Chapter 2 Part A
V8 engines

Contents

Specifications

General

Displacement

Displacement
4.3L	265 cubic inches
5.0L	305 cubic inches
5.7L	350 cubic inches

Cylinder numbers (front-to-rear)
Left (driver's) side	1-3-5-7
Right side	2-4-6-8

Firing order 1-8-4-3-6-5-7-2

Camshaft

Bearing journal diameter
1993 and earlier	1.8682 to 1.8692 inches
1994 and later	1.8677 to 1.8697 inches

Lobe lift

1991 through 1993
Intake	0.233 inch
Exhaust	0.257 inch

1994 and 1995
Intake	0.279 inch
Exhaust	0.286 inch

1996
Intake	0.276 inch
Exhaust	0.285 inch

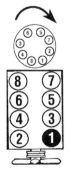

1993 and earlier

24046-1SPECS.HAYNES

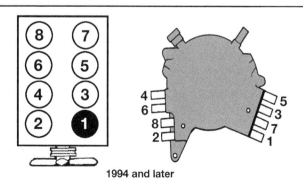

1994 and later

Cylinder location, distributor rotation and spark plug terminal locations

Torque specifications

Ft-lbs (unless otherwise indicated)

Camshaft sprocket bolts	21
Camshaft retainer bolts	108 in-lbs
Cylinder head bolts	
1991 through 1993	68
1994 and 1995	65
1996	
Step 1	22
Step 2	
Short bolts	Tighten an additional 55-degrees
Medium bolts	Tighten an additional 65-degrees
Long bolts	Tighten an additional 75-degrees
Driveplate	74
Exhaust manifold bolts	26
Intake manifold bolts	
1993 and earlier	
Step 1	124 in-lbs
Step 2	35
1994 and later	
Step 1	71 in-lbs
Step 2	35
Oil baffle nut	30
Oil pan-to-engine block	
Corner bolts/nuts	180 in-lbs
Side-rail bolts	106 in-lbs
Oil pump mounting bolt	66
Rear oil seal housing-to-block bolts	132 in-lbs
Valve cover bolts	108 in-lbs
Front cover bolts	100 in-lbs
Coolant transfer tube bolts (1994 and later)	30
Crankshaft balancer hub bolt	
1991 through 1993	70
1994 and later	75
Crankshaft balancer bolts (1994 and later)	63
Crankshaft pulley bolts	
1991 through 1993	43
1994 and later	18
Lifter guide retainer bolts	
1991 through 1993	144 in-lbs
1994 and later	180 in-lbs
Oil pump driveshaft bolt (1994 and later)	144 in-lbs

1 General information

Warning: *The models covered by this manual are equipped with airbags. Always disable the airbag system before working in the vicinity of the impact sensors, steering column or instrument panel to avoid the possibility of accidental deployment of the airbag(s), which could cause personal injury (see Chapter 12). The yellow wires and connectors routed through the instrument panel and, on 1995 and earlier models, to the front of the vehicle, are for this system. Do not use electrical test equipment on these yellow wires or tamper with them in any way.*
Caution: *On models equipped with a Delco Loc II or Theftlock audio system, be sure the lockout feature is turned off before performing any procedure which requires disconnecting the battery.*

This Part of Chapter 2 is devoted to in-vehicle repair procedures for the V8 engine. All information concerning engine removal and installation and engine block and cylinder head overhaul can be found in Part B of this Chapter.

Since the repair procedures included in this Part are based on the assumption the engine is still installed in the vehicle, if they are being used during a complete engine overhaul (with the engine already out of the vehicle and on a stand) many of the Steps included here will not apply.

The Specifications included in this Part of Chapter 2 apply only to the procedures found here. The specifications necessary for rebuilding the block and cylinder heads are included in Part B.

2 Repair operations possible with the engine in the vehicle

Many major repair operations can be accomplished without removing the engine from the vehicle.

Clean the engine compartment and the exterior of the engine with some type of pressure washer before any work is done. A clean engine will make the job easier and will help keep dirt out of the internal areas of the engine.

Depending on the components involved, it may be a good idea to remove the hood to improve access to the engine as repairs are performed (refer to Chapter 11 if necessary).

If oil or coolant leaks develop, indicating a need for gasket or seal replacement, the repairs can generally be made with the engine in the vehicle. The oil pan gasket, the cylinder head gaskets, intake and exhaust manifold gaskets, front cover gaskets and the crankshaft oil seals are accessible with the engine in place.

Exterior engine components, such as the water pump, the starter motor, the alternator, the distributor and the fuel injection unit, as well as the intake and exhaust manifolds, can be removed for repair with the engine in place.

Since the cylinder heads can be removed without pulling the engine, valve component servicing can also be accomplished with the engine in the vehicle.

Replacement of, repairs to or inspection of the timing chain and sprockets and the oil pump are all possible with the engine in place.

In extreme cases caused by a lack of necessary equipment, repair or replacement of piston rings, pistons, connecting rods and rod bearings is possible with the engine in the

4.5a Detach the brake booster vacuum hose from the plenum and (on early models) the PCV hose from the left cover

4.5b Remove the air injection fitting from each exhaust manifold and swing it out of the way

4.6 Unscrew the four mounting bolts (arrows) and detach the cover

vehicle. However, this practice is not recommended because of the cleaning and preparation work that must be done to the components involved.

3 Top Dead Center (TDC) for number one piston - locating

1   Top Dead Center (TDC) is the highest point in the cylinder that each piston reaches as it travels up-and-down when the crankshaft turns. Each piston reaches TDC on the compression stroke and again on the exhaust stroke, but TDC generally refers to piston position on the compression stroke.
2   Positioning the pistons at TDC is an essential part of many procedures such as rocker arm removal, valve adjustment, and timing chain and sprocket replacement.
3   In order to bring any piston to TDC, the crankshaft must be turned using one of the methods outlined below. When looking at the front of the engine, normal crankshaft rotation is clockwise. **Warning:** *Before beginning this procedure, be sure to place the transmission in Park or Neutral and disable the ignition system by disconnecting the primary wires from the coil.*

a) *The preferred method is to turn the crankshaft with a large socket and breaker bar attached to the balancer hub bolt that is threaded into the front of the crankshaft.*
b) *A remote starter switch, which may save some time, can also be used. Attach the switch leads to the S (switch) and B (battery) terminals on the starter solenoid. Once the piston is close to TDC, use a socket and breaker bar as described in the previous paragraph.*
c) *If an assistant is available to turn the ignition switch to the Start position in short bursts, you can get the piston close to TDC without a remote starter switch. Use a socket and breaker bar as described in Paragraph a) to complete the procedure.*

4   Insert a compression gauge (screw-in type with a hose) in the number 1 spark plug hole. Place the gauge dial where you can see it while turning the balancer hub bolt.
5   Turn the crankshaft until you see compression building up on the gauge; at that point you are on the compression stroke for number one.

a) *On 1993 and earlier models, continue rotating the crankshaft until the groove in the crankshaft balancer aligns with the 0 mark on the timing indicator. Remove the distributor cap and verify the rotor is pointing at the number one cylinder spark plug wire terminal.*
b) *On 1994 and later models, the cast-in arrow on the crankshaft balancer installed on the front of the crankshaft is referenced to the number one piston at TDC when the arrow is straight up, or at "12 o'clock" (see Section 9). Stop turning when the arrow is straight up.*

6   After the number one piston has been positioned at TDC on the compression stroke, TDC for any of the remaining cylinders can be located by turning the crankshaft 90-degrees (1/4-turn) at a time and following the firing order (refer to the Specifications). For example, turning 90-degrees past number one TDC would give you TDC for number eight cylinder, the next in the firing order.

4 Valve covers - removal and installation

Refer to illustrations 4.5a, 4.5b and 4.6

Removal

1   Disconnect the negative cable from the battery. **Caution:** *On models equipped with a Delco Loc II or Theftlock audio system, be sure the lockout feature is turned off before performing any procedure which requires disconnecting the battery.*
2   On 1993 and earlier models, remove the air cleaner assembly.
3   On 1994 and later models, remove the

serpentine drivebelt (see Chapter 1). Disconnect the transmission dipstick tube from the bellhousing and remove the alternator (see Chapter 5).
4   Remove the wiring harness clips and position the harness aside.
5   Remove the PCV hose from the left valve cover, disconnect the brake booster vacuum hose (left side) and disconnect the air injection fitting from the each exhaust manifold **(see illustrations)**. **Note:** *Use penetrating oil on the air injection fittings and allow it to soak in before trying to remove the fittings.*
6   Remove the valve cover mounting bolts **(see illustration)**.
7   Remove the valve cover. **Note:** *If the cover is stuck to the head, bump the cover with a block of wood and a hammer to release it. If it still won't come loose, try to slip a flexible putty knife between the head and cover to break the seal. Do not pry at the cover-to-head joint or damage to the sealing surface and cover flange will result and oil leaks will develop.*

Installation

8   The mating surfaces of each cylinder head and valve cover must be perfectly clean when the covers are installed. Use a gasket scraper to remove all traces of sealant or old gasket, then clean the mating surfaces with lacquer thinner or acetone. If there's sealant or oil on the mating surfaces when the cover is installed, oil leaks may develop.
9   Make sure the threaded holes are clean. Run a tap into them to remove corrosion and restore damaged threads.
10   Mate the new gaskets to the covers before the covers are installed. Apply a thin coat of RTV sealant to the cover flange, then position the gasket inside the cover lip and allow the sealant to set up so the gasket adheres to the cover (if the sealant isn't allowed to set, the gasket may fall out of the cover as it's installed on the engine).
11   Carefully position the cover on the head and install the bolts.
12   Tighten the nuts/bolts in three steps to the torque listed in this Chapter's Specifications.

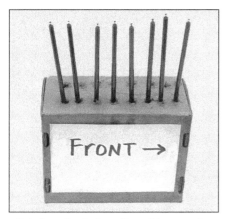

5.4 A perforated cardboard box can be used to store the pushrods to ensure that they're reinstalled in their original locations - note the label indicating the front of the engine

5.10 The ends of the pushrods and the valve stems should be lubricated with moly-base grease prior to installation of the rocker arms

5.11 Moly-base grease applied to the pivot balls will ensure adequate lubrication until oil pressure builds up when the engine is started

13   The remaining installation steps are the reverse of removal.

14   Start the engine and check carefully for oil leaks as the engine warms up.

---

5   Rocker arms and pushrods - removal, inspection and installation

---

Refer to illustrations 5.4, 5.10, 5.11 and 5.13

Removal

Note: *Any valve train components being removed (rocker arms, pivot balls, pushrods or lifters) should be stored in marked containers so they can be returned to their original locations on assembly.*

1   Refer to Section 4 and detach the valve covers from the cylinder heads.

2   Beginning at the front of one cylinder head, loosen and remove the rocker arm stud nuts. **Note:** *If the pushrods are the only items being removed, loosen each nut just enough to allow the rocker arms to be rotated to the side so the pushrods can be lifted out.*

3   Lift off the rocker arms and pivot balls and store them in the marked containers with the nuts (they must be reinstalled in their original locations).

4   Remove the pushrods and store them in order to make sure they don't get mixed up during installation **(see illustration)**.

Inspection

5   Check each rocker arm for wear, cracks and other damage, especially where the pushrods and valve stems contact the rocker arm faces.

6   Make sure the hole at the pushrod end of each rocker arm is open.

7   Check each rocker arm pivot area for wear, cracks and galling. If the rocker arms are worn or damaged, replace them with new ones and use new pivot balls as well.

8   Inspect the pushrods for cracks and excessive wear at the ends. Roll each push-

rod across a piece of plate glass to see if it's bent (if it wobbles, it's bent).

Installation

9   Lubricate the lower end of each pushrod with clean engine oil or moly-base grease and install them in their original locations. Make sure each pushrod seats completely in the lifter socket.

10   Apply moly-base grease to the ends of the valve stems and the upper ends of the pushrods before positioning the rocker arms over the studs **(see illustration)**.

11   Set the rocker arms in place, then install the pivot balls and nuts. Apply moly-base grease to the pivot balls to prevent damage to the mating surfaces before engine oil pressure builds up. Be sure to install each nut with the flat side against the pivot ball **(see illustration)**.

Valve adjustment

12   Refer to Section 3 and bring the number one piston to TDC on the compression stroke.

13   Tighten the rocker arm nuts (number one cylinder only) until all play is removed at the pushrods. This can be determined by wiggling the pushrod up-and-down as the nut is tightened **(see illustration)**. When you can no longer feel movement, the play is taken up.

5.13 When all play is removed from the pushrod, tighten the nut an additional 3/4-turn

14   Tighten each nut an additional 3/4-turn to center the lifter plungers in their travel. Valve adjustment for cylinder number one is now complete. A cylinder number illustration and the firing order is included in the Specifications.

15   Adjust the number two, five and seven intake valves and the number three, four and eight exhaust valves at this time. After adjusting the indicated valves, turn the crankshaft one complete revolution (360-degrees) and adjust the number three, four, six and eight intake valves and the number two, five, six and seven exhaust valves.

16   Refer to Section 4 and install the valve covers. Start the engine, listen for unusual valve train noises and check for oil leaks at the valve cover joints.

---

6   Valve springs, retainers and seals - replacement

Refer to illustrations 6.4, 6.7 and 6.16

Note: *Broken valve springs and defective valve stem seals can be replaced without removing the cylinder head. Two special tools and a compressed air source are normally required to perform this operation, so read through this Section carefully and rent or buy the tools before beginning the job.*

1   Refer to Section 4 and remove the valve cover from the affected cylinder head. If all of the valve stem seals are being replaced, remove both valve covers.

2   Remove the spark plug from the cylinder which has the defective component. If all of the valve stem seals are being replaced, all of the spark plugs should be removed.

3   Turn the crankshaft until the piston in the affected cylinder is at top dead center on the compression stroke (refer to Section 3 for instructions). If you're replacing all of the valve stem seals, begin with cylinder number one and work on the valves for one cylinder at a time. Move from cylinder-to-cylinder following the firing order sequence (see the specifications).

6.4 This is what the air hose adapter that threads into the spark plug hole looks like - they're commonly available from auto parts stores

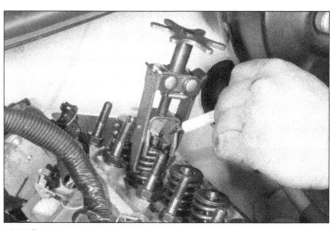

6.7 Compress the valve spring and use a magnet or small pliers to remove the keepers

4   Thread an adapter, available at most auto parts stores, into the spark plug hole and connect an air hose from a compressed air source to it **(see illustration)**. Most auto parts stores can supply an air hose adapter. **Note:** *Many cylinder compression gauges utilize a screw-in fitting that may work with your air hose quick-disconnect fitting.*
5   Remove the nut, pivot ball and rocker arm for the valve with the defective part and pull out the pushrod. If all of the valve stem seals are being replaced, all of the rocker arms and pushrods should be removed (refer to Section 5).
6   Apply compressed air to the cylinder. **Warning:** *The piston may be forced down by the compressed air! Make sure the breaker bar and socket are not attached to the balancer and you are clear of the drivebelt and pulleys before applying air pressure.*
7   Stuff shop rags into the cylinder head holes above and below the valves to prevent parts and tools from falling into the engine, then use a valve spring compressor to compress the spring/balancer assembly. Remove the keepers with a pair of small needle-nose pliers or a magnet **(see illustration)**.
8   Remove the spring retainer and valve spring assembly with the compressor still attached, then remove the valve stem seal. **Note:** *If air pressure fails to hold the valve in the closed position during this operation, the valve face or seat is probably damaged. If so, the cylinder head will have to be removed for additional repair operations.*
9   Wrap a rubber band or tape around the top of the valve stem so the valve will not fall into the combustion chamber, then release the air pressure.
10   Inspect the valve stem for damage. Rotate the valve in the guide and check the end for eccentric movement, which would indicate that the valve is bent.
11   Move the valve up-and-down in the guide and make sure it doesn't bind. If the valve stem binds, either the valve is bent or the guide is damaged. In either case, the cylinder head will have to be removed for repair.
12   Inspect the rocker arm studs for wear. Worn studs should be replaced. **Note:** *Special*

6.16 Apply a small dab of grease to each keeper as shown here before installation - it'll hold them in place on the valve stem as the spring is released

tools and expertise are required for rocker arm stud replacement, remove the cylinder heads and take them to an automotive machine shop for stud replacement, if necessary.
13   Reapply air pressure to the cylinder to retain the valve in the closed position, then remove the tape or rubber band from the valve stem.
14   Lubricate the valve stem with engine oil and install a new oil seal of the type originally used on the engine.
15   Install the spring and retainer in position over the valve.
16   Compress the spring and install the keepers in the upper groove. Apply a small dab of grease to the inside of each keeper to hold it in place if necessary **(see illustration)**. Remove the pressure from the spring tool and make sure the keepers are seated.
17   Disconnect the air hose and remove the adapter from the spark plug hole.
18   Refer to Section 5 and install the rocker arms and pushrods and adjust the valves.
19   Install the spark plugs and connect the wires.
20   Refer to Section 4 and install the valve covers.
21   Start and run the engine, then check for

7.3 Remove the alternator brace from the intake manifold, and remember the location of its stud (arrow)

oil leaks and unusual sounds coming from the valve cover area.

---

7   **Intake manifold - removal and installation**

Removal

Refer to illustrations 7.3, 7.7 and 7.8

1   Disconnect the negative cable from the battery, then refer to Chapter 1 and drain the cooling system. **Caution:** *On models equipped with a Delco Loc II or Theftlock audio system, be sure the lockout feature is turned off before performing any procedure which requires disconnecting the battery.*
2   On 1993 and earlier models, remove the air cleaner assembly and the throttle body unit. On 1994 and later models, remove the air intake duct, throttle body, fuel rails and injectors (see Chapter 4).
3   Remove the alternator rear brace **(see illustration)**. Remove the air conditioning compressor brace, if equipped.
4   Unbolt and remove the accelerator cable bracket and cable, and the cruise control cable adjuster (see Chapter 4).

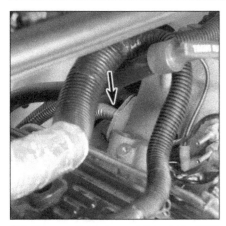

7.7 Remove the bolts and detach the EGR pipe from the back of the intake manifold

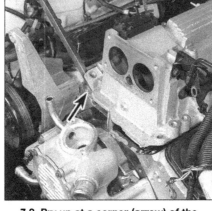

7.8 Pry up at a corner (arrow) of the manifold, being careful not to gouge the soft aluminum in a gasket sealing area

7.9 After covering the lifter valley, use a gasket scraper to remove all traces of sealant and old gasket material from the cylinder head and intake manifold mating surfaces

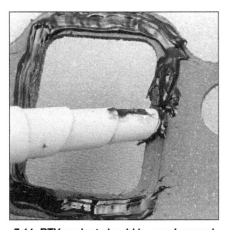

7.11 RTV sealant should be used around the coolant passage holes in the new intake manifold gaskets

7.12 Be sure to install the gaskets with the marks UP!

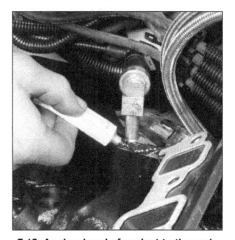

7.13 Apply a bead of sealant to the ends (ridges) of the block

5   Label and then disconnect any fuel lines, wires and vacuum hoses from the vehicle to the intake manifold. Lay the wiring harnesses on each side away from the manifold. Disconnect the coolant hoses from the intake manifold. Remove any brackets attached to the intake manifold.

6   On 1993 and earlier models, remove the distributor and ignition coil/bracket (see Chapter 5).

7   On 1994 and later models, unbolt the EGR pipe from the back of the intake manifold **(see illustration)**.

8   Loosen the manifold mounting bolts/studs in 1/4-turn increments until they can be removed by hand. **Note:** *Mark the manifold to indicate the location of the four studs.* The manifold will probably be stuck to the cylinder heads and force may be required to break the gasket seal. A large prybar can be positioned under the cast-in lug near the thermostat housing to pry up the front of the manifold **(see illustration)**. **Caution:** *Do not pry between the block and manifold or the heads and manifold or damage to the gasket sealing surfaces may result and vacuum leaks could develop.*

Installation

Refer to illustrations 7.9, 7.11, 7.12, 7.13 and 7.16

Note: *The mating surfaces of the cylinder heads, block and manifold must be perfectly clean when the manifold is installed. Gasket removal solvents are available at most auto parts stores and may be helpful when removing old gasket material that is stuck to the heads and manifold. Be sure to follow the directions printed on the container.*

9   Use a gasket scraper to remove all traces of sealant and old gasket material, then wipe the mating surfaces with a cloth saturated with lacquer thinner or acetone. If there is old sealant or oil on the mating surfaces when the manifold is installed, oil or vacuum leaks may develop. Cover the lifter valley with shop rags to keep debris out of the engine **(see illustration)**. Use a vacuum cleaner to remove any gasket material that falls into the intake ports in the heads.

10   Use a tap of the correct size to chase the threads in the bolt holes, then use compressed air (if available) to remove the debris from the holes. **Warning:** *Wear safety glasses or a face shield to protect your eyes when using compressed air.*

11   Apply a thin coat of RTV sealant **(see illustration)** around the coolant passage holes on the cylinder head side of the new intake manifold gaskets (there is normally one hole at each end).

12   Position the gaskets on the cylinder heads. Make sure all intake port openings, coolant passage holes and bolt holes are aligned correctly and the THIS SIDE UP is visible **(see illustration)**.

13   Apply a bead of RTV sealant at the ends of the block. Apply a 3/16-inch bead of sealant to the front and rear of the block as shown **(see illustration)**. Extend the bead 1/2-inch up each cylinder head to seal and retain the gaskets. Refer to the instructions with the gasket set for further information.

14   Carefully set the manifold in place. Do not disturb the gaskets and do not move the manifold fore-and-aft after it contacts the front and rear seals.

15   Apply a thin coat of a non-hardening sealant to the manifold bolt threads, then install the bolts.

16   While the sealant is still wet, tighten the bolts to the initial torque listed in this Chap-

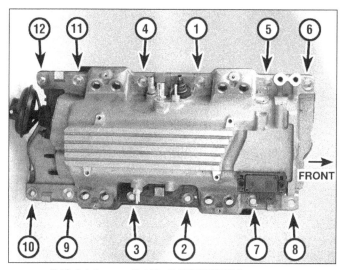

7.16 Intake manifold bolt TIGHTENING sequence

8.4 Unbolt the exhaust pipe-to-manifold flanges (arrow)

ter's Specifications following the recommended sequence **(see illustration)**, then tighten them to the final torque listed in this Chapter's Specifications, in sequence.

17 The remaining installation steps are the reverse of removal. Start the engine and check carefully for oil, vacuum and coolant leaks at the intake manifold joints.

8 Exhaust manifolds - removal and installation

Refer to illustrations 8.4 and 8.6
Warning: *Allow the engine to cool completely before following this procedure.*

Removal

1 Disconnect the negative cable from the battery. **Caution:** *On models equipped with a Delco Loc II or Theftlock audio system, be sure the lockout feature is turned off before performing any procedure which requires disconnecting the battery.*
2 Label each spark plug wire and remove them from the spark plugs. Position the wires aside.
3 Set the parking brake and block the rear wheels. Raise the front of the vehicle and support it securely on jackstands.
4 Disconnect the electrical connectors from the oxygen sensor in each manifold. Unbolt the exhaust pipe from each exhaust manifold **(see illustration)**. Lower the vehicle.

Right manifold

5 Remove the serpentine drivebelt (see Chapter 1), and remove the alternator and lower brace (see Chapter 5). Remove the air conditioning brace, if equipped.
6 Disconnect the air injection pipe from the manifold **(see illustration)**. **Note:** *Use penetrating oil on the air injection fittings and allow it to soak in before trying to remove the fittings.*

8.6 Use care not to round off the fittings when removing air injection pipes (arrow) from the manifolds

7 Remove the remaining manifold bolts/nuts and remove the manifold.

Left manifold

8 Disconnect the air injection pipe from the manifold. **Note:** *Use penetrating oil on the air injection fittings and allow it to soak in before trying to remove the fittings.*
9 Remove the brake booster vacuum hose (see Section 4).
10 Remove the remaining manifold bolts/nuts and remove the manifold.

Installation

11 Installation is the reverse of the removal procedure. Clean the manifold and head gasket surfaces to remove old gasket material, then install new gaskets. Do not use any gasket cement or sealant on exhaust system gaskets. **Note:** *The exhaust manifold gaskets are sheet metal and should only go on one way; if put on incorrectly, stamped writing on the gasket will read "installed wrong".* Apply anti-seize compound to the exhaust manifold-to-exhaust pipe studs.

12 Install all the manifold bolts and tighten them to the torque listed in this Chapter's Specifications. Work from the center out and approach the final torque in three steps.

9 Crankshaft front oil seal - replacement

Refer to illustrations 9.4, 9.6a, 9.6b, 9.7 and 9.9

1 Disconnect the negative cable from the battery. **Caution:** *On models equipped with a Delco Loc II or Theftlock audio system, be sure the lockout feature is turned off before performing any procedure which requires disconnecting the battery.*
2 Remove the serpentine drivebelt belt (see Chapter 1). On 1993 and earlier models, remove the cooling fan and shroud. Raise the vehicle and support it securely on jackstands.
3 On 1993 and earlier models, remove the crankshaft pulley. Using a puller that bolts to the center hub, remove the balancer **(see illustrations)**.

9.3a Remove the three bolts retaining the pulley to the balancer, then remove the center bolt (1993 and earlier models)

9.3b Use the recommended puller to remove the balancer or damage to the balancer may result (1993 and earlier models)

9.4 Use a two-pin spanner to hold the balancer while the three balancer bolts are removed (arrows) note the cast-in TDC arrow on the balancer hub (at top, TDC location) (1994 and later models)

9.6a The portion of the crank snout exposed inside the hub is too small for a standard puller's beveled tip to fit into, so use a long Allen bolt that fits inside the crank threads (not threaded into them, but smaller) - the puller will then press against the socket-head of the Allen bolt - arrows indicate the TDC mark on the hub and the cover rib it must align with, since the hub has no keyway (1994 and later models)

9.6b Most standard pullers will have to be adapted with four 5/16-inch bolts and nuts to attach to the balancer hub, with the puller's beveled tip fitting into the Allen bolt (1994 and later models)

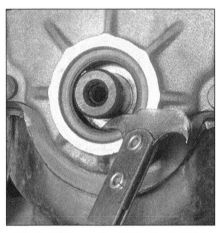

9.7 Carefully pry the old seal out of the front cover - don't damage the crankshaft surface

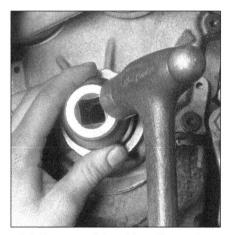

9.9 Use a special seal-installing tool or drive the seal in carefully and evenly with a large socket and hammer

4   On 1994 and later models, remove the three balancer bolts and the balancer (see illustration). A puller is not required to remove the balancer.

5   On 1994 and later models, apply matchmarks on the balancer hub and the front cover, and remove the balancer hub bolt. Caution: *Once the matchmarks have been applied, do not turn the crankshaft until the balancer has been reinstalled.* When loosening this bolt, prevent the crankshaft from turning by wedging a large screwdriver into the ring gear teeth of the driveplate.

6   On 1994 and later models, use a bolt-type puller to remove the balancer hub (see illustrations). Because the bolt holes in the hub are larger than most standard pullers will allow, you will have to use three, four-inch-long (grade 8) 5/16-inch bolts and nuts to use a standard puller, along with a four-inch-long (grade 8) 5/16-inch Allen bolt to fit into the crank snout. Because of the clearance around the distributor, it is safest to

remove the water pump (see Chapter 3) and distributor (see Chapter 5) first, which also exposes the portion of the front cover with the "TDC" rib for aligning the keyless hub when reinstalling it.

7   Carefully pry the seal out of the cover with a seal removal tool or a large screwdriver (see illustration). Be careful not to distort the cover or scratch the wall of the seal bore.

8   Clean the bore to remove any old seal material and corrosion. Position the new seal in the bore with the open end of the seal facing IN. A small amount of oil applied to the outer edge of the new seal will make installation easier - don't overdo it!

9   Drive the seal into the bore with a special tool, available at most auto parts stores, or a large socket and hammer until it's completely seated (see illustration). Select a socket that's the same outside diameter as the seal.

10   Lubricate the seal lips with engine oil and reinstall the balancer hub, lining up the matchmarks. The remaining installation steps

are the reverse of removal. Note: *On 1994 and later models, if the balancer is being replaced, be sure to transfer the bolt-on balancer weights from the old balancer to the same positions on the new balancer.*

---

10   Front cover, timing chain and sprockets - removal, inspection and installation

Removal

Refer to illustrations 10.5a, 10.5b, 10.6 and 10.8

1   Refer to Chapter 3 and remove the water pump.

2   Remove the crankshaft balancer and balancer hub (see Section 9).

10.5a On spline-type drives, paint a match mark to line up the splined shaft sticking out of the front cover for faster assembly, although it will only go back into the cover one way (1994 and later models)

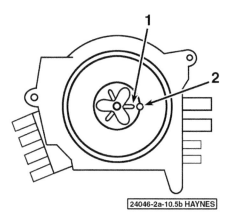

10.5b The later-type distributor drive uses a long pin on the camshaft which fits into a slot (1) on the back of the distributor - align it with the mark (2) for TDC assembly

10.6 Remove the front cover bolts and remove the cover

3   Refer to Section 3 and position the number six piston at TDC on the compression stroke on 1993 and earlier models, or position the number one piston at TDC on the compression stroke on 1994 and later models. **Caution:** *Once this has been done, do not turn the crankshaft until the timing chain and sprockets have been reinstalled.*

4   The oil pan bolts will have to be loosened and the pan lowered slightly for the front cover to be removed. If the pan has been in place for an extended period of time it's likely the pan gasket will break when the pan is lowered. In this case the pan should be removed and a new gasket installed (see Section 13).

5   On 1994 and later models, remove the distributor (see Chapter 5). There are two types of distributor drives on the V8 engines: the spline-type or "type 1", and the pin-type or "type 2" (on 1995 and later models). On the spline-type, the distributor is driven by a splined shaft coming out of the front cover and driven by the camshaft gear. On these models, the short splined shaft may come out of the front cover when the distributor is removed. It will only go back in one way and the distributor will only go back onto it one way, due to key alignment splines. These distributors should not be turned while off the engine. On the later models, the camshaft has a long pin sticking out, which meshes with a slot in the backside of the distributor. Alignment of these types is simpler **(see illustrations)**.

6   Unbolt and remove the front cover **(see illustration)**.

7   On 1996 models, remove the crankshaft position sensor reluctor ring, which is a disc over the crankshaft snout, noting which side faces the engine.

8   Remove the three bolts from the end of the camshaft, then detach the camshaft sprocket and chain as an assembly **(see illustration)**.

10.8 Unbolt and remove the camshaft sprocket, lowering it to remove the sprocket and chain as an assembly

9   Using a jaw-type puller, remove the crankshaft sprocket.

Inspection
Refer to illustration 10.11

10   The timing chain should be replaced with a new one if the engine has high mileage, the chain has visible damage, or total freeplay midway between the sprockets exceeds one-inch. Failure to replace a worn timing chain may result in erratic engine performance, loss of power and decreased fuel mileage. Loose chains can "jump" timing. In the worst case, chain "jumping" or breakage will result in severe engine damage.

11   Inspect the timing sprockets and water pump drive gear (on the back of the cam gear) for worn, non-concentric teeth or wear in the valleys between the teeth **(see illustration)**.

Installation
Refer to illustration 10.15

12   Use a gasket scraper to remove all traces of old gasket material and sealant from

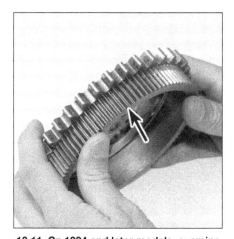

10.11 On 1994 and later models, examine the camshaft sprocket teeth and the integral water pump drive gear teeth (arrow) on the back for wear

the cover and engine block. Stuff a shop rag into the opening at the front of the oil pan to keep debris out of the engine. Clean the cover and block sealing surfaces with lacquer thinner or acetone. **Caution:** *Be careful when scraping the front cover; gouges in the soft aluminum could result in oil leaks.*

13   Replace the O-ring on the outer end of the water pump driveshaft (see Chapter 3).

14   The timing chain must be replaced as a set with the camshaft and crankshaft sprockets. Never put a new chain on old sprockets. Align the sprocket with the Woodruff key and press the sprocket onto the crankshaft with a large socket or tap it gently into place until it is completely seated. **Caution:** *If resistance is encountered, do not hammer the sprocket onto the crankshaft. It may eventually move onto the shaft, but it may be cracked in the process and fail later, causing extensive engine damage.*

15   Loop the new chain over the camshaft sprocket, then turn the sprocket until the timing mark is in the 6 o'clock position

10.15 Align the timing marks (arrows) before bolting the camshaft sprocket in place - larger arrow indicates the splined distributor driveshaft in place on the cam gear

11.3 When checking the camshaft lobe lift, the dial indicator plunger must be positioned directly above and in-line with the pushrod

11.10 Remove the three bolts retaining the lifter guide retainer to the engine block

(see illustration). Mesh the chain with the crank-shaft sprocket and position the cam-shaft sprocket on the end of the cam. If necessary, turn the camshaft so the dowel pin fits into the sprocket hole with the tim-ing mark in the 6 o'clock position. **Note:** *If necessary, turn the water pump driveshaft slightly to mesh its gear with the water pump drive gear on the back of the cam-shaft sprocket.*

16 Apply non-hardening thread locking compound to the camshaft sprocket bolt threads, then install and tighten them to the torque listed in this Chapter's Specifications. Lubricate the chain with clean engine oil.

17 The oil pan gasket should be checked for cracks and deformation before installing the front cover. If the gasket has deteriorated it must be replaced before reinstalling the cover.

18 Apply a thin layer of RTV sealant to both sides of the new cover gasket, then position it on the engine block. The dowel pins and sealant will hold it in place. **Caution:** *Wrap some plastic tape around the splines on the water pump driveshaft to protect the cover seal during installation.*

19 Install the front cover on the block, tight-ening the bolts finger-tight.

20 If the oil pan was removed, reinstall it (see Section 13). If it was only loosened, tighten the oil pan bolts, bringing the oil pan up against the front cover.

21 Tighten the front cover bolts to the torque listed in this Chapter's Specifications.

22 Lubricate the oil seal contact surface of the crankshaft balancer hub with multi-pur-pose grease or clean engine oil, then install it on the end of the crankshaft. The keyway in the balancer hub must be aligned with the Woodruff key in the crankshaft nose. If the hub cannot be seated by hand, slip a large washer over the bolt, install the bolt and tighten it to push the hub into place. Remove the large washer and tighten the bolt to the torque listed in this Chapter's Specifications.

23 If plastic tape was used to install the water pump driveshaft O-ring (see Chapter 3) remove the plastic tape from the water pump driveshaft before installing the water pump. The remaining installation steps are the reverse of removal. **Caution:** *On 1994 and later models, when rein-stalling spline-type distributors, the distributor must go on smoothly and without much effort. It is possible to install the spline-type with improper alignment and force it on and tighten the bolts, but this will cause expensive damage to the parts. Wiggle the distributor slightly until you feel the spline actually engage fully before bolting down the distributor.*

---

11 Camshaft and lifters - removal, inspection and installation

---

Camshaft lobe lift check

Refer to illustration 11.3

1 To determine the extent of cam lobe wear, the lobe lift should be checked prior to camshaft removal. Refer to Section 4 and remove the valve covers.

2 Position the number one piston at TDC on the compression stroke (see Section 3).

3 Beginning with the number one cylinder, mount a dial indicator on the engine and posi-tion the plunger against the top surface of the first rocker arm. The plunger should be directly above and in-line with the pushrod **(see illustration)**.

4 Zero the dial indicator, then very slowly turn the crankshaft in the normal direction of rotation until the indicator needle stops and begins to move in the opposite direction. The point at which it stops indicates maximum cam lobe lift.

5 Record this figure for future reference, then reposition the piston at TDC on the com-pression stroke.

6 Move the dial indicator to the other number one cylinder rocker arm and repeat the check. Be sure to record the results for each valve.

7 Repeat the check for the remaining valves. Since each piston must be at TDC on the compression stroke for this procedure, work from cylinder-to-cylinder following the firing order sequence. Turn the crankshaft 90-degrees to reach TDC for each of the cyl-inders after Number 1.

8 After the check is complete, compare the results to this Chapter's Specifications. If camshaft lobe lift is less than specified, cam lobe wear has occurred and a new camshaft should be installed.

Removal

Refer to illustrations 11.10, 11.11 and 11.12

9 Refer to the appropriate Sections and remove the intake manifold, the rocker arms, the pushrods and the timing chain and cam-shaft sprocket. The radiator must be removed as well (see Chapter 3). **Note:** *If the vehicle is equipped with air conditioning it will be nec-essary to remove the air conditioning con-denser. If the condenser must be removed, the system must first be evacuated by a dealer service department or automotive air conditioning service facility, and the refriger-ant recovered. Do not disconnect any air con-ditioning lines until the system has been prop-erly evacuated.*

10 In the engine's lifter valley, remove the bolts securing the lifter guide retainer and remove the retainer **(see illustration)**. Remove the lifter guides (one guide for every pair of lifters) and keep them in a clearly labeled box to ensure that they are reinstalled in their original locations, along with their cor-responding lifters.

11 There are several ways to extract the lift-ers from the bores. A special tool designed to grip and remove lifters is manufactured by many tool companies and is widely available, but it may not be required in every case **(see illustration)**. On newer engines without a lot of varnish buildup, the lifters can often be removed with a small magnet or even with your fingers. A machinist's scribe with a bent end can be used to pull the lifters out by posi-

11.11 A lifter removal tool may be necessary to remove the lifters on an engine with high-mileage - store the lifters in an organized container

11.12 Remove the camshaft retainer plate

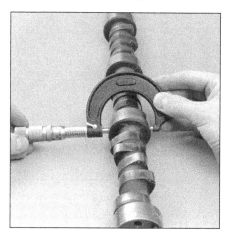

11.15 The camshaft bearing journal diameter is checked to pinpoint excessive wear and out-of-round conditions

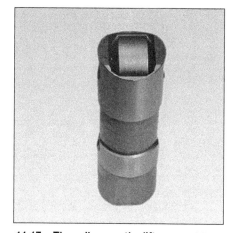

11.17a The rollers on the lifters must turn freely - check for wear and excessive play as well

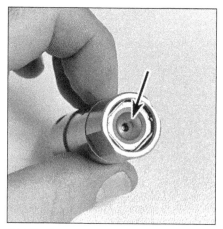

11.17b Check the pushrod seat (arrow) in the top of each lifter for wear

tioning the point under the retainer ring inside the top of each lifter. **Caution:** *Do not use pliers to remove the lifters unless you intend to replace them with new ones (along with the camshaft). The pliers will damage the precision machined and hardened lifters, rendering them useless.*

12   Remove the camshaft retainer plate **(see illustration)**. On 1994 and later models, remove the oil pump drive from inside the lifter valley at the right rear.

13   Thread six-inch long 5/16-18 bolts into the camshaft sprocket bolt holes to use as a handle when removing the camshaft from the block and carefully pull the camshaft out. Support the cam near the block so the lobes do not nick or gouge the bearings as the cam is withdrawn.

Inspection

Camshaft and bearings

Refer to illustrations 11.15, 11.17a and 11.17b

14   After the camshaft has been removed from the engine, cleaned with solvent and dried, inspect the bearing journals for uneven wear, pitting and evidence of seizure. If the

journals are damaged, the bearing inserts in the block are probably damaged as well. Both the camshaft and bearings will have to be replaced. Replacement of the camshaft bearings requires special tools and techniques which place it beyond the scope of the home mechanic. The block will have to be removed from the vehicle and taken to an automotive machine shop for this procedure.

15   Measure the bearing journals with a micrometer to determine if they are excessively worn or out-of-round **(see illustration)**.

16   Check the camshaft lobes for heat discoloration, score marks, chipped areas, pitting and uneven wear. If the lobes are in good condition and if the lobe lift measurements are as specified, the camshaft can be re-used.

17   Check the lifter rollers carefully for wear and damage and make sure they turn freely without excessive play **(see illustrations)**.

18   Used roller lifters can be reinstalled with a new camshaft and the original camshaft can be used if new lifters are installed, but the factory does recommend replacing both lifters and cam at the same time, especially if the engine has high mileage.

Installation

Refer to illustration 11.19

19   Lubricate the camshaft bearing journals and cam lobes with camshaft installation lubricant **(see illustration)**.

20   Slide the camshaft into the engine. Support the cam near the block and be careful

11.19 Coat the lobes and journals with special camshaft installation lube

12.6 Remove these two banjo bolts (arrows) holding this water transfer tube to the back of the heads (shown removed from the vehicle for clarity)

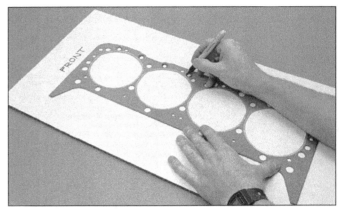

12.7 To avoid mixing up the head bolts, use a new gasket to transfer the bolt hole pattern to a piece of cardboard, then punch holes to accept the bolts

not to scrape or nick the bearings.

21   Position the camshaft retainer plate and tighten the bolts to the torque listed in this Chapter's Specifications.

22   Refer to Section 10 and install the timing chain and sprockets.

23   Lubricate the lifters with clean engine oil and install them in the block. If the original lifters are being reinstalled, be sure to return them to their original locations. If a new camshaft was installed, be sure to install new lifters as well.

24   Reinstall the oil pump drive at the back of the lifter valley and tighten it's bolt. The remaining installation steps are the reverse of removal.

25   Before starting and running the engine, change the oil and install a new oil filter (see Chapter 1).

12 Cylinder heads - removal and installation

Removal

Refer to illustrations 12.6, 12.7 and 12.17

1   Disconnect the negative cable from the battery. **Caution:** *On models equipped with a Delco Loc II or Theftlock audio system, be sure the lockout feature is turned off before performing any procedure which requires disconnecting the battery.*

2   Refer to Section 4 and remove the valve covers. Refer to Section 7 and remove the intake manifold. Note that the cooling system must be drained (see Chapter 1) to prevent coolant from getting into internal areas of the engine when the manifold and heads are removed. Refer to Section 5 and remove the rocker arms and pushrods.

3   Remove the ignition coil (see Chapter 5), spark plugs and disconnect the spark plug wire harnesses from their clips.

4   Refer to Section 8 and detach both exhaust manifolds.

5   Remove the alternator, power steering pump and air conditioning compressor, without disconnecting the hoses and position the accessories aside. Remove the brackets

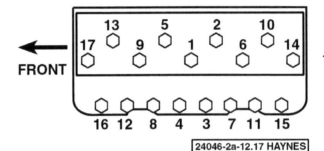

12.17 Cylinder head bolt TIGHTENING sequence

24046-2a-12.17 HAYNES

attached to the cylinder heads.

6   On 1994 and later models, at the rear of the cylinder heads there is a tubular steel water pipe connecting the back of both heads **(see illustration)**. It is retained to each head with a "banjo" bolt, which can be difficult to reach, as they are very close to the firewall. However, it is important to get a firm purchase on these bolts when removing them, to avoid rounding them off.

7   Loosen the head bolts in 1/4-turn increments until they can be removed by hand. Work from bolt-to-bolt in a pattern that's the reverse of the tightening sequence **(see illustration 12.17)**. **Note:** *Don't overlook the row of bolts on the lower edge of each head, near the spark plug holes. Store the bolts in the cardboard holder as they're removed. This will ensure the bolts are reinstalled in their original holes* **(see illustration)**.

8   Lift the heads off the engine. If resistance is felt, do not pry between the head and block as damage to the mating surfaces will result. To dislodge the head, place a block of wood against the end of it and strike the wood block with a hammer. Store the heads on blocks of wood to prevent damage to the gasket sealing surfaces.

9   Cylinder head disassembly and inspection procedures are covered in detail in Chapter 2, Part B.

Installation

10   The mating surfaces of the cylinder heads and block must be perfectly clean

when the heads are installed.

11   Use a gasket scraper to remove all traces of carbon and old gasket material, then clean the mating surfaces with lacquer thinner or acetone. If there's oil on the mating surfaces when the heads are installed, the gaskets may not seal correctly and leaks could develop. When working on the block, cover the lifter valley with shop rags to keep debris out of the engine. Use a vacuum cleaner to remove any debris that falls into the cylinders.

12   Check the block and head mating surfaces for nicks, deep scratches and other damage. If damage is slight, it can be removed with a file. If it's excessive, machining may be the only alternative.

13   Use a tap of the correct size to chase the threads in the head bolt holes in the block. Mount each bolt in a vise and run a die down the threads to remove corrosion and restore the threads. Dirt, corrosion, sealant and damaged threads will affect torque readings.

14   Position the new gaskets over the dowel pins in the block. The yellow tabs should be UP.

15   Carefully position the heads on the block without disturbing the gaskets.

16   Before installing the head bolts, coat the threads with a non-hardening sealant.

17   Install the bolts in their original locations and tighten them finger-tight. Following the recommended sequence, tighten the bolts in several steps to the torque listed in this Chap-

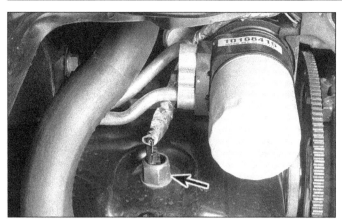

13.8 Disconnect the electrical connection from the oil level sensor (arrow), then remove the sensor - the lower bellhousing cover has been removed here also

13.13 When removing the oil pan bolts/nuts, keep track of where the studs go (two are shown here with arrows) - do not forget the metal reinforcement strips (between the bolts/nuts and the pan) during reassembly

ter's Specifications **(see illustration)**.
18   The remaining installation steps are the reverse of removal.

13   Oil pan - removal and installation

Refer to illustrations 13.8 and 13.13

Removal

1   Disconnect the negative cable from the battery. **Caution:** *On models equipped with a Delco Loc II or Theftlock audio system, be sure the lockout feature is turned off before performing any procedure which requires disconnecting the battery.*
2   Remove the air cleaner assembly/resonator. Remove the cooling fan and shroud. On 1993 and earlier models, remove the distributor cap. Disconnect the electrical connector from the windshield wiper motor.
3   Raise the vehicle and support it securely on jackstands. Drain the engine oil and remove the oil filter (see Chapter 1) and the oil dipstick.
4   On 1993 and earlier models, remove the exhaust crossover pipe. On 1994 and later models, Disconnect the pre-catalytic converters from the exhaust manifolds and the support under the transmission bellhousing.
5   Remove the lower bellhousing cover.
6   Remove the starter, if necessary for clearance (see Chapter 5).
7   Remove the transmission fluid lines from the clip at the oil pan.
8   Disconnect the electrical connector from the oil level sensor in the pan, and remove the sensor **(see illustration)**. **Note:** *The sensor can be damaged if left in the pan during pan removal.*
9   Attach an engine hoist to the engine from above and raise the engine slightly to allow removal of the engine mount through-bolts.
10   Raise the engine approximately three inches. **Caution:** *When lifting the engine, check that no upper engine components or wire harnesses are hitting the edge of the cowl.*

11   Place blocks of wood between the crossmember and the engine block in the area of the engine mounts to hold the engine in the raised position.
12   Turn the crankshaft until the timing mark on the balancer is at the 6 o'clock position.
13   Remove the oil pan bolts **(see illustration)** and reinforcements. Note that studs and nuts are used in some positions, and mark their locations.
14   Remove the pan by tilting the rear down and working it away from the crankshaft throws, oil pump pick-up and front crossmember.

Installation

15   Clean the sealing surfaces with lacquer thinner or acetone. Make sure the bolt holes in the block are clean.
16   Check the oil pan flange for distortion, particularly around the bolt holes. If necessary, place the pan on a block of wood and use a hammer to flatten and restore the gasket surface.
17   Apply a one-inch long bead of RTV sealant to the corners of the timing cover-to-block and rear oil seal housing-to-block junctions.
18   The one-piece rubber gasket should be checked carefully and replaced with a new

one if damage is noted. Apply a small amount of RTV sealant to the corners of the semi-circular cutouts at both ends of the pan, then attach the rubber gasket to the pan.
19   Carefully position the pan against the block and install the bolts/nuts finger-tight (don't forget the reinforcement strips). Make sure the seals and gaskets haven't shifted, then tighten the bolts/nuts in three steps to the torque listed in this Chapter's Specifications. Start at the center of the pan and work out toward the ends in a spiral pattern.
20   The remaining steps are the reverse of removal. **Caution:** *Don't forget to refill the engine with oil before starting it (see Chapter 1).*

14   Oil pump - removal and installation

Refer to illustrations 14.2 and 14.3

1   Remove the oil pan as described in Section 13.
2   While supporting the oil pump, remove the pump-to-rear main bearing cap bolt **(see illustration)**.
3   Lower the pump and remove it along with the pump driveshaft. Note that on most

14.2 Remove the single bolt holding the oil pump to the rear main cap and remove the oil pump

14.3 Make sure the nylon sleeve is in place between the oil pump and driveshaft

15.4 The driveplate only goes on one way because the crankshaft has a long locator pin (arrow) corresponding to a locator hole in the driveplate

16.2 The rear main oil seal retainer has notches where a screwdriver can be used to pry out the old seal

models a hard nylon sleeve is used to align the oil pump driveshaft and the oil pump shaft. Make sure this sleeve is in place on the oil pump driveshaft **(see illustration)**. If it is not there, check the oil pan for the pieces of the sleeve, clean them out of the pan, then get a new sleeve for the oil pump driveshaft.

4   If a new oil pump is installed, make sure the pump driveshaft is mated with the shaft inside the pump.

5   Position the pump on the engine and make sure the slot in the upper end of the driveshaft is aligned with the tang on the lower end of the distributor/oil pump drive. It is absolutely essential that the components mate properly.

6   Install the mounting bolt and tighten it to the torque listed in this Chapter's Specifications.

7   Install the oil pan.

15   Driveplate - removal and installation

Refer to illustration 15.4

1   Raise the vehicle and support it securely on jackstands.

2   Refer to Chapter 7 and remove the transmission.

3   Use paint or a center-punch to make alignment marks on the driveplate and crankshaft to ensure correct alignment during reinstallation.

4   Remove the bolts that secure the driveplate to the crankshaft **(see illustration)**. If the crankshaft turns, hold the driveplate with a pry bar or wedge a screwdriver into the ring gear teeth to jam the driveplate.

5   Remove the driveplate from the crankshaft. Since the driveplate is fairly heavy, be sure to support it while removing the last bolt.

6   Clean the driveplate to remove grease and oil. Inspect the surface for cracks. Check for cracked and broken ring gear teeth. Lay the driveplate on a flat surface and use a straightedge to check for warpage.

7   Clean and inspect the mating surfaces of the driveplate and the crankshaft. If the crankshaft rear seal is leaking, replace it before reinstalling the driveplate.

8   Position the driveplate against the crankshaft. Be sure to align the marks made during removal. Note that some engines have an alignment dowel or staggered bolt holes to ensure correct installation. Before installing the bolts, apply thread locking compound to the threads.

9   Wedge a screwdriver into the ring gear teeth to keep the driveplate from turning. Using a criss-cross tightening sequence, torque the bolts to the Specifications listed in this Chapter.

10   The remainder of installation is the reverse of the removal procedure.

16   Rear main oil seal - replacement

Refer to illustration 16.2

1   The one-piece rear main oil seal is installed in a bolt-on housing. Replacing this seal requires removal of the transmission and driveplate. Refer to Chapter 7 for the transmission removal procedures and Section 15 of this Chapter for driveplate removal.

2   Insert a screwdriver blade into the notches in the seal housing and pry out the old seal **(see illustration)**. Be sure to note how far it's recessed into the housing bore before removal so the new seal can be installed to the same depth. Although the seal can be removed this way, installation with the housing still mounted on the block requires the use of a special tool, which attaches to the threaded holes in the crankshaft flange and then presses the new seal into place.

3   If the special installation tool is not available, remove the oil pan (see Section 13) and the bolts securing the housing to the block, then detach the housing and gasket. Whenever the housing is removed from the block a new seal and gasket must be installed. **Note:**

The oil pan must be removed because the two rearmost oil pan fasteners are studs that are part of the rear seal retainer, and the pan must come away enough to clear these studs to allow retainer removal and installation.

4   Clean the housing thoroughly, then apply a thin coat of engine oil to the new seal. Set the seal squarely into the recess in the housing, then, using two pieces of wood, one on each side of the housing, use a hammer to press the seal into place.

5   Carefully slide the seal over the crankshaft and bolt the seal housing to the block. Be sure to use a new gasket, but don't use any gasket sealant.

6   The remainder of installation is the reverse of the removal procedure.

17   Engine mounts - check and replacement

1   Engine mounts seldom require attention, but broken or deteriorated mounts should be replaced immediately or the added strain placed on the driveline components may cause damage or wear.

Check

2   During the check, the engine must be raised slightly to remove the weight from the mounts.

3   Raise the vehicle and support it securely on jackstands, then position a jack under the engine oil pan. Place a large block of wood between the jack head and the oil pan, then carefully raise the engine just enough to take the weight off the mounts. **Warning:** *DO NOT place any part of your body under the engine when it's supported only by a jack!*

4   Check the mounts to see if the rubber is cracked, hardened or separated from the metal plates. Sometimes the rubber will split right down the center.

5   Check for relative movement between the mount plates and the engine or frame (use a large screwdriver or pry bar to attempt to move the mounts). If movement is noted, lower the engine and tighten the mount fasteners.

6   Rubber preservative should be applied to the mounts to slow deterioration.

Replacement

7   Disconnect the negative battery cable from the battery, then raise the vehicle and support it securely on jackstands (if not already done).

8   Loosen the nut on the through bolt and remove the bolt and nut that secure the mount to the frame bracket.

9   Raise the engine slightly with a jack or hoist (make sure the fan doesn't hit the radiator or shroud). Remove the through bolt and nut and detach the mount.

10   Installation is the reverse of removal. Use thread locking compound on the mount bolts and be sure to tighten them securely.

Chapter 2 Part B
General engine overhaul procedures

Contents

Specifications

General

Displacement
4.3L	265 cubic inches
5.0L	305 cubic inches
5.7L	350 cubic inches
Cylinder compression pressure	100 psi minimum
Maximum variation between cylinders	30-percent

Oil pressure
@1000 RPM	6 psi minimum
@2000 RPM	18 psi minimum
@4000 RPM	24 psi minimum

Cylinder head

Warpage limit	0.003 inch per 6 inch span/0.006 inch overall
Maximum allowable cut	0.010 inch

Valves and related components

Seat angle	46-degrees
Face angle	45-degrees

Stem-to-guide clearance
Standard	0.0010 to 0.0027 inch

Service limit
Intake	0.0037 inch
Exhaust	0.0047 inch
Spring free length	2.02 inches
Spring installed height	1.70 inches

Crankshaft and connecting rods

Connecting rod journal

Diameter
1993 and earlier	2.0893 to 2.0998 inches
1994 and later	2.0978 to 2.0998 inches
Bearing oil clearance	0.0013 to 0.0035 inch
Taper limit	0.0010 inch
Out-of-round limit	0.0010 inch
Connecting rod side clearance	0.006 to 0.014 inch

Crankshaft and connecting rods (continued)

Main bearing journal
 Diameter
  1993 and earlier
   Journal no. 1 ....   2.4488 to 2.4493 inches
   Journals no. 2, 3 and 4 ....   2.4481 to 2.4490 inches
   Journal no. 5 ....   2.4481 to 2.4488 inches
  1994 and later ....   2.4485 to 2.4491 inches
 Bearing oil clearance
  Journal no. 1 ....   0.0010 to 0.0015 inch
  Journals no. 2, 3 and 4 ....   0.0010 to 0.0025 inch
  Journal no. 5 ....   0.0025 to 0.0035 inch
 Taper limit ....   0.0010 inch
 Out-of-round limit ....   0.0010 inch
Crankshaft endplay ....   0.0010 to 0.0070 inch

Engine block

Cylinder bore
 Diameter
  4.3L ....   3.7355 to 3.7385 inches
  5.0L ....   3.7350 to 3.7385 inches
  5.7L ....   4.0007 to 4.0017 inches
 Taper limit ....   0.0010 inch
 Out-of-round limit ....   0.0020 inch

Pistons and rings

Piston-to-bore clearance
 Standard
  1993 and earlier ....   0.0005 to 0.0022 inch
  1994 and later ....   0.0010 to 0.0027 inch
 Service limit ....   0.0027 inch
Piston ring end gap
 Compression rings
  Standard
   Top ring ....   0.010 to 0.020 inch
   Second ring ....   0.018 to 0.026 inch
  Service limit ....   0.035 inch
 Oil control ring
  Standard
   1993 and earlier ....   0.015 to 0.055 inch
   1994 and later ....   0.010 to 0.030 inch
  Service limit ....   0.065 inch
Piston ring side clearance
 Compression rings
  Standard
   1993 and earlier ....   0.0012 to 0.0032 inch
   1994 and later ....   0.0012 to 0.0029 inch
  Service limit ....   0.0042 inch
 Oil control ring
  Standard ....   0.002 to 0.007 inch
  Service limit ....   0.008 inch

Torque specifications*   **Ft-lbs**

Connecting rod cap nuts
 1993 and earlier   44
 1994   47
 1995 and 1996
  Step 1   20
  Step 2 ....   Rotate an additional 55-degrees
Main bearing cap bolts/nuts   77

* **Note:** *Refer to Chapter 2 Part A for additional torque specifications.*

---

1   General information

Included in this portion of Chapter 2 are the general overhaul procedures for the cylinder heads and internal engine components.

The information ranges from advice concerning preparation for an overhaul and the purchase of replacement parts to detailed, step-by-step procedures covering removal and installation of internal engine components and the inspection of parts.

The following Sections have been written based on the assumption the engine has been removed from the vehicle. For information concerning in-vehicle engine repair, as well as removal and installation of the external components necessary for the overhaul, see Part A of this Chapter and Section 6 of this Part.

The Specifications included in this Part are only those necessary for the inspection and overhaul procedures which follow. Refer to Chapter 2 Part A for additional Specifications.

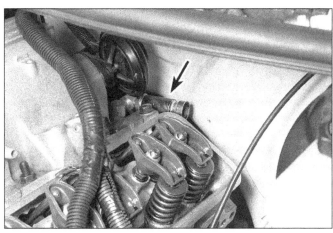

2.4 The oil pressure sending unit (arrow) is located behind the intake manifold

3.6 A compression gauge with a threaded fitting for the spark plug hole is preferred over the type that requires hand pressure to maintain the seal - be sure to open the throttle valve as far as possible during the compression check

2 Engine overhaul - general information

Refer to illustration 2.4

It's not always easy to determine when, or if, an engine should be completely overhauled, as a number of factors must be considered.

High mileage isn't necessarily an indication an overhaul is needed, while low mileage doesn't preclude the need for an overhaul. Frequency of servicing is probably the most important consideration. An engine that's had regular and frequent oil and filter changes, as well as other required maintenance, will most likely give many thousands of miles of reliable service. Conversely, a neglected engine may require an overhaul very early in its life.

Excessive oil consumption is an indication that piston rings, valve seals and/or valve guides are in need of attention. Make sure oil leaks aren't responsible before deciding the rings and/or guides are bad. Perform a cylinder compression check to determine the extent of the work required (see Section 3).

Remove the oil pressure sending unit **(see illustration)** and check the oil pressure with a gauge installed in its place. Compare the results to this Chapter's Specifications. As a general rule, engines should have ten psi oil pressure for every 1,000 rpm. If the pressure is extremely low, the bearings and/or oil pump are probably worn out.

Loss of power, rough running, knocking or metallic engine noises, excessive valve-train noise and high fuel consumption rates may also point to the need for an overhaul, especially if they're all present at the same time. If a complete tune-up doesn't remedy the situation, major mechanical work is the only solution.

An engine overhaul involves restoring the internal parts to the specifications of a new engine. During an overhaul, the piston rings are replaced and the cylinder walls are reconditioned (rebored and/or honed). If a rebore is done by an automotive machine shop, new oversize pistons will also be installed. The

main bearings and connecting rod bearings are generally replaced with new ones and, if necessary, the crankshaft may be reground to restore the journals. Generally, the valves are serviced as well, since they're usually in less-than-perfect condition at this point. While the engine is being overhauled, other components, such as the starter and alternator, can be rebuilt as well. The end result should be a like-new engine that will give many trouble free miles. **Note:** *Critical cooling system components such as the hoses, drivebelts, thermostat and water pump MUST be replaced with new parts when an engine is overhauled. The radiator should be checked carefully to ensure it isn't clogged or leaking (see Chapter 3). Also, we don't recommend overhauling the oil pump - always install a new one when an engine is rebuilt.*

Before beginning the engine overhaul, read through the entire procedure to familiarize yourself with the scope and requirements of the job. Overhauling an engine isn't particularly difficult, if you follow all of the instructions carefully, have the necessary tools and equipment and pay close attention to all specifications; however, it can be time consuming. Plan on the vehicle being tied up for a minimum of two weeks, especially if parts must be taken to an automotive machine shop for repair or reconditioning. Check on availability of parts and make sure any necessary special tools and equipment are obtained in advance. Most work can be done with typical hand tools, although a number of precision measuring tools are required for inspecting parts to determine if they must be replaced. Often an automotive machine shop will handle the inspection of parts and offer advice concerning reconditioning and replacement. **Note:** *Always wait until the engine has been completely disassembled and all components, especially the engine block, have been inspected before deciding what service and repair operations must be performed by an automotive machine shop. Since the block's condition will be the major factor to consider when determining whether to overhaul the original engine or buy a rebuilt*

one, *never purchase parts or have machine work done on other components until the block has been thoroughly inspected. As a general rule, time is the primary cost of an overhaul, so it doesn't pay to install worn or substandard parts.*

As a final note, to ensure maximum life and minimum trouble from a rebuilt engine, everything must be assembled with care in a spotlessly clean environment.

3 Cylinder compression check

Refer to illustration 3.6

1 A compression check will tell you what mechanical condition the upper end (pistons, rings, valves, head gaskets) of the engine is in. Specifically, it can tell you if the compression is down due to leakage caused by worn piston rings, defective valves and seats or a blown head gasket. **Note:** *The engine must be at normal operating temperature and the battery must be fully charged for this check.*

2 Begin by cleaning the area around the spark plugs before you remove them. Compressed air should be used, if available, otherwise a small brush or even a bicycle tire pump will work. The idea is to prevent dirt from getting into the cylinders as the compression check is being done.

3 Remove all of the spark plugs from the engine (see Chapter 1).

4 Block the throttle wide open.

5 Disable the fuel and ignition systems by removing the ECM IGN fuse from the instrument panel fuse block and the IGNITION fuse from the underhood fuse block (1993 and earlier models), the PCM IGN fuse from the instrument panel fuse block and the IGNITION fuse from the underhood fuse block (1994 and 1995 models), or the PCM BATT fuse from the instrument panel fuse block and the IGNITION fuse from the underhood fuse block (1996 models).

6 Install the compression gauge in the number one spark plug hole **(see illustration)**.

7   Crank the engine over at least seven compression strokes and watch the gauge. The compression should build up quickly in a healthy engine. Low compression on the first stroke, followed by gradually increasing pressure on successive strokes, indicates worn piston rings. A low compression reading on the first stroke, which doesn't build up during successive strokes, indicates leaking valves or a blown head gasket (a cracked head could also be the cause). Deposits on the undersides of the valve heads can also cause low compression. Record the highest gauge reading obtained.

8   Repeat the procedure for the remaining cylinders, turning the engine over for the same length of time for each cylinder, and compare the results to this Chapter's Specifications.

9   If the readings are below normal, add some engine oil (about three squirts from a plunger-type oil can) to each cylinder, through the spark plug hole, and repeat the test.

10   If the compression increases after the oil is added, the piston rings are definitely worn. If the compression doesn't increase significantly, the leakage is occurring at the valves or head gasket. Leakage past the valves may be caused by burned valve seats and/or faces or warped, cracked or bent valves.

11   If two adjacent cylinders have equally low compression, there's a strong possibility the head gasket between them is blown. The appearance of coolant in the combustion chambers or the crankcase would verify this condition.

12   If one cylinder is about 20-percent lower than the others, and the engine has a slightly rough idle, a worn exhaust lobe on the camshaft could be the cause.

13   If the compression is unusually high, the combustion chambers are probably coated with carbon deposits. If that's the case, the cylinder heads should be removed and decarbonized.

14   If compression is way down or varies greatly between cylinders, it would be a good idea to have a leak-down test performed by an automotive repair shop. This test will pinpoint exactly where the leakage is occurring and how severe it is.

15   Install the fuses and drive the vehicle to restore the "block learn" memory.

4   Vacuum gauge diagnostic checks

A vacuum gauge provides valuable information about what is going on in the engine at a low cost. You can check for worn rings or cylinder walls, leaking head or intake manifold gaskets, incorrect carburetor adjustments, restricted exhaust, stuck or burned valves, weak valve springs, improper ignition or valve timing and ignition problems.

Unfortunately, vacuum gauge readings are easy to misinterpret, so they should be used in conjunction with other tests to confirm the diagnosis.

Both the gauge readings and the rate of needle movement are important for accurate interpretation. Most gauges measure vacuum in inches of mercury (in-Hg). As vacuum increases (or atmospheric pressure decreases), the reading will increase. Also, for every 1,000-foot increase in elevation above sea level, the gauge readings will decrease about one inch of mercury.

Connect the vacuum gauge directly to intake manifold vacuum, not to ported (carburetor) vacuum. Be sure no hoses are left disconnected during the test or false readings will result.

Before you begin the test, allow the engine to warm up completely. Block the wheels and set the parking brake. With the transmission in Park, start the engine and allow it to run at normal idle speed.

Read the vacuum gauge; an average, healthy engine should normally produce about 17 to 22 inches of vacuum with a fairly steady needle. Refer to the following vacuum gauge readings and what they indicate about the engine's condition:

1   A low, steady reading usually indicates a leaking gasket between the intake manifold and carburetor or throttle body, a leaky vacuum hose, late ignition timing or incorrect camshaft timing. Eliminate all other possible causes, utilizing the tests provided in this Chapter before you remove the timing chain cover to check the timing marks.

2   If the reading is three to eight inches below normal and it fluctuates at that low reading, suspect an intake manifold gasket leak at an intake port.

3   If the needle has regular drops of about two to four inches at a steady rate, the valves are probably leaking. Perform a compression or leak-down test to confirm this.

4   An irregular drop or down-flick of the needle can be caused by a sticking valve or an ignition misfire. Perform a compression or leak-down test and read the spark plugs.

5   A rapid vibration of about four inches-Hg vibration at idle combined with exhaust smoke indicates worn valve guides. Perform a leak-down test to confirm this. If the rapid vibration occurs with an increase in engine speed, check for a leaking intake manifold gasket or head gasket, weak valve springs, burned valves or ignition misfire.

6   A slight fluctuation, say one inch up and down, may mean ignition problems. Check all the usual tune-up items and, if necessary, run the engine on an ignition analyzer.

7   If there is a large fluctuation, perform a compression or leak-down test to look for a weak or dead cylinder or a blown head gasket.

8   If the needle moves slowly through a wide range, check for a clogged PCV system, incorrect idle fuel mixture, throttle body or intake manifold gasket leaks.

9   Check for a slow return after revving the engine by quickly snapping the throttle open until the engine reaches about 2,500 rpm and let it shut. Normally the reading should drop to near zero, rise above normal idle reading (about 5 in-Hg over) and then return to the previous idle reading. If the vacuum returns slowly and doesn't peak when the throttle is snapped shut, the rings may be worn. If there is a long delay, look for a restricted exhaust system (often the muffler or catalytic converter). An easy way to check this is to temporarily disconnect the exhaust ahead of the suspected part and re-test.

5   Engine removal - methods and precautions

If you've decided the engine must be removed for overhaul or major repair work, several preliminary steps should be taken. Locating a suitable place to work is extremely important. Adequate work space, along with storage space for the vehicle, will be needed.

Cleaning the engine compartment and engine before beginning the removal procedure will help keep tools clean and organized. An engine hoist will also be necessary. Safety is of primary importance, considering the potential hazards involved in removing the engine from this vehicle.

If the engine is being removed by a novice, a helper should be available. Advice and aid from someone more experienced would also be helpful. There are many instances when one person cannot simultaneously perform all of the operations required when lifting the engine out of the vehicle.

Plan the operation ahead of time. Arrange for or obtain all of the tools and equipment you'll need prior to beginning the job. Some of the equipment necessary to perform engine removal and installation safely and with relative ease in addition to a hydraulic jack, jack stands and an engine hoist) are a complete sets of wrenches and sockets as described in the front of this manual, wooden blocks and plenty of rags and cleaning solvent for mopping up spilled oil, coolant and gasoline.

Plan for the vehicle to be out of use for quite a while. A machine shop will be required to perform some of the work which the do-it-yourselfer can't accomplish without special equipment. These shops often have a busy schedule, so it would be a good idea to consult them before removing the engine in order to accurately estimate the amount of time required to rebuild or repair components that may need work.

Always be extremely careful when removing and installing the engine. Serious injury can result from careless actions. Plan ahead, take your time and a job of this nature, although major, can be accomplished successfully. **Note:** *Because it may be some time before you reinstall the engine, it is very helpful to make sketches or take photos of various accessory mountings and wiring hookups before removing the engine.*

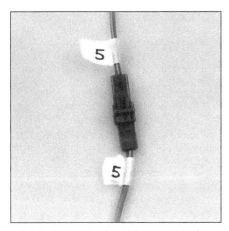

6.4 Label each wire before unplugging
the connector

6 Engine - removal and installation

Warning 1: *The models covered by this manual are equipped with airbags. Always disable the airbag system before working in the vicinity of the impact sensors, steering column or instrument panel to avoid the possibility of accidental deployment of the airbag(s), which could cause personal injury (see Chapter 12). The yellow wires and connectors routed through the instrument panel and, on 1995 and earlier models, to the front of the vehicle, are for this system. Do not use electrical test equipment on these yellow wires or tamper with them in any way.*

Warning 2: *Gasoline is extremely flammable, so take extra precautions when you work on any part of the fuel system. Don't smoke or allow open flames or bare light bulbs near the work area, and don't work in a garage where a natural gas-type appliance (such as a water heater or a clothes dryer) with a pilot light is present. Since gasoline is carcinogenic, wear latex gloves when there's a possibility of being exposed to fuel, and, if you spill any fuel on your skin, rinse it off immediately with soap and water. Mop up any spills immediately and do not store fuel-soaked rags where they could ignite. The fuel system is under constant pressure, so, if any fuel lines are to be disconnected, the fuel pressure in the system must be relieved first. When you perform any kind of work on the fuel system, wear safety glasses and have a Class B type fire extinguisher on hand.*

Warning 3: *The air conditioning system is under high pressure - have a dealer service department or service station evacuate the system and recapture the refrigerant before disconnecting any of the hoses or fittings.*

Caution: *On models equipped with a Delco Loc II or Theftlock audio system, be sure the lockout feature is turned off before performing any procedure which requires disconnecting the battery.*

Removal
Refer to illustrations 6.4, 6.8, 6.11 and 6.20
1   Relieve the fuel system pressure (see

6.8 While the vehicle is raised, disconnect
any wiring harnesses attached to
the block

Chapter 4), then disconnect the negative cable from the battery.
2   Cover the fenders and cowl. Special pads are available to protect the fenders, but an old bedspread or blanket will also work. Remove the hood (see Chapter 11).
3   Remove the air intake duct assembly (see Chapter 4).
4   To ensure correct reassembly, label each vacuum line, emission system hose, electrical connector, ground strap and fuel line. Pieces of masking tape with numbers or letters written on them prevent confusion at assembly time **(see illustration)**. Or sketch the engine compartment routing of lines, hoses and wires.
5   Drain the cooling system (see Chapter 1) and label and detach all coolant hoses from the engine.
6   Disconnect the throttle and cruise control cables (see Chapter 4) and TV cable (see Chapter 7).
7   Disconnect the air conditioning hoses from the compressor, hang them out of the way, detach the air conditioning compressor from its mounting bracket and remove it from the engine compartment (see Chapter 3).
8   Raise the vehicle and suitably support it on jackstands. While raised, perform the disassembly procedures that can be done only from underneath: exhaust system (see Chapter 4), transmission cooler lines (see Chapter 7) and wiring harnesses attached to the block **(see illustration)**.
9   Drain the engine oil and remove the filter (see Chapter 1).
10   Remove the starter (see Chapter 5). Remove the driveplate access cover.
11   Make an alignment mark between the driveplate and the torque converter **(see illustration)**, then rotate the crankshaft and remove the driveplate bolts.
12   Remove the lower bellhousing-to-engine bolts.
13   Lower the vehicle and place a floor jack under the transmission, then support the engine with an engine hoist from above. Place a block of wood on the head of the jack to protect the fluid pan.

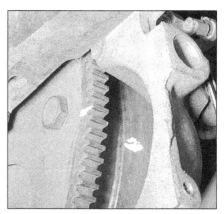

6.11 Paint or scribe alignment marks on
the driveplate and the torque converter to
ensure that the two components are still
in balance when they're reassembled

14   Remove the radiator and air conditioning condenser (see Chapter 3).
15   Unbolt the power steering pump and bracket and tie them out of the way.
16   Remove the intake manifold (see Chapter 4).
17   Raise the engine enough to take the weight off the engine mounts and remove the engine mount through-bolts (see Chapter 2A).
18   Working at the back of the engine, remove the remaining bellhousing bolts.
19   Raise the engine and pull it forward to free it from the torque converter snout.
20   Tie the wiring harnesses out of the way, raise the engine with the hoist, and with a combination of tilting, twisting and raising, move the engine around any obstructions and pull it out of the vehicle, raising it high enough to clear the front of the body **(see illustration)**.
21   Remove the driveplate (see Chapter 2A) and mount the engine on an engine stand.

Installation
22   While the engine is out, check the engine mounts and the transmission mount (see Chapter 7). If they're worn or damaged, replace them.

6.20 With the accessories tied out of the
way and the engine mount through-bolts
removed, tilt, twist and raise the engine
with the hoist until it clears everything

23   Carefully lower the engine, twisting it to clear any harnesses or obstructions, until the converter snout lines up and the bellhousing-to-engine bolts can be inserted. **Caution:** *DO NOT use the transmission-to-engine bolts to force the transmission and engine together. Take great care when mating the torque converter to the driveplate, following the procedure outlined in Chapter 7. Make sure the alignment marks you made on the driveplate and the torque converter during removal are lined up.*
24   Install the driveplate-to-torque converter bolts and tighten them to the torque listed in the Chapter 7 Specifications.
25   Reinstall the remaining components in the reverse order of removal. Double-check to make sure everything is hooked up right, using the sketches or photos taken earlier to go by. Make sure the metal tubing that connects the air pump ports from one side of the engine to the other is installed right after the engine is installed, and before the other accessories are bolted on.
26   Add coolant, oil, power steering and transmission fluid as needed.
27   Run the engine and check for leaks and proper operation of all accessories, then install the hood and test drive the vehicle.

7   Engine rebuilding alternatives

The home mechanic is faced with a number of options when performing an engine overhaul. The decision to replace the engine block, piston/connecting rod assemblies and crankshaft depends on a number of factors, with the number one consideration being the condition of the block. Other considerations are cost, access to machine shop facilities, parts availability, time required to complete the project and the extent of prior mechanical experience.

Some of the rebuilding alternatives include:

Individual parts - If the inspection procedures reveal the engine block and most engine components are in reusable condition, purchasing individual parts may be the most economical alternative. The block, crankshaft and piston/connecting rod assemblies should all be inspected carefully. Even if the block shows little wear, the cylinder bores should be surface-honed.

Short-block - A short-block consists of an engine block with a crankshaft and piston/connecting rod assemblies already installed. All new bearings are incorporated and all clearances will be correct. The existing camshaft, valve train components, cylinder heads and external parts can be bolted to the short block with little or no machine shop work necessary. Some rebuilding companies include a new timing chain, camshaft and lifters with their short-block assemblies.

Long-block - A long-block consists of a short block plus an oil pump, oil pan, cylinder heads, rocker arm covers, camshaft and

valve train components, timing sprockets and chain and timing chain cover. All components are installed with new bearings, seals and gaskets incorporated throughout. The installation of manifolds and external parts is all that's necessary. Give careful thought to which alternative is best for you and discuss the situation with local automotive machine shops, auto parts dealers and experienced rebuilders before ordering or purchasing replacement parts.

8   Engine overhaul - disassembly sequence

1   It's much easier to disassemble and work on the engine if it's mounted on a portable engine stand. A stand can often be rented quite cheaply from an equipment rental yard. Before it's mounted on a stand, the flywheel/driveplate should be removed from the engine (see Chapter 2A).
2   If a stand isn't available, it's possible to disassemble the engine with it blocked up on the floor. Be extra careful not to tip or drop the engine when working without a stand.
3   If you're going to obtain a rebuilt engine, all external components must come off first, to be transferred to the replacement engine, just as they will if you're doing a complete engine overhaul yourself. These include:

Alternator and brackets
Emissions control components
Ignition coil, spark plug wires and spark plugs
Thermostat and housing cover
Water pump
Distributor
Engine front cover
Fuel injection components
Intake/exhaust manifolds
Oil filter
Engine mounts
Driveplate
Rear main oil seal housing

Note: *When removing the external components from the engine, pay close attention to details that may be helpful or important during installation. Note the installed position of gaskets, seals, spacers, pins, brackets, washers, bolts and other small items.*
4   If you're obtaining a short-block, then the cylinder heads, oil pan and oil pump will have to be removed as well. See *Engine rebuilding alternatives* for additional information regarding the different possibilities to be considered.
5   If you're planning a complete overhaul, the engine must be disassembled and the internal components removed in the following general order:

Intake and exhaust manifolds
Valve covers
Rocker arms and pushrods
Valve lifters
Cylinder heads
Timing chain and sprockets

Oil pump drive assembly (1994 and later models)
Camshaft
Water pump driveshaft (1994 and later models)
Oil pan
Oil pump
Piston/connecting rod assemblies
Crankshaft and main bearings

6   Before beginning the disassembly and overhaul procedures, make sure the following items are available. Also, refer to *Engine overhaul - reassembly sequence* for a list of tools and materials needed for engine reassembly.

Common hand tools
Small cardboard boxes or plastic bags for storing parts
Gasket scraper
Ridge reamer
Engine balancer puller
Micrometers
Telescoping gauges
Dial indicator set
Valve spring compressor
Cylinder surfacing hone
Piston ring groove-cleaning tool
Electric drill motor
Tap and die set
Wire brushes
Oil gallery brushes
Cleaning solvent

9   Cylinder head - disassembly

Refer to illustrations 9.2, 9.3 and 9.4
Note: *New and rebuilt cylinder heads are commonly available for most engines at dealerships and auto parts stores. Due to the fact that some specialized tools are necessary for the disassembly and inspection procedures, and replacement parts aren't always readily available, it may be more practical and economical for the home mechanic to purchase replacement heads rather than taking the time to disassemble, inspect and recondition the originals.*
1   Cylinder head disassembly involves removal of the intake and exhaust valves and related components. Remove the rocker arm bolts, pivots and rocker arms from the cylinder heads. Label the parts or store them separately so they can be reinstalled in their original locations.
2   Before the valves are removed, arrange to label and store them, along with their related components, so they can be kept separate and reinstalled in their original locations **(see illustration)**.
3   Compress the springs on the first valve with a spring compressor and remove the keepers **(see illustration)**. Carefully release the valve spring compressor and remove the retainer, the spring and the spring seat (if used).
4   Pull the valve out of the head, then remove the oil seal from the guide. If the valve binds in the guide (won't pull through), push it back into the head and deburr the area

9.2 A small plastic bag, with an appropriate label, can be used to store the valve train components so they can be kept together and reinstalled in the original positions

9.3 Use a valve spring compressor to compress the spring, then remove the keepers from the valve stem

9.4 If the valve won't pull through the guide, deburr the edge of the stem end and the area around the top of the keeper groove with a file or whetstone

around the keeper groove with a fine file or whetstone **(see illustration)**.

5   Repeat the procedure for the remaining valves. Remember to keep all the parts for each valve together so they can be reinstalled in the same locations.

6   Once the valves and related components have been removed and stored in an organized manner, the heads should be thoroughly cleaned and inspected. If a complete engine overhaul is being done, finish the engine disassembly procedures before beginning the cylinder head cleaning and inspection process.

10   Cylinder head - cleaning and inspection

1   Thorough cleaning of the cylinder heads and related valve train components, followed by a detailed inspection, will enable you to decide how much valve service work must be done during the engine overhaul. **Note:** *If the engine was severely overheated, the cylinder head is probably warped* (see Step 12).

Cleaning

2   Scrape all traces of old gasket material and sealant off the head gasket, intake manifold and exhaust manifold mating surfaces. Be very careful not to gouge the cylinder head, especially on engines with aluminum heads. Special gasket removal solvents that soften gaskets and make removal much easier are available at auto parts stores.

3   Remove all built-up scale from the coolant passages.

4   Run a stiff wire brush through the various holes to remove deposits that may have formed in them.

5   Run an appropriate-size tap into each of the threaded holes to remove corrosion and thread sealant that may be present. If compressed air is available, use it to clear the holes of debris produced by this operation. **Warning:** *Wear eye protection when using compressed air!*

6   Clean the rocker arm pivot stud or bolt threads with a wire brush.

10.12 Check the cylinder head gasket surface for warpage by trying to slip a feeler gauge under the straightedge (see this Chapter's Specifications for the maximum warpage allowed and use a feeler gauge of that thickness)

7   Clean the cylinder head with solvent and dry it thoroughly. Compressed air will speed the drying process and ensure that all holes and recessed areas are clean. **Note:** *Decarbonizing chemicals are available and may prove very useful when cleaning cylinder heads and valve train components. They're very caustic and should be used with caution. Be sure to follow the instructions on the container.*

8   Clean the rocker arms, pivots, bolts and pushrods with solvent and dry them thoroughly (don't mix them up during the cleaning process). Compressed air will speed the drying process and can be used to clean out the oil passages.

9   Clean all the valve springs, keepers and retainers with solvent and dry them thoroughly. Do the components from one valve at a time to avoid mixing up the parts.

10   Scrape off any heavy deposits that may have formed on the valves, then use a motorized wire brush to remove deposits from the valve heads and stems. Again, make sure the valves don't get mixed up.

10.14 A dial indicator can be used to determine the valve stem-to-guide clearance (move the valve stem as indicated by the arrows)

Inspection

Note: *Be sure to perform all of the following inspection procedures before concluding machine shop work is required. Make a list of the items that need attention.*

Cylinder head

Refer to illustrations 10.12 and 10.14

11   Inspect the head very carefully for cracks, evidence of coolant leakage and other damage. If cracks are found, check with an automotive machine shop concerning repair. If repair isn't possible, a new cylinder head must be obtained.

12   Using a straightedge and feeler gauge, check the head gasket mating surface for warpage **(see illustration)**. If the warpage exceeds the limit in this Chapter's Specifications, it can be resurfaced at an automotive machine shop. **Note:** *If the heads are resurfaced, the intake manifold flanges will also require machining.*

13   Examine the valve seats in each of the combustion chambers. If they're pitted, cracked or burned, the head will require valve service that's beyond the scope of the home mechanic.

14   Check the valve stem-to-guide clearance

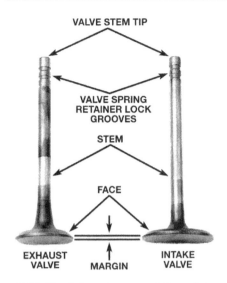

10.15 Check for valve wear at the points shown here

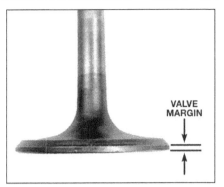

10.16 The margin width on the valve must be as specified (if no margin exists, the valve cannot be re-used)

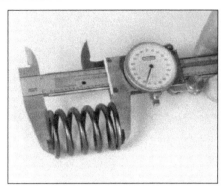

10.17 Measure the free length of each valve spring with a dial or vernier caliper

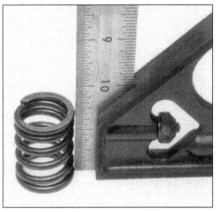

10.18 Check each valve spring for squareness

by measuring the lateral movement of the valve stem with a dial indicator attached securely to the head **(see illustration)**. The valve must be in the guide and approximately 1/16-inch off the seat. The total valve stem movement indicated by the gauge needle must be divided by two to obtain the actual clearance. After this is done, if there's still some doubt regarding the condition of the valve guides, they should be checked by an automotive machine shop (the cost should be minimal).

Valves

Refer to illustrations 10.15 and 10.16

15   Carefully inspect each valve face for uneven wear, deformation, cracks, pits and burned areas. Check the valve stem for scuffing and galling and the neck for cracks. Rotate the valve and check for any obvious indication that it's bent. Look for pits and excessive wear on the end of the stem. The presence of any of these conditions **(see illustration)** indicates the need for valve service by an automotive machine shop.

16   Measure the margin width on each valve **(see illustration)**. Any valve with a margin narrower than 1/32-inch should be replaced with a new one.

Valve components

Refer to illustrations 10.17 and 10.18

17   Check each valve spring for wear (on the ends) and pits. Measure the free length and compare it to this Chapter's Specifications **(see illustration)**. Any springs that are shorter than specified have sagged and shouldn't be re-used. The tension of all springs should be checked with a special fixture before deciding they're suitable for use in a rebuilt engine (take the springs to an automotive machine shop for this check).

18   Stand each spring on a flat surface and check it for squareness **(see illustration)**. If any of the springs are distorted or sagged, replace all of them with new parts.

19   Check the spring retainers and keepers for obvious wear and cracks. Any questionable parts should be replaced with new ones, as extensive damage will occur if they fail during engine operation.

Rocker arm components

20   Check the rocker arm faces (the areas that contact the pushrod ends and valve stems) for pits, wear, galling, score marks and rough spots. Check the rocker arm pivot contact areas and pivots as well. Look for cracks in each rocker arm and bolt.

21   Inspect the pushrod ends for scuffing and excessive wear. Roll each pushrod on a flat surface, like a piece of plate glass, to determine if it's bent. Check the pushrod guide plates for signs of excessive wear.

22   Check the rocker arm bolt holes or studs in the cylinder heads for damaged threads.

23   Any damaged or excessively worn parts must be replaced with new ones.

All components

24   If the inspection process indicates the valve components are in generally poor condition and worn beyond the limits specified, which is usually the case in an engine that's being overhauled, reassemble the valves in the cylinder head (see Section 11 for valve servicing recommendations).

---

11   Valves - servicing

---

1   Because of the complex nature of the job and the special tools and equipment needed, servicing of the valves, the valve seats and the valve guides, commonly known as a valve job, should be done by a professional.

2   The home mechanic can remove and disassemble the head, do the initial cleaning and inspection, then reassemble and deliver it to a dealer service department or an automotive machine shop for the actual service work. Doing the inspection will enable you to see what condition the head and valvetrain components are in and will ensure that you know what work and new parts are required when dealing with an automotive machine shop.

3   The dealer service department, or automotive machine shop, will remove the valves and springs, recondition or replace the valves and valve seats, recondition the valve guides, check and replace the valve springs, rotators, spring retainers and keepers (as necessary), replace the valve seals with new ones, reassemble the valve components and make sure the installed spring height is correct. The cylinder head gasket surface will also be resurfaced if it's warped.

4   After the valve job has been performed by a professional, the head will be in like new condition. When the head is returned, be sure to clean it again before installation on the engine to remove any metal particles and abrasive grit that may still be present from the valve service or head resurfacing operations. Use compressed air, if available, to blow out all the oil holes and passages.

---

12   Cylinder head - reassembly

Refer to illustrations 12.6 and 12.7

1   Regardless of whether or not the head was sent to an automotive repair shop for valve servicing, make sure it's clean before beginning reassembly.

2   If the head was sent out for valve servicing, the valves and related components will already be in place. Begin the reassembly procedure with Step 8.

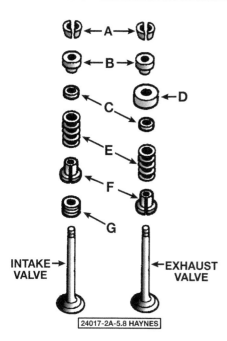

12.6 Typical valve components

A Keepers
B Retainers
C O-ring seals
D Oil shield
E Springs
F Spring dampers
G Valve stem oil seal

3 Beginning at one end of the head, lubricate and install the first valve. Apply moly-base grease or clean engine oil to the valve stem.
4 Install the shims, if originally installed, before the valve seals.
5 Install new seals on each of the valve guides. Gently tap each seal into place until it's completely seated on the guide. Many seal sets come with a plastic installer, but use hand pressure. Do not hammer on the seals or they could be driven down too far and subsequently leak. Don't twist or cock the seals during installation or they won't seal properly on the valve stems.
6 Install the components in the following order (see illustration):

Shims (if necessary)
Valve, followed by the stem seals
Springs
Retainers
Keepers

7 Compress the springs with a valve spring compressor and carefully install the keepers in the groove, then slowly release the compressor and make sure the keepers seat properly. Apply a small dab of grease to each keeper to hold it in place if necessary (see illustration). Tap the valve stem tips with a plastic hammer to seat the keepers, if necessary.
8 Repeat the procedure for the remaining valves. Be sure to return the components to their original locations - don't mix them up!
9 Check the installed valve spring height with a ruler graduated in 1/32-inch increments or a dial caliper. If the head was sent out for service work, the installed height

12.7 Apply a small dab of grease to each keeper as shown here before installation - it'll hold them in place on the valve stem as the spring is released

should be correct (but don't automatically assume it is). The measurement is taken from the top of each spring seat or top shim to the bottom of the retainer. If the height is greater than specified in this Chapter, shims can be added under the springs to correct it. **Caution:** *Do not, under any circumstances, shim the springs to the point where the installed height is less than specified.*
10 If the rocker arm studs had been previously removed from the head, install the guide plates, coat the bottom threads of the rocker studs with non-hardening sealant and torque them to Specifications.
11 Apply moly-base grease to the rocker arm faces and the pivots/balls, then install the rocker arms and pivots/balls on the cylinder heads. Tighten the bolts/nuts finger-tight.

13 Pistons and connecting rods - removal

Refer to illustrations 13.1, 13.3, 13.4 and 13.6
Note: *Prior to removing the piston/connecting rod assemblies, remove the cylinder heads, the oil pan and the oil pump by referring to the appropriate Sections in Chapter 2A.*
1 Use your fingernail to feel if a ridge has formed at the upper limit of ring travel (about 1/4-inch down from the top of each cylinder). If carbon deposits or cylinder wear have produced ridges, they must be completely removed with a special tool (see illustration). Follow the manufacturer's instructions provided with the tool. Failure to remove the ridges before attempting to remove the piston/connecting rod assemblies may result in piston breakage.
2 After the cylinder ridges have been removed, turn the engine upside-down so the crankshaft is facing up.
3 Before the connecting rods are removed, check the endplay with feeler gauges. Slide them between the first connecting rod and the crankshaft throw until the play is removed (see illustration). The endplay is equal to the thickness of the feeler gauge(s). If the endplay exceeds the service limit, new connecting rods will be required. If new rods (or a new crankshaft) are installed, the endplay may fall under the minimum specified in this Chapter (if it does, the rods will have to be machined

13.1 A ridge reamer is required to remove the ridge from the top of each cylinder - do this before removing the pistons!

13.3 Check the connecting rod side clearance with a feeler gauge as shown

13.4 Mark the rod bearing caps in order from the front of the engine to the rear (one mark for the front cap, two for the second one and so on)

to restore it - consult an automotive machine shop for advice if necessary). Repeat the procedure for the remaining connecting rods.
4 Check the connecting rods and caps for identification marks. If they aren't plainly marked, use a small center-punch (see illustration) to make the appropriate number of indentations on each rod and cap (1, 2, 3, etc., depending on the cylinder they're associated with).

13.6 To prevent damage to the crankshaft journals and cylinder walls, slip sections of rubber or plastic hose over the rod bolts before removing the pistons/rods

14.3 Checking crankshaft endplay with a feeler gauge

14.4 The arrow on the main bearing cap indicates the front of the engine

5   Loosen each of the connecting rod cap nuts 1/2-turn at a time until they can be removed by hand. Remove the number one connecting rod cap and bearing insert. Don't drop the bearing insert out of the cap.

6   Slip a short length of plastic or rubber hose over each connecting rod cap bolt to protect the crankshaft journal and cylinder wall as the piston is removed **(see illustration)**.

7   Remove the bearing insert and push the connecting rod/piston assembly out through the top of the engine. Use a wooden or plastic hammer handle to push on the upper bearing surface in the connecting rod. If resistance is felt, double-check to make sure all of the ridge was removed from the cylinder.

8   Repeat the procedure for the remaining cylinders.

9   After removal, reassemble the connecting rod caps and bearing inserts in their respective connecting rods and install the cap nuts finger tight. Leaving the old bearing inserts in place until reassembly will help prevent the connecting rod bearing surfaces from being accidentally nicked or gouged.

10   Don't separate the pistons from the connecting rods.

---

14 Crankshaft - removal

Refer to illustrations 14.3 and 14.4
Note: *The crankshaft can be removed only after the engine has been removed from the vehicle. It's assumed the flywheel/driveplate, timing chain, oil pan, oil pump and piston/connecting rod assemblies have already been removed. The rear main oil seal retainer must also be removed first.*

1   Before the crankshaft is removed, check the endplay. Mount a dial indicator with the stem in line with the crankshaft and touching one of the crank throws.

2   Push the crankshaft all the way to the rear and zero the dial indicator. Next, pry the crankshaft to the front as far as possible and check the reading on the dial indicator. The distance it moves is the endplay. If it's greater

15.4a A hammer and large punch can be used to knock the core plugs sideways in their bores

than listed in this Chapter's Specifications, check the crankshaft thrust surfaces for wear. If no wear is evident, new main bearings should correct the endplay.

3   If a dial indicator isn't available, feeler gauges can be used. Gently pry or push the crankshaft all the way to the front of the engine. Slip feeler gauges between the crankshaft and the front face of the thrust main bearing to determine the clearance **(see illustration)**. **Note:** *The thrust bearing is located at the number five main bearing cap.*

4   Check the main bearing caps to see if they're marked to indicate their locations. They should be numbered consecutively from the front of the engine to the rear. If they aren't, mark them with number stamping dies or a center-punch. Main bearing caps generally have a cast-in arrow, which points to the front of the engine **(see illustration)**. Loosen the main bearing cap bolts 1/4-turn at a time each, until they can be removed by hand. Note if any stud bolts are used and make sure they're returned to their original locations when the crankshaft is reinstalled.

5   Gently tap the caps with a soft-face hammer, then separate them from the engine block. If necessary, use the bolts as levers to remove the caps. Try not to drop the bearing inserts if they come out with the caps.

15.4b Pull the core plugs from the block with pliers

6   Carefully lift the crankshaft straight out of the engine. It may be a good idea to have an assistant available, since the crankshaft is quite heavy. With the bearing inserts in place in the engine block and main bearing caps, return the caps to their respective locations on the engine block and tighten the bolts finger tight.

---

15 Engine block - cleaning

Refer to illustrations 15.4a, 15.4b, 15.8 and 15.10

1   Remove the main bearing caps and separate the bearing inserts from the caps and the engine block. Tag the bearings, indicating which cylinder they were removed from and whether they were in the cap or the block, then set them aside.

2   Using a gasket scraper, remove all traces of gasket material from the engine block. Be very careful not to nick or gouge the gasket sealing surfaces.

3   Remove all of the covers and threaded oil gallery plugs from the block. The plugs are usually very tight - they may have to be drilled out and the holes retapped. Use new plugs when the engine is reassembled.

4   Remove the core plugs from the engine block. To do this, knock one side of the plugs

15.8 All bolt holes in the block - particularly the main bearing cap and head bolt holes - should be cleaned and restored with a tap (be sure to remove debris from the holes after this is done)

15.10 A large socket on an extension can be used to drive the new core plugs into the bores

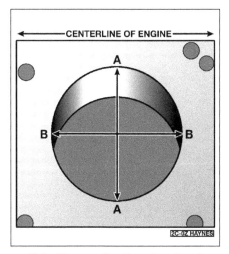

16.4a Measure the diameter of each cylinder at a right angle to engine centerline (A), and parallel to engine centerline (B) - out-of-round is the difference between A and B; taper is the difference between the diameter at the top of the cylinder and the diameter at the bottom of the cylinder

into the block with a hammer and punch, then grasp them with large pliers and pull them out **(see illustrations)**.

5   If the engine is extremely dirty, it should be taken to an automotive machine shop to be cleaned. **Note:** *If the block is cleaned in a caustic-solution hot tank, this will ruin any camshaft bearings left in the block. If the engine is being rebuilt, these bearings should be replaced anyway.*

6   After the block is returned, clean all oil holes and oil galleries one more time. Brushes specifically designed for this purpose are available at most auto parts stores. Flush the passages with warm water until the water runs clear, dry the block thoroughly and wipe all machined surfaces with a light, rust preventive oil. If you have access to compressed air, use it to speed the drying process and blow out all the oil holes and galleries. **Warning:** *Wear eye protection when using compressed air!*

7   If the block isn't extremely dirty or sludged up, you can do an adequate cleaning job with hot soapy water and a stiff brush. Take plenty of time and do a thorough job. Regardless of the cleaning method used, be sure to clean all oil holes and galleries very thoroughly, dry the block completely and coat all machined surfaces with light oil.

8   The threaded holes in the block must be clean to ensure accurate torque readings during reassembly. Run the proper size tap into each of the holes to remove rust, corrosion, thread sealant or sludge and restore damaged threads **(see illustration)**. If possible, use compressed air to clear the holes of debris produced by this operation. Now is a good time to clean the threads on the head bolts and the main bearing cap bolts as well.

9   Reinstall the main bearing caps and tighten the bolts finger tight.

10   After coating the sealing surfaces of the new core plugs with a non-hardening sealant (such as Permatex no. 2), install them in the engine block **(see illustration)**. Make sure they're driven in straight and seated properly

or leakage could result. Special tools are available for this purpose, but a large socket, with an outside diameter that will just slip into the core plug, a 1/2-inch drive extension and a hammer will work just as well.

11   Apply non-hardening sealant (such as Permatex no. 2 or Teflon pipe sealant) to the new oil gallery plugs and thread them into the holes in the block. Make sure they're tightened securely.

12   If the engine isn't going to be reassembled right away, cover it with a large plastic trash bag to keep it clean.

16   Engine block - inspection

Refer to illustrations 16.4a, 16.4b and 16.4c
Note: *The manufacturer recommends checking the block deck for warpage and the main bearing bore concentricity and alignment. Since special measuring tools are needed, the checks should be done by an automotive machine shop.*

1   Before the block is inspected, it should be cleaned as described in Section 15.

2   Visually check the block for cracks, rust and corrosion. Look for stripped threads in the threaded holes. It's also a good idea to have the block checked for hidden cracks by an automotive machine shop that has the special equipment to do this type of work. If defects are found, have the block repaired, if possible, or replaced.

3   Check the cylinder bores for scuffing and scoring.

4   Check the cylinders for taper and out-of-round conditions as follows **(see illustrations)**:

5   Measure the diameter of each cylinder at the top (just under the ridge area), center and bottom of the cylinder bore, parallel to the crankshaft axis.

6   Next, measure each cylinder's diameter at the same three locations perpendicular to the crankshaft axis.

7   The taper of each cylinder is the difference between the bore diameter at the top of

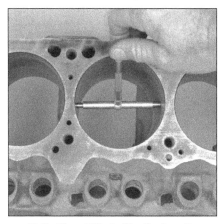

16.4b The ability to "feel" when the telescoping gauge is at the correct point will be developed over time, so work slowly and repeat the check until you're satisfied the bore measurement is accurate

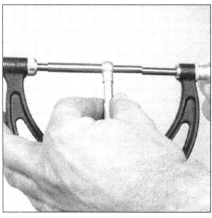

16.4c The gauge is then measured with a micrometer to determine the bore size

17.3a A "bottle brush" hone will produce better results if you've never honed cylinders before

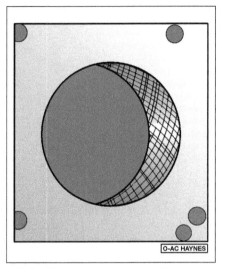

17.3b The cylinder hone should leave a smooth, crosshatch pattern with the lines intersecting at approximately a 60-degree angle

18.4a The piston ring grooves can be cleaned with a special tool, as shown here . . .

the cylinder and the diameter at the bottom. The out-of-round specification of the cylinder bore is the difference between the parallel and perpendicular readings. Compare your results to this Chapter's Specifications.

8   If the cylinder walls are badly scuffed or scored, or if they're out-of-round or tapered beyond the limits given in this Chapter's Specifications, have the engine block rebored and honed at an automotive machine shop.

9   If a rebore is done, oversize pistons and rings will be required.

10   Using a precision straightedge and feeler gauge, check the block deck (the surface the cylinder heads mate with) for distortion as you did with the cylinder heads (see Section 10). If it's distorted beyond the specified limit, the block decks can be resurfaced by an automotive machine shop.

11   If the cylinders are in reasonably good condition and not worn to the outside of the limits, and if the piston-to-cylinder clearances can be maintained properly, they don't have to be rebored. Honing is all that's necessary (see Section 17).

17 Cylinder honing

Refer to illustrations 17.3a and 17.3b

1   Prior to engine reassembly, the cylinder bores must be honed so the new piston rings will seat correctly and provide the best possible combustion chamber seal. **Note:** *If you don't have the tools or don't want to tackle the honing operation, most automotive machine shops will do it for a reasonable fee.*

2   Before honing the cylinders, install the main bearing caps and tighten the bolts to the torque listed in this Chapter's Specifications.

3   Two types of cylinder hones are commonly available - the flex hone or "bottle brush" type and the more traditional surfacing hone with spring-loaded stones. Both will do the job, but for the less experienced mechanic the "bottle brush" hone will probably be easier to use. You'll also need some

honing oil (kerosene will work if honing oil isn't available), rags and an electric drill motor. Proceed as follows:

a)   *Mount the hone in the drill motor, compress the stones and slip it into the first cylinder* **(see illustration)**. *Be sure to wear safety goggles or a face shield!*

b)   *Lubricate the cylinder with plenty of honing oil, turn on the drill and move the hone up-and-down in the cylinder at a pace that will produce a fine crosshatch pattern on the cylinder walls, and with the drill square and centered with the bore. Ideally, the crosshatch lines should intersect at approximately a 45-60-degree angle* **(see illustration)**. *Be sure to use plenty of lubricant and don't take off any more material than is absolutely necessary to produce the desired finish.* **Note:** *Piston ring manufacturers may specify a different crosshatch angle - read and follow any instructions included with the new rings.*

c)   *Don't withdraw the hone from the cylinder while it's running. Instead, shut off the drill and continue moving the hone up-and-down in the cylinder until it comes to a complete stop, then compress the stones and withdraw the hone. If you're using a "bottle brush" type hone, stop the drill motor, then turn the chuck in the normal direction of rotation while withdrawing the hone from the cylinder.*

d)   *Wipe the oil out of the cylinder and repeat the procedure for the remaining cylinders.*

4   After the honing job is complete, chamfer the top edges of the cylinder bores with a small file so the rings won't catch when the pistons are installed. Be very careful not to nick the cylinder walls with the end of the file.

5   The entire engine block must be washed again very thoroughly with warm, soapy water

to remove all traces of the abrasive grit produced during the honing operation. **Note:** *The bores can be considered clean when a lint-free white cloth - dampened with clean engine oil - used to wipe them out doesn't pick up any more honing residue, which will show up as gray areas on the cloth. Be sure to run a brush through all oil holes and galleries and flush them with running water.*

6   After rinsing, dry the block and apply a coat of light rust preventive oil to all machined surfaces. Wrap the block in a plastic trash bag to keep it clean and set it aside until reassembly.

18 Pistons and connecting rods - inspection

Refer to illustrations 18.4a, 18.4b, 18.10 and 18.11

1   Before the inspection process can be carried out, the piston/connecting rod assemblies must be cleaned and the original piston rings removed from the pistons. **Note:** *Always use new piston rings when the engine is reassembled.*

2   Using a piston ring installation tool, carefully remove the rings from the pistons. Be careful not to nick or gouge the pistons in the process.

3   Scrape all traces of carbon from the top of the piston. A hand-held wire brush or a piece of fine emery cloth can be used once the majority of the deposits have been scraped away. Do not, under any circumstances, use a wire brush mounted in a drill motor to remove deposits from the pistons. The piston material is soft and may be eroded away by the wire brush.

4   Use a piston ring groove-cleaning tool to remove carbon deposits from the ring grooves. If a tool isn't available, a piece broken off the old ring will do the job. Be very careful to remove only the carbon deposits - don't remove any metal and do not nick or scratch the sides of the ring grooves **(see illustrations)**.

18.4b ... or a section of broken ring

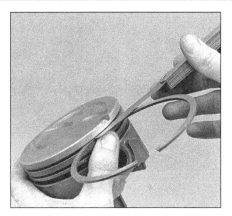

18.10 Check the ring side clearance with a feeler gauge at several points around the groove

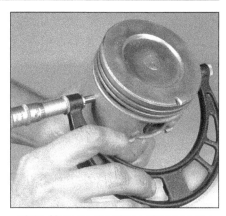

18.11 Measure the piston diameter at a 90-degree angle to the piston pin and in line with it

5   Once the deposits have been removed, clean the piston/rod assemblies with solvent and dry them with compressed air (if available). **Warning:** *Wear eye protection. Make sure the oil return holes in the back sides of the ring grooves are clear.*

6   If the pistons and cylinder walls aren't damaged or worn excessively, and if the engine block isn't rebored, new pistons won't be necessary. Normal piston wear appears as even vertical wear on the piston thrust surfaces and slight looseness of the top ring in its groove. New piston rings, however, should always be used when an engine is rebuilt.

7   Carefully inspect each piston for cracks around the skirt, at the pin bosses and at the ring lands.

8   Look for scoring and scuffing on the thrust faces of the skirt, holes in the piston crown and burned areas at the edge of the crown. If the skirt is scored or scuffed, the engine may have been suffering from overheating and/or abnormal combustion, which caused excessively high operating temperatures. The cooling and lubrication systems should be checked thoroughly. A hole in the piston crown is an indication that abnormal combustion (preignition) was occurring. Burned areas at the edge of the piston crown are usually evidence of spark knock (detonation). If any of the above problems exist, the causes must be corrected or the damage will occur again. The causes may include intake air leaks, incorrect fuel/air mixture, low octane fuel, ignition timing and EGR system malfunctions.

9   Corrosion of the piston, in the form of small pits, indicates coolant is leaking into the combustion chamber and/or the crankcase. Again, the cause must be corrected or the problem may persist in the rebuilt engine.

10   Measure the piston ring side clearance by laying a new piston ring in each ring groove and slipping a feeler gauge in beside it **(see illustration)**. Check the clearance at three or four locations around each groove. Be sure to use the correct ring for each groove - they are different. If the side clearance is greater than specified in this Chapter, new pistons will have to be used.

11   Check the piston-to-bore clearance by measuring the bore (see Section 16) and the piston diameter. Make sure the pistons and bores are correctly matched. Measure the piston across the skirt, at a 90-degree angle to the piston pin **(see illustration)**. The measurement must be taken at a specific point to be accurate: The pistons are measured 0.45-inch from the bottom of the skirt, at right angles to the piston pin. Measure the cylinder bore 2.5-inches from the top for comparison with the piston measurement.

12   Subtract the piston diameter from the bore diameter to obtain the clearance. If it's greater than specified, the block will have to be rebored and new pistons and rings installed.

13   Check the piston-to-rod clearance by twisting the piston and rod in opposite directions. Any noticeable play indicates excessive wear, which must be corrected. The piston/connecting rod assemblies should be taken to an automotive machine shop to have the pistons and rods re-sized and new pins installed.

14   If the pistons must be removed from the connecting rods for any reason, they should be taken to an automotive machine shop. While they are there, have the connecting rods checked for bend and twist, since auto-

motive machine shops have special equipment for this purpose. **Note:** *Unless new pistons and/or connecting rods must be installed, do not disassemble the pistons and connecting rods.*

15   Check the connecting rods for cracks and other damage. Temporarily remove the rod caps, lift out the old bearing inserts, wipe the rod and cap bearing surfaces clean and inspect them for nicks, gouges and scratches. After checking the rods, replace the old bearings, slip the caps into place and tighten the nuts finger tight. **Note:** *If the engine is being rebuilt because of a connecting rod knock, be sure to install new rods.*

19   Crankshaft - inspection

Refer to illustrations 19.1, 19.2, 19.5 and 19.7

1   Remove all burrs from the crankshaft oil holes with a stone, file or scraper **(see illustration)**.

2   Clean the crankshaft with solvent and dry it with compressed air (if available). **Warning:** *Wear eye protection when using compressed air.* Be sure to clean the oil holes with a stiff brush **(see illustration)** and flush them with solvent.

19.1 The oil holes should be chamfered so sharp edges don't gouge or scratch the new bearings

19.2 Use a wire or stiff plastic bristle brush to clean the oil passages in the crankshaft

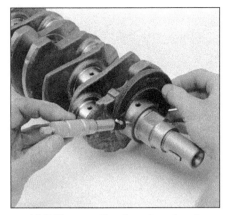

19.5 Measure the diameter of each crankshaft journal at several points to detect taper and out-of-round conditions

19.7 If the seals have worn grooves in the crankshaft journals, or if the seal contact surfaces are nicked or scratched, the new seals will leak

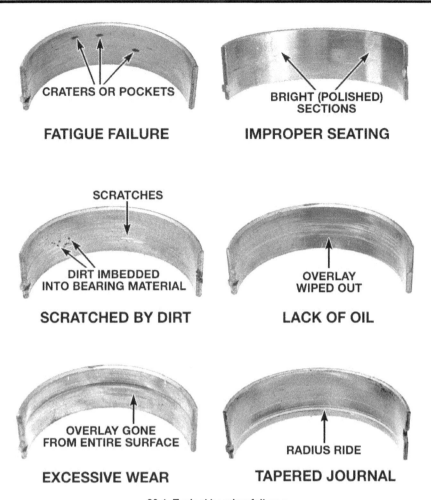

20.1 Typical bearing failures

3   Check the main and connecting rod bearing journals for uneven wear, scoring, pits and cracks.

4   Check the rest of the crankshaft for cracks and other damage. It should be Magnafluxed to reveal hidden cracks - an automotive machine shop will handle the procedure.

5   Using a micrometer, measure the diameter of the main and connecting rod journals and compare the results to this Chapter's Specifications **(see illustration)**. By measuring the diameter at a number of points around each journal's circumference, you'll be able to determine whether or not the journal is out-of-round. Take the measurement at each end of the journal, near the crank throws, to determine if the journal is tapered.

6   If the crankshaft journals are damaged, tapered, out-of-round or worn beyond the limits given in the Specifications, have the crankshaft reground by an automotive machine shop. Be sure to use the correct-size bearing inserts if the crankshaft is reconditioned.

7   Check the oil seal journals at each end of the crankshaft for wear and damage. If the seal has worn a groove in the journal, or if it's nicked or scratched **(see illustration)**, the

new seal may leak when the engine is reassembled. In some cases, an automotive machine shop may be able to repair the journal by pressing on a thin sleeve. If repair isn't feasible, a new or different crankshaft should be installed.

8   Examine the main and rod bearing inserts (see Section 20).

20 Main and connecting rod bearings - inspection

Refer to illustration 20.1

1   Even though the main and connecting rod bearings should be replaced with new ones during the engine overhaul, the old bearings should be retained for close examination, as they may reveal valuable information about the condition of the engine **(see illustration)**.

2   Bearing failure occurs because of lack of lubrication, the presence of dirt or other foreign particles, overloading the engine and corrosion. Regardless of the cause of bearing failure, it must be corrected before the engine is reassembled to prevent it from happening again.

3   When examining the bearings, remove them from the engine block, the main bearing caps, the connecting rods and the rod caps and lay them out on a clean surface in the same general position as their location in the engine. This will enable you to match any bearing problems with the corresponding crankshaft journal.

4   Dirt and other foreign particles get into the engine in a variety of ways. It may be left in the engine during assembly, or it may pass through filters or the PCV system. It may get into the oil, and from there into the bearings. Metal chips from machining operations and normal engine wear are often present. Abrasives are sometimes left in engine components after reconditioning, especially when parts aren't thoroughly cleaned using the proper cleaning methods. Whatever the source, these foreign objects often end up embedded in the soft bearing material and are easily recognized. Large particles won't embed in the bearing and will score or gouge the bearing and journal. The best prevention for this cause of bearing failure is to clean all parts thoroughly and keep everything spotlessly clean during engine assembly. Frequent and regular engine oil and filter changes are also recommended.

5   Lack of lubrication (or lubrication break-down) has a number of interrelated causes. Excessive heat (which thins the oil), overload-ing (which squeezes the oil from the bearing face) and oil leakage or throw off (from exces-sive bearing clearances, worn oil pump or high engine speeds) all contribute to lubrica-tion breakdown. Blocked oil passages, which usually are the result of misaligned oil holes in a bearing shell, will also oil starve a bearing and destroy it. When lack of lubrication is the cause of bearing failure, the bearing material is wiped or extruded from the steel backing of the bearing. Temperatures may increase to the point where the steel backing turns blue from overheating.

6   Driving habits can have a definite effect on bearing life. Low speed operation in too high a gear (lugging the engine) puts very high loads on bearings, which tends to squeeze out the oil film. These loads cause the bearings to flex, which produces fine cracks in the bearing face (fatigue failure). Eventually the bearing material will loosen in pieces and tear away from the steel backing. Short trip driving leads to corrosion of bearings because insufficient engine heat is produced to drive off the con-densed water and corrosive gases. These products collect in the engine oil, forming acid and sludge. As the oil is carried to the engine bearings, the acid attacks and corrodes the bearing material.

7   Incorrect bearing installation during engine assembly will lead to bearing failure as well. Tight-fitting bearings leave insuffi-cient oil clearance and will result in oil starva-tion. Dirt or foreign particles trapped behind a bearing insert result in high spots on the bearing which lead to failure.

21   Engine overhaul - reassembly sequence

1   Before beginning engine reassembly, make sure you have all the necessary new parts, gaskets and seals as well as the fol-lowing items on hand:

Common hand tools
Torque wrench (1/2-inch drive) with
*   angle-torque gauge*
Piston ring installation tool
Piston ring compressor
Crankshaft balancer installation tool
Short lengths of rubber or plastic hose to
*   fit over connecting rod bolts*
Plastigage
Feeler gauges
Fine-tooth file
New engine oil
Engine assembly lube or moly-base
*   grease*
Gasket sealant
Thread locking compound

2   In order to save time and avoid prob-lems, engine reassembly must be done in the following general order:

Crankshaft and main bearings
Piston/connecting rod assemblies

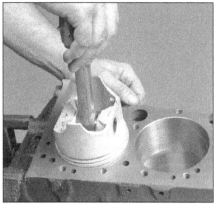

22.3 When checking piston ring end gap, the ring must be square in the cylinder bore (this is done by pushing the ring down with the top of a piston as shown)

Water pump drive gear (1994 and later
*   models)*
Camshaft
Oil pump drive assembly (1994 and later
*   models)*
Rear main oil seal
Water pump driveshaft (1994 and later
*   models)*
Timing chain and sprockets
Timing chain cover
Oil pump
Oil pan
Cylinder heads
Valve lifters
Rocker arms and pushrods
Driveplate

Assembled after engine installation

Intake and exhaust manifolds
Valve covers

22   Piston rings - installation

Refer to illustrations 22.3, 22.4, 22.5, 22.9a, 22.9b and 22.12

1   Before installing the new piston rings, the ring end gaps must be checked. It's assumed the piston ring side clearance has been checked and verified correct (see Sec-tion 18).

2   Lay out the piston/connecting rod assemblies and the new ring sets so the ring sets will be matched with the same piston and cylinder during the end gap measure-ment and engine assembly.

3   Insert the top (number one) ring into the first cylinder and square it up with the cylin-der walls by pushing it in with the top of the piston **(see illustration)**. The ring should be near the bottom of the cylinder, at the lower limit of ring travel.

4   To measure the end gap, slip feeler gauges between the ends of the ring until a gauge equal to the gap width is found **(see illustration)**. The feeler gauge should slide between the ring ends with a slight amount of

22.4 With the ring square in the cylinder, measure the end gap with a feeler gauge

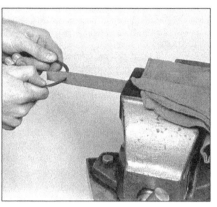

22.5 If the end gap is too small, clamp a file in a vise and file the ring ends (from the outside in only) to enlarge the gap slightly

drag. Compare the measurement to this Chapter's Specifications. If the gap is larger or smaller than specified, double-check to make sure you have the correct rings before proceeding.

5   If the gap is too small, it must be enlarged or the ring ends may come in con-tact with each other during engine operation, which can cause serious engine damage. The end gap can be increased by filing the ring ends very carefully with a fine file. Mount the file in a vise equipped with soft jaws, slip the ring over the file with the ends contacting the file teeth and slowly move the ring to remove material from the ends. When performing this operation, file only from the outside in **(see illustration)**. **Note:** *When you have the end gap correct, remove any burrs from the filed ends of the rings with a whetstone.*

6   Excess end gap isn't critical unless it's greater than 0.040-inch. Again, double-check to make sure you have the correct rings for the engine. If the engine block has been bored oversize, necessitating oversize pis-tons, matching oversize rings are required.

7   Repeat the procedure for each ring that will be installed in the first cylinder and for each ring in the remaining cylinders. Remem-ber to keep rings, pistons and cylinders matched up.

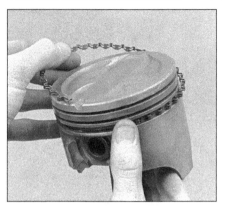

22.9a Installing the spacer/expander in the oil control ring groove

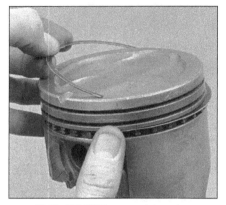

22.9b DO NOT use a piston ring installation tool when installing the oil ring side rails

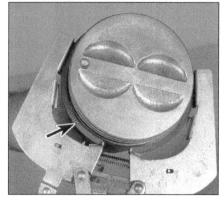

22.12 Installing the compression rings with a ring expander - the mark (arrow) must face up

8   Once the ring end gaps have been checked/corrected, the rings can be installed on the pistons.

9   The oil control ring (lowest one on the piston) is usually installed first. It's composed of three separate components. Slip the spacer/expander into the groove **(see illustration)**. If an anti-rotation tang is used, make sure it's inserted into the drilled hole in the ring groove. Next, install the lower side rail. Don't use a piston ring installation tool on the oil ring side rails, as they may be damaged. Instead, place one end of the side rail into the groove between the spacer/expander and the ring land, hold it firmly in place and slide a finger around the piston while pushing the rail into the groove **(see illustration)**. Next, install the upper side rail in the same manner.

10   After the three oil ring components have been installed, check to make sure both the upper and lower side rails can be turned smoothly in the ring groove.

11   The number two (middle) ring is installed next. It's usually stamped with a mark, which must face up, toward the top of the piston. **Note:** *Always follow the instructions printed on the ring package or box - different manufacturers may require different approaches. Don't mix up the top and middle rings, as they have different cross-sections.*

12   Use a piston ring installation tool and make sure the identification mark is facing the top of the piston, then slip the ring into the middle groove on the piston **(see illustration)**. Don't expand the ring any more than necessary to slide it over the piston.

13   Install the number one (top) ring in the same manner. Make sure the mark is facing up. Be careful not to confuse the number one and number two rings.

14   Repeat the procedure for the remaining pistons and rings.

---

23   Crankshaft - installation and main bearing oil clearance check

---

1   Crankshaft installation is the first step in engine reassembly. It's assumed at this point that the engine block and crankshaft have

been cleaned, inspected and repaired or reconditioned.

2   Position the engine with the bottom facing up.

3   Remove the main bearing cap bolts and lift out the caps. Lay them out in the proper order to ensure correct installation.

4   If they're still in place, remove the original bearing inserts from the block and the main bearing caps. Wipe the bearing surfaces of the block and caps with a clean, lint-free cloth. They must be kept spotlessly clean.

Main bearing oil clearance check

Refer to illustrations 23.11 and 23.15
Note: *Don't touch the faces of the new bearing inserts with your fingers. Oil and acids from your skin can etch the bearings.*

5   Clean the back sides of the new main bearing inserts and lay one in each main bearing saddle in the block. If one of the bearing inserts from each set has a large groove in it, make sure the grooved insert is installed in the block. Lay the other bearing from each set in the corresponding main bearing cap. Make sure the tab on the bearing insert fits into the recess in the block or cap, neither higher than the cap's edge nor lower. **Caution:** *The oil holes in the block must line up with the oil holes in the bearing inserts. Do not hammer the bearing into place and don't nick or gouge the bearing faces. No lubrication should be used at this time.*

6   The flanged thrust bearing must be installed in the last main cap. **Caution:** *Some engines have a 0.008-inch oversize rear main bearing. Check your crankshaft for a marking on the last counterweight, and check the backside of the old bearing insert for a similar marking. If they are marked oversize, an oversize rear bearing will be required.*

7   Clean the faces of the bearings in the block and the crankshaft main bearing journals with a clean, lint-free cloth.

8   Check or clean the oil holes in the crankshaft, as any dirt here can go only one way - straight through the new bearings.

9   Once you're certain the crankshaft is clean, carefully lay it in position in the main bearings.

23.11 Lay the Plastigage strips (arrow) on the main bearing journals, parallel to the crankshaft centerline

10   Before the crankshaft can be permanently installed, the main bearing oil clearance must be checked.

11   Cut several pieces of the appropriate size Plastigage (they should be slightly shorter than the width of the main bearings) and place one piece on each crankshaft main bearing journal, parallel with the journal axis **(see illustration)**.

12   Clean the faces of the bearings in the caps and install the caps in their original locations (don't mix them up) with the arrows pointing toward the front of the engine. Don't disturb the Plastigage.

13   Starting with the center main and working out toward the ends, tighten the main bearing cap bolts to the torque listed in this Chapter's Specifications in three steps. Don't rotate the crankshaft at any time during this operation, and do not tighten one cap completely - tighten all caps equally. The main caps can be seated using light taps with a brass or plastic mallet.

14   Remove the bolts/studs and carefully lift off the main bearing caps. Keep them in order. Don't disturb the Plastigage or rotate the crankshaft. If any of the main bearing caps are difficult to remove, tap them gently from side-to-side with a soft-face hammer to loosen them.

15   Compare the width of the crushed

23.15 Measuring the width of the crushed Plastigage to determine the main bearing oil clearance (be sure to use the correct scale - standard and metric ones are included)

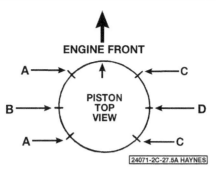

25.5 Piston ring end gap positions

A   *Oil ring rail gaps*
B   *Second compression ring gap*
C   *Oil ring spacer gap (position between marks)*
D   *Top compression ring gap*

25.9 The notch or arrow on each piston must face the front end of the engine as the pistons are installed

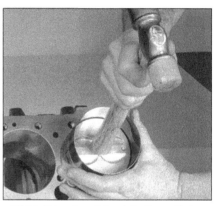

25.11 Drive the piston into the cylinder bore with the end of a wooden or plastic hammer handle

Plastigage on each journal to the scale printed on the Plastigage envelope to obtain the main bearing oil clearance **(see illustration)**. Check the Specifications to make sure it's correct.

16   If the clearance is not as specified, the bearing inserts may be the wrong size (which means different ones will be required). Before deciding different inserts are needed, make sure no dirt or oil was between the bearing inserts and the caps or block when the clearance was measured. If the Plastigage was wider at one end than the other, the journal may be tapered (see Section 19).

17   Carefully scrape all traces of the Plastigage material off the main bearing journals and/or the bearing faces. Use your fingernail or the edge of a credit card - don't nick or scratch the bearing faces.

Final crankshaft installation

18   Carefully lift the crankshaft out of the engine.

19   Clean the bearing faces in the block, then apply a thin, uniform layer of moly-base grease or engine assembly lube to each of the bearing surfaces. Be sure to coat the thrust faces as well as the journal face of the thrust bearing.

20   Make sure the crankshaft journals are clean, then lay the crankshaft back in place in the block.

21   Clean the faces of the bearings in the caps, then apply lubricant to them.

22   Install the caps in their original locations with the arrows pointing toward the front of the engine.

23   With all caps in place and bolts just started, tap the ends of the crankshaft forward and backward with a lead or brass hammer to line up the main bearing and crankshaft thrust surfaces.

24   Following the procedures outlined in Step 13, retighten all main bearing cap bolts to the torque listed in this Chapter's Specifications, starting with the center main and working out toward the ends.

25   Rotate the crankshaft a number of times by hand to check for any obvious binding.

26   The final step is to check the crankshaft endplay with feeler gauges or a dial indicator as described in Section 14. The endplay should be correct if the crankshaft thrust faces aren't worn or damaged and new bearings have been installed.

24   Rear main oil seal - replacement

Refer to Chapter 2 part A for the rear main seal replacement procedure.

25   Pistons and connecting rods - installation and rod bearing oil clearance check

1   Before installing the piston/connecting rod assemblies, the cylinder walls must be perfectly clean, the top edge of each cylinder must be chamfered, and the crankshaft must be in place.

2   Remove the cap from the end of the number one connecting rod (check the marks made during removal). Remove the original bearing inserts and wipe the bearing surfaces of the connecting rod and cap with a clean, lint-free cloth. They must be kept spotlessly clean.

Piston installation and rod bearing oil clearance check

Refer to illustrations 25.5, 25.9, 25.11, 25.13 and 25.17

Note: *Don't touch the faces of the new bearing inserts with your fingers. Oil and acids from your skin can etch the bearings.*

3   Clean the back side of the new upper bearing insert, then lay it in place in the connecting rod. Make sure the tab on the bearing fits into the recess in the rod. Don't hammer the bearing insert into place and be very careful not to nick or gouge the bearing face. Don't lubricate the bearing at this time.

4   Clean the back side of the other bearing insert and install it in the rod cap. Again, make

sure the tab on the bearing fits into the recess in the cap, and don't apply any lubricant. It's critically important that the mating surfaces of the bearing and connecting rod are perfectly clean and oil free when they're assembled.

5   Stagger the piston ring gaps around the piston **(see illustration)**.

6   Slip a section of plastic or rubber hose over each connecting rod cap bolt.

7   Lubricate the piston and rings with clean engine oil and attach a piston ring compressor to the piston. Leave the skirt protruding about 1/4-inch to guide the piston into the cylinder. The rings must be compressed until they're flush with the piston.

8   Rotate the crankshaft until the number one connecting rod journal is at BDC (bottom dead center) and apply a coat of engine oil to the cylinder walls.

9   With the mark or notch on top of the piston **(see illustration)** facing the front of the engine, gently insert the piston/connecting rod assembly into the number one cylinder bore and rest the bottom edge of the ring compressor on the engine block.

10   Tap the top edge of the ring compressor to make sure it's contacting the block around its entire circumference.

11   Gently tap on the top of the piston with the end of a wooden or plastic hammer handle **(see illustration)** while guiding the end of

25.13 Lay the Plastigage strips on each rod bearing journal, parallel to the crankshaft centerline

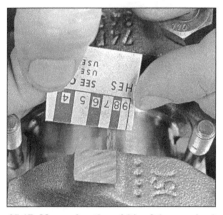

25.17 Measuring the width of the crushed Plastigage to determine the rod bearing oil clearance (be sure to use the correct scale - standard and metric ones are included)

the connecting rod into place on the crankshaft journal. The piston rings may try to pop out of the ring compressor just before entering the cylinder bore, so keep some pressure down on the ring compressor. Work slowly, and if any resistance is felt as the piston enters the cylinder, stop immediately. Find out what's hanging up and fix it before proceeding. Do not, for any reason, force the piston into the cylinder - you might break a ring and/or the piston.

12   Once the piston/connecting rod assembly is installed, the connecting rod bearing oil clearance must be checked before the rod cap is permanently bolted in place.

13   Cut a piece of the appropriate size Plastigage slightly shorter than the width of the connecting rod bearing and lay it in place on the number one connecting rod journal, parallel with the journal axis **(see illustration)**.

14   Clean the connecting rod cap bearing face, remove the protective hoses from the connecting rod bolts and install the rod cap. Make sure the mating mark on the cap is on the same side as the mark on the connecting rod.

15   Install the nuts and tighten them to the torque listed in this Chapter's Specifications. Work up to it in three steps. **Note:** *Use a thin-wall socket to avoid erroneous torque readings that can result if the socket is wedged between the rod cap and nut. If the socket tends to wedge itself between the nut and the cap, lift up on it slightly until it no longer contacts the cap.* Do not rotate the crankshaft at any time during this operation.

16   Remove the nuts and detach the rod cap, being very careful not to disturb the Plastigage.

17   Compare the width of the crushed Plastigage to the scale printed on the Plastigage envelope to obtain the oil clearance **(see illustration)**. Compare it to this Chapter's Specifications to make sure the clearance is correct.

18   If the clearance is not as specified, the bearing inserts may be the wrong size (which means different ones will be required). Before deciding different inserts are needed, make sure no dirt or oil was between the bearing

inserts and the connecting rod or cap when the clearance was measured. Also, recheck the journal diameter. If the Plastigage was wider at one end than the other, the journal may be tapered (see Section 19).

Final connecting rod installation

19   Carefully scrape all traces of the Plastigage material off the rod journal and/or bearing face. Be very careful not to scratch the bearing - use your fingernail or the edge of a credit card.

20   Make sure the bearing faces are perfectly clean, then apply a uniform layer of clean moly-base grease or engine assembly lube to both of them. You'll have to push the piston into the cylinder to expose the face of the bearing insert in the connecting rod - be sure to slip the protective hoses over the rod bolts first.

21   Slide the connecting rod back into place on the journal, remove the protective hoses from the rod cap bolts, install the rod cap and tighten the nuts to the torque listed in this Chapter's Specifications. Again, work up to the torque in three steps.

22   Repeat the entire procedure for the remaining pistons/connecting rods.

23   The important points to remember are ...

   a) *Keep the back sides of the bearing inserts and the insides of the connecting rods and caps perfectly clean when assembling them.*

   b) *Make sure you have the correct piston/rod assembly for each cylinder.*

   c) *The arrow or mark on the piston must face the front of the engine.*

   d) *Lubricate the cylinder walls with clean oil.*

   e) *Lubricate the bearing faces when installing the rod caps after the oil clearance has been checked.*

24   After all the piston/connecting rod assemblies have been properly installed, rotate the crankshaft a number of times by hand to check for any obvious binding.

25   As a final step, the connecting rod end-play must be checked (see Section 14).

26   Compare the measured endplay to this Chapter's Specifications to make sure it's correct. If it was correct before disassembly and the original crankshaft and rods were reinstalled, it should still be right. If new rods or a new crankshaft were installed, the end-play may be inadequate. If so, the rods will have to be removed and taken to an automotive machine shop for re-sizing.

---

26 Initial start-up and break-in after overhaul

Warning: *Have a fire extinguisher handy when starting the engine for the first time.*

1   Once the engine has been installed in the vehicle, double-check the oil and coolant levels.

2   With the spark plugs out of the engine, remove ECM IGN fuse from the instrument panel fuse block and the IGNITION fuse from the underhood fuse block (1993 and earlier models), the PCM IGN fuse from the instrument panel fuse block and the IGNITION fuse from the underhood fuse block (1994 and 1995 models), or the PCM BATT fuse from the instrument panel fuse block and the IGNITION fuse from the underhood fuse block (1996 models). Crank the engine until oil pressure registers on the gauge or the light goes out.

3   Install the spark plugs, hook up the plug wires and install the fuses.

4   Start the engine. It may take a few moments for the fuel system to build up pressure, but the engine should start without a great deal of effort. **Note:** *If the engine keeps backfiring, recheck the valve timing and spark plug wire routing.*

5   After the engine starts, it should be allowed to warm up to normal operating temperature. While the engine is warming up, make a thorough check for fuel, oil and coolant leaks.

6   Shut the engine off and recheck the engine oil and coolant levels.

7   Drive the vehicle to an area with no traffic, accelerate from 30 to 50 mph, then allow the vehicle to slow to 30 mph with the throttle closed. Repeat the procedure 10 or 12 times. This will load the piston rings and cause them to seat properly against the cylinder walls. Check again for oil and coolant leaks.

8   Drive the vehicle gently for the first 500 miles (no sustained high speeds) and keep a constant check on the oil level. It isn't unusual for an engine to use oil during the break-in period.

9   At approximately 500 to 600 miles, change the oil and filter.

10   For the next few hundred miles, drive the vehicle normally. Don't pamper it or abuse it.

11   After 2000 miles, change the oil and filter again and consider the engine broken in.

Chapter 3
Cooling, heating and air conditioning systems

Contents

Specifications

General

Coolant capacity	See Chapter 1
Radiator pressure cap rating	15 psi
Thermostat opening temperature	180-degrees F
Refrigerant type	
1993 and earlier	R-12
1994 and later	R-134a
Refrigerant capacity	
1993 and earlier	3.125 pounds
1994 and later	1.75 pounds

Torque specifications

	Ft-lbs (unless otherwise indicated)
Oil cooler lines	18
Thermostat housing bolts	21
Water pump attaching bolts	30
Water pump pulley bolts	18

1 General information

All vehicles covered by this manual employ a pressurized engine cooling system with thermostatically controlled coolant circulation. Coolant is drawn from the radiator by an impeller-type water pump mounted at the front of the block. The coolant is then circulated through the engine block, intake manifold and the cylinder heads before it's redirected back into the radiator.

A wax pellet type thermostat is located in the thermostat housing near the front of the engine. During warm up, the closed thermostat prevents coolant from circulating through the radiator. When the engine reaches normal operating temperature, the thermostat opens and allows hot coolant to travel through the radiator, where it is cooled before returning to the engine.

The cooling system is pressurized by a spring-loaded radiator cap, which, by maintaining pressure, increases the boiling point of the coolant. If the coolant temperature goes above this increased boiling point, the extra pressure in the system forces the radiator cap valve off its seat and allows the coolant to escape through the overflow tube into a coolant reservoir. When the system cools, the excess coolant is automatically drawn from the reservoir tank back into the radiator.

The coolant reservoir serves as both the point at which fresh coolant is added to the cooling system to maintain the proper fluid level and as a holding tank for overheated coolant.

The heating system works by directing air through the heater core mounted in the dash and then to the interior of the vehicle by a system of ducts. Temperature is controlled by mixing heated air with fresh air, using a system of doors in the ducts, and a blower motor.

Air conditioning is an optional accessory, consisting of an evaporator core located under the dash, a condenser in front of the radiator, an accumulator in the engine compartment and a belt-driven compressor mounted at the front of the engine.

2.4 An inexpensive hydrometer can be used to test the condition of your coolant

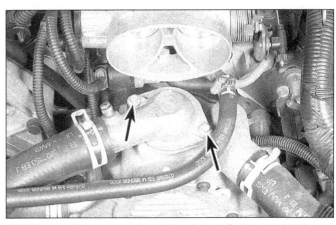

3.9 Thermostat cover bolt locations (arrows) - late model shown

2   Antifreeze - general information

Refer to illustration 2.4

Warning: *Do not allow antifreeze to come in contact with your skin or painted surfaces of the vehicle. Rinse off spills immediately with plenty of water. Antifreeze is highly toxic if ingested. Never leave antifreeze lying around in an open container or in puddles on the floor; children and pets are attracted by it's sweet smell and may drink it. Check with local authorities about disposing of used antifreeze. Many communities have collection centers which will see that antifreeze is disposed of safely. Never dump used anti-freeze on the ground or pour it into drains.*

Note: *Non-toxic antifreeze is now available at most auto parts stores, but even these types should be disposed of properly.*

The cooling system should be filled with a water/ethylene glycol based antifreeze solution which will prevent freezing down to at least -20-degrees F (even lower in cold climates). It also provides protection against corrosion and increases the coolant boiling point.

The cooling system should be drained, flushed and refilled at least every other year (see Chapter 1). The use of antifreeze solutions for periods of longer than two years is likely to cause damage and encourage the formation of rust and scale in the system. However, 1996 models are filled with a new, long-life coolant called "Dex-Cool", which the factory claims is good for five years.

Before adding antifreeze to the system, check all hose connections. Antifreeze can leak through very minute openings.

The exact mixture of antifreeze to water which you should use depends on the relative weather conditions. The mixture should contain at least 50-percent antifreeze, but should never contain more than 70-percent anti-freeze. Consult the mixture ratio chart on the antifreeze container before adding coolant. Hydrometers are available at most auto parts stores to test the coolant **(see illustration)**. Use antifreeze which meets the vehicle manufacturer's specifications.

3   Thermostat - check and replacement

Refer to illustrations 3.9, 3.10 and 3.13

Warning: *The engine must be completely cool when this procedure is performed.*

Note: *Don't drive the vehicle without a thermostat! The computer may stay in open loop mode and emissions and fuel economy will suffer.*

Check

1   Before condemning the thermostat, check the coolant level, drivebelt tension and temperature gauge (or light) operation.

2   If the engine takes a long time to warm up, the thermostat is probably stuck open. Replace the thermostat.

3   If the engine runs hot, check the temperature of the upper radiator hose. If the hose isn't hot, the thermostat is probably stuck shut. Replace the thermostat.

4   If the upper radiator hose is hot, it means the coolant is circulating and the thermostat is open. Refer to the *Troubleshooting* at the front of this manual for the cause of overheating.

5   If an engine has been overheated, you may find damage such as leaking head gas-

kets, scuffed pistons and warped or cracked cylinder heads.

Replacement

6   Drain coolant (about 1 gallon) from the radiator, until the coolant level is below the thermostat housing.

7   Remove the air intake duct (see Chapter 4).

8   Disconnect the radiator hose from the thermostat cover. **Note:** *Later models do not have a thermostat housing gasket, and it isn't necessary to remove the hose unless it is being replaced.*

9   Remove the bolts and lift the cover off **(see illustration)**. It may be necessary to tap the cover with a soft-face hammer to break the gasket seal.

10   Note how it's installed, then remove the thermostat **(see illustration)**. Be sure to use a replacement thermostat with the correct opening temperature (see this Chapter's Specifications).

11   If a gasket was used, use a scraper or putty knife to remove all traces of old gasket material and sealant from the mating surfaces. **Caution:** *Be careful not to gouge or damage the gasket surfaces, because a leak could develop after assembly. Make sure no*

3.10 Note how it's installed, then remove the thermostat (the spring end points toward the engine)

3.13 On later models, install a new rubber seal around the thermostat

4.9 Disconnect the electrical connector from the fan and apply fused battery power and ground to test the fan

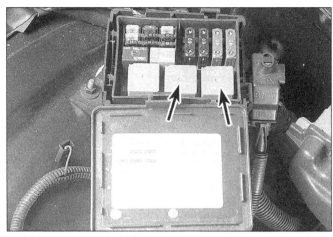

4.10 The cooling fan relays (arrows) are easily identifiable by viewing the decal on the inside of the underhood fuse/relay box cover

gasket material falls into the coolant passage; it's a good idea to stuff a rag in the passage. Wipe the mating surfaces with a rag saturated with lacquer thinner or acetone.

12   Install the thermostat and make sure the correct end faces out - the spring is directed toward the engine **(see illustration 3.10).**

13   Later models will not have a traditional gasket, but rather a rubber ring around the thermostat. If so, replace this ring and install the thermostat cover without gasket sealant **(see illustration)**. **Note:** *If a gasket was used, apply a thin coat of RTV sealant to both sides of the new gasket and position it on the engine side, over the thermostat, and make sure the gasket holes line up with the bolt holes in the housing.*

14   Carefully position the cover and install the bolts. Tighten them to the torque listed in this Chapter's Specifications - do not over-tighten the bolts or the cover may crack or become distorted.

15   Reattach the radiator hose to the cover and tighten the clamp - now may be a good time to check and replace the hoses and clamps (see Chapter 1).

16   Refer to Chapter 1 and refill the system, then run the engine and check carefully for leaks.

17   Repeat steps 1 through 5 to be sure the repairs corrected the previous problem(s).

---

4   Engine cooling fan(s) - check

Warning: *Keep hands, tools and clothing away from the fan. To avoid injury or damage DO NOT operate the engine with a damaged fan. Do not attempt to repair fan blades - replace a damaged fan with a new one.*

Mechanical fan

1   Early model vehicles covered by this manual are equipped with a thermostatically controlled fan clutch.

2   Begin the fan clutch check with a luke-warm engine (start the engine when cold and let it run for two minutes only).

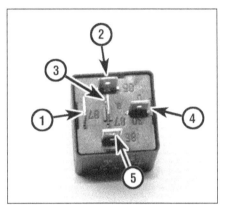

4.11 Test for continuity between relay terminals 1 and 4 - there should be no continuity until power is applied to terminal 5 and ground to terminal 2

3   Remove the key from the ignition switch for safety purposes.

4   Turn the fan blades and note the resistance. There should be moderate resistance, depending on temperature. If the fan spins freely by hand, the clutch should be replaced.

5   Drive the vehicle until the engine is warmed up. Shut it off and remove the key.

6   Turn the fan blades by hand and again note the resistance. There should be a notice-able increase in resistance.

7   If the fan clutch fails this check or is locked up, replacement is necessary. If excessive fluid is leaking from the hub or lateral play over 1/4-inch is noted, replace the fan clutch. **Note:** *Excessive noise from the fan assembly at all times when the engine is above high idle indicates the clutch may be locked up.*

8   If any fan blades are bent, DO NOT attempt to straighten them! The metal will become fatigued and could cause the blades to break or fly off during operation. Replace the fan with a new one.

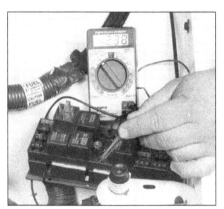

4.13 Probing with a grounded test light or voltmeter, check terminal no.4 on the relay panel - there should be power at all times, and only with the key On at terminal no. 2

Electric fan

Refer to illustrations 4.9 4.10, 4.11 and 4.13

Note: *Later model vehicles are equipped with two electric fans, except late models equipped with a heavy-duty cooling system option which incorporates a mechanical fan and an electric fan. The following procedures apply to all electric fans.*

9   To test a fan motor, unplug the electrical connector at the motor and use fused jumper wires to connect the fan directly to the battery **(see illustration)**. If the fan still doesn't work, replace the motor.

10   If the motor tests OK, check the cooling fan relays, located in the underhood fuse/relay panel **(see illustration)**.

11   Remove cooling fan relay no. 1 and test for continuity between terminals 1 and 4 **(see illustration)**. There should be no continuity.

12   Apply power to terminal 5, and ground to terminal 2, and test for continuity again between 1 and 4. There should now be continuity; if not, replace the relay. This test applies to both relays.

13   Test the number 4 terminal (on the fuse/relay panel) for each relay **(see illustration)**. There should be power at all times,

5.3a Detach the bolts (arrows) securing the upper fan shroud to the radiator support . . .

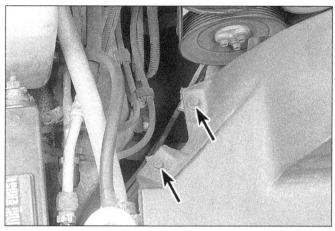

5.3b . . . then remove the bolts (arrows) securing each side of the upper fan shroud to the lower fan shroud

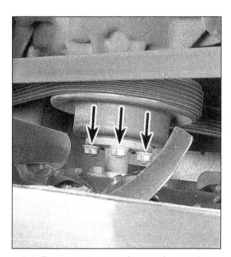

5.4 Remove the nuts (arrows) securing the fan assembly to the engine pulley

5.6 Remove the four bolts securing the fan clutch to the fan blade assembly

5.12a On later models equipped with electric fans, detach the retaining bolts (arrows) securing the top of the fan . . .

5.12b . . . and the outer edges of the fan assembly

probing with a grounded test light. The number 2 terminal should have power only with the key ON.

14   If the relays and the fan motor are good, check the wiring from the relays to the PCM (computer) for open or short circuits.

15   If the circuit checks OK but the fan(s) still don't come on, check the engine coolant temperature sensor.

---

5   Engine cooling fan(s) - removal and installation

Warning: *Keep hands, tools and clothing away from the fan. To avoid injury or damage DO NOT operate the engine with a damaged fan. Do not attempt to repair fan blades - replace a damaged fan with a new one.*
Caution: *On models equipped with a Delco Loc II or Theftlock audio system, be sure the lockout feature is turned off before performing any procedure which requires disconnecting the battery.*

Mechanical fan

Refer to illustrations 5.3a, 5.3b, 5.4 and 5.6

1   Disconnect the cable from the negative terminal of the battery.
2   Remove the air intake duct and air cleaner on later models equipped with a heavy-duty cooling system (mechanical fan and electric) (see Chapter 4).
3   Remove the bolts and detach the upper fan shroud from the radiator **(see illustrations)**.
4   Remove the fasteners securing the fan assembly to the water pump hub **(see illustration)**.
5   Detach the fan and clutch assembly from the vehicle as a unit.
6   To replace the fan clutch, simply unbolt it from the fan blade assembly **(see illustration)**.
7   Installation is the reverse order of removal. Tighten all bolts and fasteners securely.

Electric fan(s)

Refer to illustrations 5.12a and 5.12b
Note: *This procedure applies to either fan.*
8   Disconnect the cable from the negative

terminal of the battery.
9   Remove the air intake duct and air cleaner (see Chapter 4).
10   Remove the bolts and detach upper fan shroud from the radiator on later models equipped with a heavy-duty cooling system **(see illustrations 5.3a and 5.3b)**.

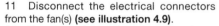

6.6a Disconnect the transmission cooler lines (arrows) from the right side of the radiator

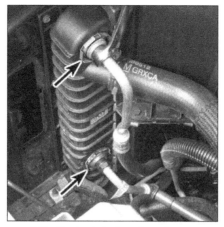

6.6b On the left side, if equipped, disconnect the oil cooler lines (arrows)

11 Disconnect the electrical connectors from the fan(s) **(see illustration 4.9)**.
12 Detach the fan retaining bolts **(see illustrations)**. **Note:** *It is extremely important to mark the installed position of each fan (left side or right side) because each fan motor has a different wattage rating and insufficient cooling could occur if they are not installed back into their original position.*
13 Pull the fan assembly up slightly to dislodge its tabs from the radiator, then guide the fan assembly out from the engine compartment, making sure that all wiring clips are disconnected. Be careful not to contact the radiator cooling fins
14 Installation is the reverse of removal.

---

6 Radiator and coolant reservoir - removal and installation

Warning: *The engine must be completely cool when this procedure is performed.*
Caution: *On models equipped with a Delco*

Loc II or Theftlock audio system, be sure the lockout feature is turned off before performing any procedure which requires disconnecting the battery.

Radiator

Refer to illustrations 6.6a, 6.6b and 6.7
1 Disconnect the cable from the negative terminal of the battery.
2 Drain the cooling system as described in Chapter 1. Refer to the coolant **Warning** in Section 2.
3 Remove the air intake duct and air cleaner (see Chapter 4).
4 Remove the cooling fan assembly (see Section 5).
5 Disconnect the upper and lower radiator hoses and the heater hose (early models only) from the radiator, then disconnect the coolant recovery hoses at the radiator neck.
6 Remove the automatic transmission cooler lines from the right tank of the radiator and the engine oil cooler lines (if equipped) from the left tank of the radiator **(see illustrations)**. Be careful not to damage the lines or

fittings. Plug the ends of the disconnected lines to prevent leakage and stop dirt from entering the system. Have a drip pan ready to catch any spills.
7 Remove the upper radiator support panel on vehicles equipped with electric cooling fans **(see illustration)**.
8 The radiator can be removed from the top with some wiggling to clear various components surrounding the radiator.
9 Prior to installation of the radiator, replace any damaged hose clamps, radiator hoses and radiator mounts.
10 If leaks have been noticed or there have been cooling problems, have the radiator cleaned and tested at a radiator shop.
11 Radiator installation is the reverse of removal. Be sure to seat the radiator securely in the lower mounts before installing the upper support panel or upper fan shroud.
12 After installation, fill the system with the proper mixture of antifreeze, bleed the air from the cooling system as described in Chapter 1, and check the automatic transmission fluid level. Also check the engine oil level, if equipped with an oil cooler (see Chapter 1).

Coolant reservoir

Refer to illustration 6.16
13 Disconnect the cable from the negative terminal of the battery.
14 Drain the cooling system as described in Chapter 1 until the coolant reservoir is empty. Refer to the coolant **Warning** in Section 2.
15 Remove the coolant recovery hoses from the coolant reservoir.
16 Disconnect the connector from the low coolant warning sensor **(see illustration)**.
17 Detach the reservoir mounting bolts and remove it from the vehicle.
18 Prior to installation make sure the reservoir is clean and free of debris which could be drawn into the radiator (wash it with soapy water and a brush if necessary, then rinse thoroughly).
19 Installation is the reverse of removal.

6.7 Remove the upper radiator support bolts (arrows)

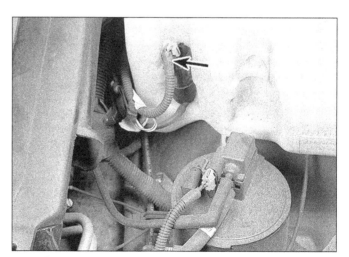

6.16 Disconnect the wiring harness from low coolant warning sensor

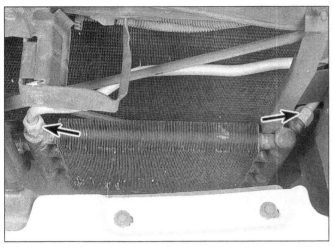

8.3 Detach the transmission cooler lines (arrows), then plug them to prevent fluid leakage and dirt from entering the system

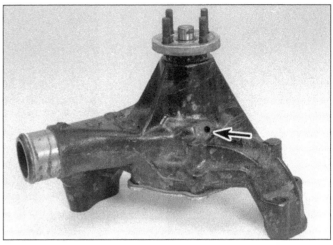

9.2a On early models the weep hole (arrow) is located on the bottom of the water pump (pump removed for clarity)

7 Coolant temperature sending unit - check and replacement

Warning: *Wait until the engine is completely cool before beginning this procedure.*

Check

1   The coolant temperature indicator system is composed of a temperature gauge or warning light mounted in the dash and a coolant temperature sending unit mounted on the engine. The sending unit is mounted on the left cylinder head, below the exhaust manifold.

2   If an overheating indication occurs, check the coolant level in the system and then make sure the wiring between the gauge and the sending unit is secure and all fuses are intact.

3   Check the circuit and gauge operation by grounding the wire to the sending unit while the ignition is On (engine NOT running). If the gauge deflects full scale, the circuit and gauge are OK. The problem lies in the sending unit.

4   To confirm the sending unit is defective, check the resistance of the unit when the engine is cool. Resistance should be high - approximately 1365 ohms at 100-degrees F. Next, run the engine until it is fully warmed up and check the resistance of the sending unit again. The resistance should now be low - approximately 55 ohms at 260-degrees F. If it doesn't respond at close to these numbers, replace the sending unit.

Replacement

5   Make sure the engine is cool before removing the defective sending unit. There will be some coolant loss as the unit is removed, so be prepared to catch it. Refer to the coolant **Warning** in Section 2.

6   Prepare the new sending unit by wrapping the threads with Teflon tape or by coating them with sealant. Disconnect the electrical connector and unscrew the sensor. Install

9.2b On later models the weep hole (arrow) is located on the top of the water pump (pump removed for clarity)

the new sensor as quickly as possible to minimize coolant loss.

7   Check the coolant level after the replacement unit has been installed and fill up the system, if necessary (see Chapter 1). Check now for proper operation of the gauge and sending unit.

8 Engine oil cooler and transmission oil cooler - general information and replacement

Refer to illustration 8.3

1   Several types of oil coolers were used during the production period covered by this manual. The engine oil cooler (if equipped) is internally mounted in the left radiator tank and it works in the same fashion as the transmission oil cooler which is mounted in the right tank of the radiator. Police and taxi vehicles are equipped with transmission oil coolers that are externally mounted in front of the radiator and have a greater cooling capacity.

2   Internally mounted oil coolers can not be serviced and should be taken to a radiator repair facility for further inspection.

3   To replace an externally mounted transmission cooler, simply detach the cooler lines **(see illustration)**. Be careful not to damage the lines or fittings. Plug the ends of the disconnected lines to prevent leakage and stop dirt from entering the system. Have a drip pan ready to catch any spills.

4   Detach the cooler mounting bolts and remove the cooler from the vehicle.

5   Installation is the reverse of removal

9 Water pump - check

Refer to illustrations 9.2a, 9.2b and 9.4

1   Water pump failure can cause overheating and serious damage to the engine. There are three ways to check the operation of the water pump while it is installed on the engine. If any one of the following quick-checks indicates water pump problems, it should be replaced immediately.

2   A seal protects the water pump impeller shaft bearing from contamination by engine coolant. If this seal fails, a weep hole in the water pump snout will leak coolant **(see illustrations)** (an inspection mirror can be used to look at the underside of the pump if the hole

9.4 On early models, check the pump for loose or rough bearings (this can be done with the belt removed and the fan and the pulley in place)

10.5 Use locking pliers to squeeze the hose clamp (arrow), then detach the lower radiator hose from the water pump

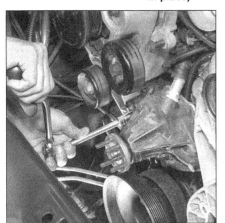

10.6 Unscrew the water pump retaining bolts and remove the water pump

10.9 Disconnect the electrical connector from the coolant temperature sensor (arrow)

10.11 After removing the air injection pump, remove these two bolts and the pump bracket - it covers a water pump stud

isn't on top). If the weep hole is leaking, shaft bearing failure will follow. Replace the water pump immediately.

3 The water pump impeller shaft bearing on early model vehicles can also prematurely wear out by an improperly tensioned drivebelt. When the bearing wears out, it emits a high-pitched squealing sound. If such a noise is coming from the water pump during engine operation, the shaft bearing has failed - replace the water pump immediately. **Note:** *Do not confuse belt noise with bearing noise.*

4 To identify excessive bearing wear on early models equipped with belt driven water pumps, grasp the water pump pulley and try to force it up-and-down or from side-to-side. If the pulley can be moved either horizontally or vertically, the bearing is nearing the end of its service life **(see illustration)**. Replace the water pump. Don't mistake drivebelt slippage, which causes a squealing sound, for water pump bearing failure.

5 It is possible for a water pump to be bad, even if it doesn't howl or leak water. Sometimes the fins on the back of the impeller can corrode away until the pump is no longer effective. The only way to check for this is to remove the pump for examination.

10 Water pump - removal and installation

Warning: *Wait until the engine is completely cool before starting this procedure.*

Removal

1 Disconnect the cable from the negative terminal of the battery. **Caution:** *On models equipped with a Delco Loc II or Theftlock audio system, be sure the lockout feature is turned off before performing any procedure which requires disconnecting the battery.*
2 Drain the coolant (see Chapter 1).
3 Remove the cooling fan (see Section 5) then remove the serpentine drivebelt (see Chapter 1).

1993 and earlier models

Refer to illustrations 10.5 and 10.6
4 Remove the water pump pulley from the front of the water pump.
5 Remove the lower radiator hose from the water pump **(see illustration)**.
6 Detach the water pump retaining bolts **(see illustration)** and remove the pump from

the vehicle. If necessary, strike the pump with a soft-face hammer or a block of wood or wooden hammer handle to break the gasket seal. Do not pry between the pump and the block.

1994 and later models

Refer to illustrations 10.9, 10.11 and 10.12
7 If equipped with a mechanical fan, remove the cooling fan idler pulley and bracket assembly located in front of the water pump.
8 Remove the upper and lower radiator hoses and heater hoses from the water pump.
9 Disconnect the electrical connector from the coolant temperature sensor **(see illustration)**, then unclip its wire harness from the two clips attached to the face of the water pump.
10 Unbolt and move the air injection pump (see Chapter 6).
11 Remove the two bolts holding the air injection pump bracket **(see illustration)**.
12 Remove the water pump mounting bolts and detach it. If necessary, strike the pump

10.12 Use a soft-faced hammer to break the gasket seal on the pump - arrows here indicate hoses to be disconnected on later models

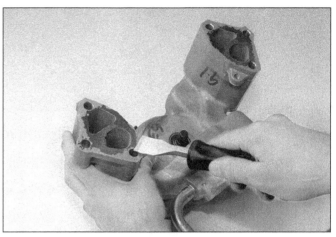

10.13 Remove all traces of old gasket material - use care to avoid gouging the soft aluminum on later models

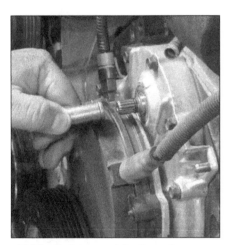

10.15a On later models, the water pump is mechanically driven by this splined shaft from the timing cover - held here is the splined coupler that joins the shaft to the splines on the back of the water pump

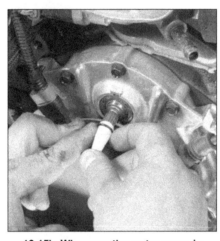

10.15b Whenever the water pump is removed on models, replace the O-ring seals - one at the pump splines, one at the drive splines - use something tapered to slip the seal evenly around the shafts

11 Blower motor and circuit - check

Refer to illustrations 11.4, 11.8a, 11.8b and 11.8c

Warning: *The models covered by this manual are equipped with airbags. Always disable the airbag system before working in the vicinity of the impact sensors, steering column or instrument panel to avoid the possibility of accidental deployment of the airbag(s), which could cause personal injury (see Chapter 12). The yellow wires and connectors routed through the instrument panel are for this system. Do not use electrical test equipment on these yellow wires or tamper with them in any way while working under the instrument panel.*

1   Check the fuse (marked HVAC) and all connections in the circuit for looseness and corrosion. Make sure the battery is fully charged.

2   With the transmission in Park, the parking brake securely set, turn the ignition switch to the Run position. It isn't necessary to start the vehicle.

3   Remove the lower right dash insulator panel (below the glove box) for access to the blower motor.

4   Backprobe the blower motor electrical connector with two small paper clips (straightened out) and connect a voltmeter to the blower motor connector and ground (**see illustration**).

5   Move the blower switch through each of its positions and note the voltage readings. Changes in voltage indicate that the motor speeds will also vary as the switch is moved to the different positions.

6   If there is voltage present, but the blower motor does not operate, the blower motor is probably faulty. Disconnect the blower motor connector, then hook one side of the blower motor terminals to a chassis ground and the other to a fused source of battery voltage. If the blower doesn't operate, it is faulty.

with a soft-face hammer or a block of wood or wooden hammer handle to break the gasket seal (**see illustration**). Do not pry between the pump and the block.

Installation

Refer to illustrations 10.13, 10.15a and 10.15b

13   Clean the sealing surfaces of all gasket material on both the water pump and block (**see illustration**). Wipe the mating surfaces with a rag saturated with lacquer thinner or acetone.

14   Apply a thin layer of RTV sealant to both sides of the new gasket and install the gasket on the water pump.

15   On later model vehicles, pull the water pump driveshaft coupling (**see illustration**) from the driveshaft (the short, splined shaft coming out of the block above the distributor). Use something tapered like a marking-pen top and slip lightly-oiled, new O-ring

seals over the splines at the driven-gear end and the splines at the back of the water pump (**see illustration**).

16   Place the water pump in position and install the bolts finger tight. Use caution to ensure that the gasket doesn't slip out of position. Remember to replace any mounting brackets secured by the water pump mounting bolts/studs. Tighten the bolts to the torque listed in this Chapter's Specifications. **Note:** *On later model vehicles, install the water pump driveshaft coupling over the driveshaft and align the pump's splines with the coupling, when placing the pump on the engine.*

17   The remainder of the installation procedure is the reverse of removal.

18   Add coolant to the specified level (see Chapter 1). Start the engine and check for the proper coolant level and the water pump and hoses for leaks. Bleed the cooling system of air as described in Chapter 1.

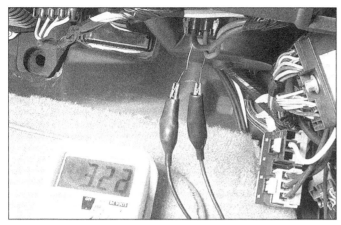

11.4 Connect a voltmeter to the heater blower motor connector (arrows) by backprobing, and check the running voltage at each blower switch position

11.8a The blower motor resistor (arrow) is located just to the left of the blower motor

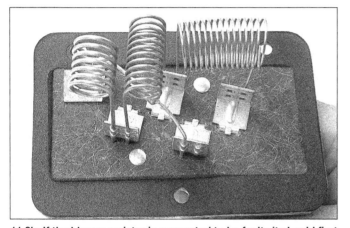

11.8b If the blower resistor is suspected to be faulty it should first be visually inspected for an open circuit in the resistor windings . . .

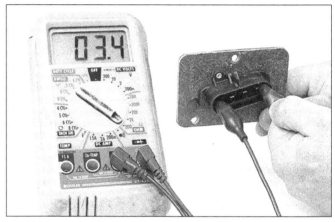

11.8c . . . then it should be checked for continuity between the terminals

7   If there was no voltage present at the blower motor at one or more speeds, and the motor itself tested OK, check the blower motor resistor.

8   Disconnect the electrical connector from the blower motor resistor **(see illustrations)**. With the ignition on, check for voltage at each of the terminals in the connector as the blower speed switch is moved to the different positions. If the voltmeter responds correctly to the switch then the resistor is probably faulty. If there is no voltage present from the switch, then the switch, control panel or related wiring is probably faulty.

---

12 Blower motor - removal and installation

Refer to illustrations 12.2 and 12.3
Warning: *The models covered by this manual are equipped with airbags. Always disable the airbag system before working in the vicinity of the impact sensors, steering column or instrument panel to avoid the possibility of accidental deployment of the airbag(s), which could cause personal injury (see Chapter 12). The yellow wires and connectors routed through*

12.2 Remove the three screws retaining the blower motor to the housing

the instrument panel are for this system. Do not use electrical test equipment on these yellow wires or tamper with them in any way while working under the instrument panel.

1   Remove the lower right dash insulator panel (below the glove box) for access to the blower motor.

12.3 Lower the blower motor and fan assembly straight down to remove it from the vehicle

2   Disconnect the electrical connector from the blower motor and remove the three screws from the blower housing **(see illustration)**.

3   Pull the blower motor and fan straight down **(see illustration)**.

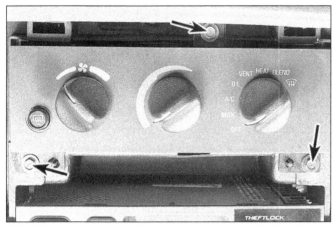

13.3a Remove the screws (arrows) to allow removal of the heater/air conditioning control panel

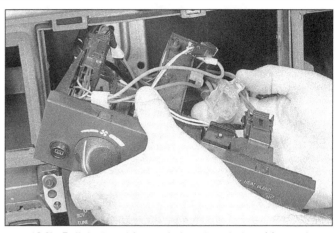

13.3b Pull the bezel forward, then detach the wiring and vacuum harness

4   To remove the fan from the blower motor, squeeze the spring clip together and slip the fan off the shaft. **Note:** *On some models, there is no clip or nut retaining the fan to the shaft; in this case the fan can be pried off with two screwdrivers.*

5   Install the fan onto the motor and install the blower motor into the heater housing.

---

13   Heater and air conditioning control assembly - removal and installation

Warning: *The models covered by this manual are equipped with airbags. Always disable the airbag system before working in the vicinity of the impact sensors, steering column or instrument panel to avoid the possibility of accidental deployment of the airbag(s), which could cause personal injury (see Chapter 12). The yellow wires and connectors routed through the instrument panel are for this system. Do not use electrical test equipment on these yellow wires or tamper with them in any way while working under the instrument panel.*

Removal

Refer to illustrations 13.3a, 13.3b and 13.3c

1   Disconnect the battery cable from the negative battery terminal. **Caution:** *On models equipped with a Delco Loc II or Theftlock audio system, be sure the lockout feature is turned off before performing any procedure which requires disconnecting the battery.*

2   Remove the main instrument panel bezel to allow access to the heater/air conditioning control mounting screws (see Chap-ter 11).

3   Remove the control assembly retaining screws and pull the unit from the dash **(see illustration)**. It can be pulled out just far enough to allow disconnecting the control cable end, electrical connections and vacuum harness (on air-conditioned models) from the control head. **Note:** *Use a small screwdriver to release the clips holding the control cable* **(see illustrations)**.

13.3c If equipped, use a small screwdriver to pry apart the snaps (arrow) that retain the control cable (upper arrow)

Installation

4   To install the control assembly, reverse the removal procedure. **Caution:** *When reconnecting vacuum harness to the control assembly, do not use any lubricant to make them slip on easier; it can affect vacuum operation. If necessary, use a drop of plain water to make reconnection easier.*

---

14   Heater core - removal and installation

Warning 1: *The models covered by this manual are equipped with airbags. Always disable the airbag system before working in the vicinity of the impact sensors, steering column or instrument panel to avoid the possibility of accidental deployment of the airbag(s), which could cause personal injury (see Chapter 12). The yellow wires and connectors routed through the instrument panel are for this system. Do not use electrical test equipment on these yellow wires or tamper with them in any way while working under the instrument panel.*

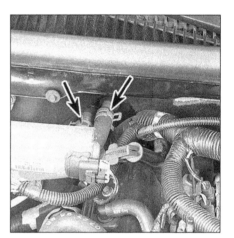

14.3 Loosen the hose clamps (arrows) and disconnect the heater hoses from the heater core tubes at the firewall

Warning 2: *The air conditioning system is under high pressure. DO NOT loosen any fittings or remove any components until after the system has been discharged. Air conditioning refrigerant should be properly discharged into an EPA-approved container at a dealership service department or an automotive air conditioning facility. Always wear eye protection when disconnecting air conditioning system fittings.*

Removal

Refer to illustrations 14.3, 14.5, 14.6, 14.7a and 14.7b

1   Disconnect the battery cable at the negative battery terminal. **Caution:** *On models equipped with a Delco Loc II or Theftlock audio system, be sure the lockout feature is turned off before performing any procedure which requires disconnecting the battery.*

2   Drain the cooling system (see Chapter 1).

3   Disconnect the heater hoses at the heater core inlet and outlet on the engine side of the firewall (passenger side) and plug the open fittings **(see illustration)**. If the hoses are stuck to the pipes, cut them off.

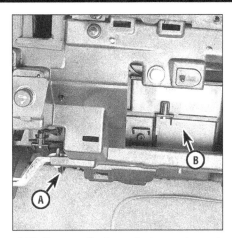

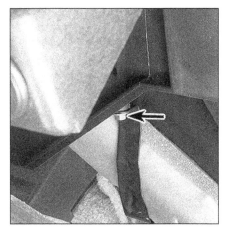

14.5 After the lower insulator panel and the glove box have been removed, disconnect any vacuum lines (arrows) attached to the heater core cover

14.6 Remove the lower instrument panel brace (A) and the heater core cover (B)

14.7a Remove the retaining screw (arrow) and the heater core retaining plate . . .

4   From the inside of the car, remove the lower right dash insulator panel (below the glove box) and the glove box (see Chapter 11).

5   Detach any vacuum lines which would interfere with the removal of the heater core cover **(see illustration)**.

6   Remove the lower instrument panel brace and heater core cover **(see illustration)**.

7   Remove the heater core clamp bolt and clamp **(see illustration)**, then slide the heater core out carefully **(see illustration)**.

Installation

Refer to illustration 14.8

8   Installation is the reverse of removal. **Note:** *When reinstalling the heater core, make sure any original insulating/sealing materials are in place around the heater core pipes and around the core* **(see illustration)**.

9   Refill the cooling system (see Chapter 1).

10   Start the engine and check for proper operation.

---

15   Air conditioning and heating system - check and maintenance

Warning: *The air conditioning system is under high pressure. DO NOT loosen any fittings or remove any components until after the system has been discharged. Air conditioning refrigerant should be properly discharged into an EPA-approved recovery container at a dealership service department or an automotive air conditioning repair facility. Always wear eye protection when disconnecting air conditioning system fittings.*

1   The following maintenance steps should be performed on a regular basis to ensure that the air conditioner continues to operate at peak efficiency:

a) *Check the drivebelt (see Chapter 1).*

b) *Check the condition of the hoses. Look*

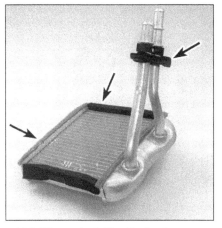

14.7b . . . then pull the heater core down and out to remove it

14.8 When reinstalling the heater core, make sure all of the sealing material (arrows) is in place

for cracks, hardening and deterioration. Look at potential leak areas (hoses and fittings) for signs of refrigerant oil leaking out. **Warning:** *Do not replace air conditioning hoses until the system has been discharged by a dealership or air conditioning repair facility.*

c) *Check the fins of the condenser for leaves, bugs and other foreign material. A soft brush and compressed air can be used to remove them.*

d) *Check the wire harness for correct routing, broken wires, damaged insulation, etc. Make sure the electrical connectors are clean and tight.*

e) *Maintain the correct refrigerant charge.*

2   The system should be run for about 10 minutes at least once a month. This is particularly important during the winter months because long-term non-use can cause hardening of the internal seals.

3   Because of the complexity of the air conditioning system and the special equipment required to effectively work on it, accurate troubleshooting of the system should be left to a certified air conditioning technician.

4   If the air conditioning system doesn't operate at all, check the fuse panel. Check

the HVAC fuse and the air conditioning compressor relay.

5   The most common cause of poor cooling is simply a low system refrigerant charge. If a noticeable drop in cool air output occurs, the following quick check will help you determine if the refrigerant level is low. For more complete information on the air conditioning system, refer to the *Haynes Automotive Heating and Air Conditioning Manual*.

Checking the refrigerant charge

6   Warm the engine up to normal operating temperature.

7   Place the air conditioning temperature selector at the coldest setting and the blower at the highest setting. Open the doors (to make sure the air conditioning system doesn't cycle off as soon as it cools the passenger compartment).

8   With the compressor engaged - the clutch will make an audible click and the center of the clutch will rotate - feel the surface of the accumulator and the evaporator inlet pipe. If there's no perceptible difference between the inlet pipe and the accumulator,

15.10 A basic charging kit for R-134a systems is available at most auto parts stores - 1993 and earlier models use R-12 refrigerant (which is no longer available to the consumer) and 1994 and later models use R-134a refrigerant

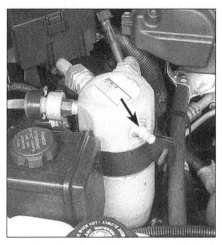

15.13 Add refrigerant to the low-side port only (arrow) - the procedure is easier if you wrap the can with a warm, wet towel to prevent icing

the system is properly charged. If there's a difference, there's something wrong with the system. It might be low charge, but it might be something else. To be sure, take the vehicle to a dealer service department or other qualified repair facility for further diagnosis.

9   Place a thermometer in the dashboard vent nearest the evaporator and add refrigerant to the system until the indicated temperature is around 40 to 45-degrees F. If the ambient (outside) air temperature is very high, say 110-degrees F, the duct air temperature will probably be higher, but generally the air conditioning is 30 to 40-degrees F cooler than the ambient air. **Note:** *Humidity of the ambient air also affects the cooling capacity of the system. Higher ambient humidity lowers the effectiveness of the air conditioning system.*

Adding refrigerant

Refer to illustrations 15.10 and 15.13
Caution: *There are two types of refrigerant; R-12, used on 1993 and earlier models, and the more environmentally friendly R-134a used on 1994 and later models. These two refrigerants (and their appropriate refrigerant oils) are not compatible and must never be mixed or components will be damaged.*
Note: *R-12 refrigerant is no longer available to the consumer. On 1993 and earlier models, the system must be serviced by a licensed automotive air conditioning technician.*
10   Buy an automotive charging kit at an auto parts store. A charging kit includes a 14-ounce can of refrigerant, a tap valve and a short section of hose that can be attached between the tap valve and the system low side service valve **(see illustration)**. Because one can of refrigerant may not be sufficient to bring the system charge up to the proper level, it's a good idea to buy an extra can. Make sure that one of the cans contains red refrigerant dye. If the system is leaking, the

red dye will leak out with the refrigerant and help you pinpoint the location of the leak.
11   Hook up the charging kit by following the manufacturer's instructions. **Warning:** *DO NOT hook the charging kit hose to the system high side!* The fittings on the charging kit are designed to fit **only** on the low side of the system.
12   Back off the valve handle on the charging kit and screw the kit onto the refrigerant can, making sure first that the O-ring or rubber seal inside the threaded portion of the kit is in place. **Warning:** *Wear protective eyewear when dealing with pressurized refrigerant cans.*
13   Remove the dust cap from the low-side charging connection and attach the quick-connect fitting on the kit hose **(see illustration)**.
14   Warm up the engine and turn on the air conditioner. Keep the charging kit hose away from the fan and other moving parts. **Note:** *The charging process requires the compressor to be running. Your compressor may cycle off if the pressure is low due to a low charge. If the clutch cycles off, you can pull the low-pressure cycling switch plug from the evaporator inlet line and attach a jumper wire across the terminals. This will keep the compressor ON.*
15   Turn the valve handle on the kit until the stem pierces the can, then back the handle out to release the refrigerant. You should be able to hear the rush of gas. Add refrigerant to the low side of the system until both the accumulator surface and the evaporator inlet pipe feel about the same temperature . Allow stabilization time between each addition.
16   If you have an accurate thermometer, you can place it in the center air conditioning duct inside the vehicle and keep track of the "conditioned" air temperature. A charged system that is working properly should put out air that is 40-degrees F. If the ambient (outside) air temperature is very high, say 110-degrees F, the duct air temperature will

probably be higher, but generally the air conditioning is 30 to 40-degrees-F cooler than the ambient air.
17   When the can is empty, turn the valve handle to the closed position and release the connection from the low-side port. Replace the dust cap. **Warning:** *Never add more than one can of refrigerant to the system (if more than one can is required, the system should be evacuated and leak tested).*
18   Remove the charging kit from the can and store the kit for future use with the piercing valve in the UP position, to prevent inadvertently piercing the can on the next use.

Heating systems

19   If the carpet under the heater core is damp, or if antifreeze vapor or steam is coming through the vents, the heater core is leaking. Remove it (see Section 14) and install a new unit (most radiator shops will not repair a leaking heater core).
20   If the air coming out of the heater vents isn't hot, the problem could stem from any of the following causes:

  a) *The thermostat is stuck open, preventing the engine coolant from warming up enough to carry heat to the heater core. Replace the thermostat (see Section 3).*
  b) *A heater hose is blocked, preventing the flow of coolant through the heater core. Feel both heater hoses at the firewall. They should be hot. If one of them is cold, there is an obstruction in one of the hoses or in the heater core, or the heater control valve is shut. Detach the hoses and back flush the heater core with a water hose. If the heater core is clear but circulation is impeded, remove the two hoses and flush them out with a water hose.*
  c) *If flushing fails to remove the blockage from the heater core, the core must be replaced (see Section 14).*

16   Air conditioning accumulator/ drier - removal and installation

Removal

Refer to illustration 16.2
Warning: *The air conditioning system is under high pressure. DO NOT loosen any fittings or remove any components until after the system has been discharged. Air conditioning refrigerant should be properly discharged into an EPA-approved container at a dealership service department or an automotive air conditioning repair facility. Always wear eye protection when disconnecting air conditioning system fittings.*
1   Have the air conditioning system discharged (see **Warning** above). Disconnect the cable from the negative terminal of the battery. **Caution:** *On models equipped with a Delco Loc II or Theftlock audio system, be sure the lockout feature is turned off before performing any procedure which requires disconnecting the battery.*

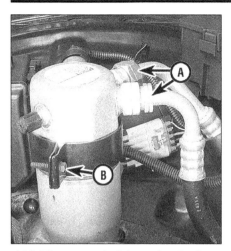

16.2 Disconnect the inlet and outlet lines (A) then remove the mounting bracket bolt (B)

17.5 Disconnect the electrical connector (arrow) from the air conditioning compressor

17.6 Remove the retaining bolt (arrow) securing the refrigerant lines to back of the compressor (early model shown)

2   Disconnect the refrigerant inlet and outlet lines **(see illustration)**, using back-up wrenches. Cap or plug the open lines immediately to prevent the entry of dirt or moisture.

3   Loosen the clamp bolt on the mounting bracket and slide the accumulator/drier assembly up and out of the compartment.

Installation

4   If you are replacing the accumu-lator/drier with a new one, add one ounce of fresh refrigerant oil to the new unit (oil must be R-12 or R-134a compatible depending on the model year of the vehicle).

5   Place the new accumulator/drier into position in the bracket.

6   Install the inlet and outlet lines, using clean refrigerant oil on the new O-rings. Tighten the mounting bolt securely.

7   Connect the cable to the negative terminal of the battery.

8   Have the system evacuated, recharged and leak tested by a dealership service department or an automotive air conditioning repair facility.

17   Air conditioning compressor - removal and installation

Refer to illustrations 17.5 and 17.6

Removal

Warning: *The air conditioning system is under high pressure. DO NOT loosen any fittings or remove any components until after the system has been discharged. Air conditioning refrigerant should be properly discharged into an EPA-approved container at a dealership service department or an automotive air conditioning repair facility. Always wear eye protection when disconnecting air conditioning system fittings.*

Note: *The accumulator/drier (see Section 16) should be replaced whenever the compressor is replaced.*

1   Have the air conditioning system discharged (see **Warning** above). Disconnect the cable from the negative terminal of the battery. **Caution:** *On models equipped with a Delco Loc II or Theftlock audio system, be sure the lockout feature is turned off before performing any procedure which requires disconnecting the battery.*

2   Clean the compressor thoroughly around the refrigerant line fittings.

3   Remove the serpentine drivebelt (see Chapter 1).

4   On later model vehicles equipped with heavy-duty cooling systems, remove the cooling fan (see Section 5).

5   Disconnect the electrical connector from the air conditioning compressor **(see illustration)**. On later model vehicles, raise the vehicle and support it securely on jackstands.

6   Disconnect the suction and discharge lines from the compressor. Both lines are mounted to the back of the compressor with a plate secured by one bolt. Plug the open fittings to prevent the entry of dirt and moisture, and discard the seals between the plate and compressor **(see illustration)**.

7   Unbolt and remove the rear compressor mount. **Note:** *On later models it may necessary to remove the right side engine mount bolt and raise the engine to access the rear compressor mount bolts.*

8   Remove the compressor-to-front-bracket bolts and nuts and remove the compressor from the engine compartment.

Installation

9   If a new compressor is being installed, pour the oil from the old compressor into a graduated container and add that exact amount of new refrigerant oil to the new compressor. Also follow any directions included with the new compressor. **Note:** *Some replacement compressors come with refrigerant oil in them. Follow the directions with the compressor regarding the draining*

of excess oil prior to installation. **Caution:** *The oil used must be labeled as compatible with R-12 for 1993 and earlier systems and labeled as compatible with R-134a for 1994 and later refrigerant systems.*

10   Installation is the reverse of the disassembly. When installing the line fitting bolt to the compressor, use new seals lubricated with clean refrigerant oil, and tighten the bolt securely.

11   Reconnect the battery cable to the negative battery terminal.

12   Have the system evacuated, recharged and leak tested by a dealership service department or an automotive air conditioning repair facility.

18   Air conditioning condenser - removal and installation

Warning: *The air conditioning system is under high pressure. DO NOT loosen any fittings or remove any components until after the system has been discharged. Air conditioning refrigerant should be properly discharged into an EPA-approved container at a dealership service department or an automotive air conditioning repair facility. Always wear eye protection when disconnecting air conditioning system fittings.*

Removal

Refer to illustration 18.2

1   Have the air conditioning system discharged (see **Warning** above). Disconnect the cable from the negative terminal of the battery. **Caution:** *On models equipped with a Delco Loc II or Theftlock audio system, be sure the lockout feature is turned off before performing any procedure which requires disconnecting the battery.*

2   Disconnect the refrigerant line fittings from the right side of the condenser and cap the open fittings to prevent the entry of dirt

18.2 The condenser lines (arrows) are located just to the right of the radiator

and moisture. **(see illustration)**.

3   Remove the cooling fan(s) (see Section 5).

4   If equipped with electric cooling fans, refer to Section 6 and remove the upper radiator support panel.

5   Tilt the upper half of the radiator forward and dislodge the condenser insulator mounts from the radiator support.

6   Pull the condenser up between the radiator and the radiator support to remove it from the vehicle. **Caution:** *The condenser is made of aluminum - be careful not to damage it during removal.*

Installation

7   Installation is the reverse of removal. Be sure to use new, compatible O-rings on the

refrigerant line fittings (lubricate the O-rings with clean refrigerant oil. If a new condenser is installed, add 1 ounce of new refrigerant oil to the system (oil must be R-12 or R-134a compatible depending on the model year of the vehicle).

8   Have the system evacuated, recharged and leak tested by a dealership service department or an automotive air conditioning repair facility.

---

19   Air conditioning evaporator - removal and installation

Warning 1: *The models covered by this manual are equipped with airbags. Always disable the airbag system before working in the vicinity of the impact sensors, steering column or instrument panel to avoid the possibility of accidental deployment of the airbag(s), which could cause personal injury (see Chapter 12). The yellow wires and connectors routed through the instrument panel are for this system. Do not use electrical test equipment on these yellow wires or tamper with them in any way while working under the instrument panel.*

Warning 2: *The air conditioning system is under high pressure. DO NOT loosen any fittings or remove any components until after the system has been discharged. Air conditioning refrigerant should be properly discharged into an EPA-approved container at a dealership service department or an automotive air conditioning repair facility. Always wear eye protection when disconnecting air conditioning system fittings.*

Removal

Refer to illustrations 19.3 and 19.5

1   Have the air conditioning system discharged (see **Warning** above). Disconnect the cable from the negative terminal of the battery. **Caution:** *On models equipped with a Delco Loc II or Theftlock audio system, be sure the lockout feature is turned off before performing any procedure which requires disconnecting the battery.*

2   Drain the cooling system (see Chapter 1).

3   Disconnect the air conditioning lines at the passenger side of the firewall **(see illustration)**.

4   Follow the procedures in Section 14 for removing the heater core cover.

5   Remove the evaporator core clamp bolt and clamp **(see illustration)**, then slide the evaporator core out carefully.

6   Check the core over carefully for signs of leaks.

7   If a new evaporator core is to be installed, save all of the sealing gaskets from the original unit and transfer them.

Installation

8   Installation is the reverse of the removal procedure. Lubricate all O-rings with clean refrigerant oil.

9   If a new evaporator has been installed, add 1 ounce of refrigerant oil (oil must be R-12 or R-134a compatible depending on the model year of the vehicle). Have the system evacuated, recharged and leak tested by a dealership service department or an automotive air conditioning repair facility.

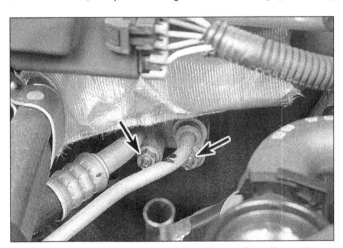

19.3 Disconnect the refrigerant lines (arrows) leading to the evaporator core

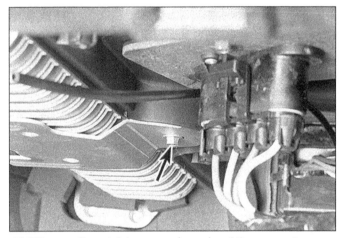

19.5 Remove the retaining screw (arrow) and the evaporator core retaining plate - pull the evaporator core down and out to remove it

Chapter 4
Fuel and exhaust systems

Contents

Specifications

General

Fuel pressure (key On, engine not running)	
Throttle Body Injection (TBI) system	9 to 13 psi
Multiport Fuel Injection (MFI) system	41 to 47 psi
Injector resistance	
TBI system	1.16 to 1.36 ohms
MFI system	11.6 to 12.4 ohms

Torque specifications

	Ft-lbs (unless otherwise indicated)
Exhaust pipe-to-manifold nuts	15 to 22
Fuel meter cover screws (TBI)	27 in-lbs
Fuel meter body-to-throttle body screws (TBI)	35 in-lbs
Fuel inlet nut (TBI)	30
Fuel outlet nut (TBI)	21
Fuel rail resonator bracket bolts (MFI)	89 in-lbs
Fuel rail mounting bolts (MFI)	89 in-lbs
IAC valve	
TBI system	13
MFI system (screws)	27 in-lbs
Throttle body bolts/nuts	16

1   General information

These models are equipped with either the Throttle Body Injection (TBI) system (1993 and earlier models) or the Multiport Fuel Injection (MFI) system (1994 and later models). 1996 models are equipped with the updated OBD II self diagnosis system (see Chapter 6).

The fuel system consists of a fuel tank, an electric fuel pump and fuel pump relay, an air cleaner assembly and a fuel injection system.

Throttle Body Injection (TBI) system

The throttle body system utilizes two injectors, centrally mounted in a carburetor-like housing. The injector is an electrical solenoid, with fuel delivered to the injector at a constant pressure level. To maintain the fuel pressure at a constant level, fuel is supplied at a greater pressure than required and controlled by a fuel pressure regulator, which returns excess fuel to the fuel tank.

A signal from the PCM opens the solenoid, allowing fuel to spray through the injec-tor into the throttle body. The amount of time the injector is held open by the PCM determines the fuel/air mixture ratio.

Multiport Fuel Injection (MFI) system

This system utilizes injectors of a different type than the throttle body injection system. Instead of an injector mounted in a centrally located throttle body as in the TBI system, one injector is installed above each intake port. The throttle body on the MFI system serves only to control the amount of air

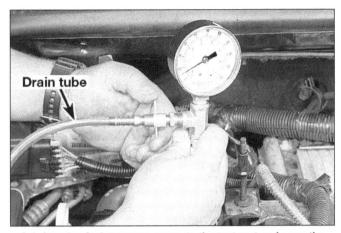

2.4 Attach a fuel pressure gauge to the test port and open the valve to drain the excess fuel out of the drain tube (arrow) and into an approved fuel container

3.2 Attach the fuel pressure gauge to the test port and check the pressure with the ignition key ON (engine not running) (MFI system shown)

passing into the system. Because each cylinder is equipped with an injector mounted immediately adjacent to the intake valve, much better control of the fuel/air mixture ratio is possible.

Fuel pump and lines

Fuel is circulated from the fuel tank to the fuel injection system, and back to the fuel tank, through a pair of lines running along the underside of the vehicle. An electric fuel pump is attached to the fuel sending unit inside the fuel tank. A return system routes all vapors and excess fuel back to the fuel tank through separate return lines.

Exhaust system

The exhaust system includes an exhaust manifold fitted with an exhaust oxygen sensor, a catalytic converter, an exhaust pipe, and a muffler. The catalytic converter is an emission control device added to the exhaust system to reduce pollutants. A single-bed converter is used in combination with a three-way (reduction) catalyst. Refer to Chapter 6 for more information regarding the catalytic converter.

---

2   Fuel pressure relief procedure

Warning: *Gasoline is extremely flammable, so take extra precautions when you work on any part of the fuel system. Don't smoke or allow open flames or bare light bulbs near the work area, and don't work in a garage where a natural gas-type appliance (such as a water heater or a clothes dryer) with a pilot light is present. Since gasoline is carcinogenic, wear latex gloves when there's a possibility of being exposed to fuel, and, if you spill any fuel on your skin, rinse it off immediately with soap and water. Mop up any spills immediately and do not store fuel-soaked rags where they could ignite. The fuel system is under constant pressure, so, if any fuel lines are to be disconnected, the fuel pressure in the system must be relieved first. When you*

perform any kind of work on the fuel system, wear safety glasses and have a Class B type fire extinguisher on hand.
Note: *After the fuel pressure has been relieved, it's a good idea to lay a shop towel over any fuel connection to be disassembled, to absorb the residual fuel that may leak out when servicing the fuel system.*
1   Before servicing any fuel system component, you must relieve the fuel pressure to minimize the risk of fire or personal injury.
2   Remove the fuel filler cap - this will relieve any pressure built up in the tank.

TBI systems

Note: *Although the Model 220 TBI units have an automatic internal bleed system, follow the procedure described below to relieve the fuel pressure on all TBI engines.*
3   TBI systems have an internal bleed feature that allows the fuel system to automatically depressurize while not running (ignition key OFF). It is a good idea to make sure all the fuel pressure has been bled out of the fuel lines before working on the fuel system. Position a metal pan under the fuel filter. Use shop rags or towels to cover the inlet fuel line at the fuel filter (see Chapter 1) and then disconnect the fuel line to allow any residual pressure to dissipate.

MFI systems

Refer to illustration 2.4
4   MFI systems are equipped with a test port located on the fuel rail. The test port is mounted near the fuel pressure regulator near the rear of the cylinder head. Install a fuel pressure gauge onto the test port and relieve the fuel by bleeding the fuel through the bleed-off valve and into a metal can **(see illustration)**.
5   If the specialized fuel pressure gauge is not available, follow this alternate method that will easily bleed the fuel pressure. Cover the test port fitting with shop rags or towels, then carefully depress the Schrader valve with a small screwdriver or awl. Dispose of the rags in a covered, marked container.

6   Unless this procedure is followed before servicing fuel lines or connections, fuel spray (and possible injury) may occur.

---

3   Fuel pump/fuel pressure - check

Warning: *Gasoline is extremely flammable, so take extra precautions when you work on any part of the fuel system. Don't smoke or allow open flames or bare light bulbs near the work area, and don't work in a garage where a natural gas-type appliance (such as a water heater or a clothes dryer) with a pilot light is present. Since gasoline is carcinogenic, wear latex gloves when there's a possibility of being exposed to fuel, and, if you spill any fuel on your skin, rinse it off immediately with soap and water. Mop up any spills immediately and do not store fuel-soaked rags where they could ignite. The fuel system is under constant pressure, so, if any fuel lines are to be disconnected, the fuel pressure in the system must be relieved first (see Section 2). When you perform any kind of work on the fuel system, wear safety glasses and have a Class B type fire extinguisher on hand.*
Note 1: *The On Board Diagnostic (OBD) system may set a trouble code 44 or 45 while performing this procedure. Refer to the code extracting procedure in Chapter 6 for additional fuel system diagnostic information.*
Note 2: *The following checks assume the fuel filter is in good condition. If you doubt its condition, install a new one (see Chapter 1).*
1   Check that there is adequate fuel in the fuel tank. Relieve the fuel pressure (see Section 2).

Fuel pump output and pressure check

Refer to illustrations 3.2, 3.3, 3.5 and 3.6
2   Install a fuel pressure gauge onto the fuel lines or the fuel rail. Each type of system has a different location on the fuel line for the fuel gauge test connection. The MFI system is equipped with a fuel pressure test port **(see illustration)**; you will need a fuel pressure

3.3 The fuel pump can be activated by attaching a jumper wire to the battery positive post (+) and touch the jumper to the fuel pump test terminal (red wire) located near the receiver drier in the engine compartment

3.5 Connect a vacuum pump to the fuel pressure regulator, apply vacuum to the fuel pressure regulator and check the fuel pressure - the fuel pressure should decrease as the vacuum increases (MFI system shown)

gage capable of measuring high fuel pressure and equipped with the correct fitting to thread onto the test port. On the TBI system the fuel pressure gauge must be connected inline with the fuel inlet line using a T-fitting. Since the inlet fuel line to the throttle body on TBI system is equipped with quick-connect fittings, it will be necessary to install the correct fuel line adapters between the throttle body and inlet fuel line. Consult a specialty tool company for the correct fuel pressure gauge and adapter set.

3   Turn the ignition switch ON (engine not running). The fuel pump should run for about two seconds - note the reading on the gauge. After the pump stops running the pressure should hold steady. It should be within the range listed in this Chapter's Specifications. **Note:** *If there is no response from the fuel pump, use a jumper wire attached to the positive terminal of the battery and apply battery voltage to the test terminal (red wire) located on the driver's side of the engine compartment* **(see illustration)**.

4   Start the engine and let it idle at normal operating temperature. The pressure should be lower by 3 to 10 psi. If all the pressure readings are within the limits listed in this Chapter's Specifications, the system is operating properly.

5   If the pressure did not drop by 3 to 10 psi after starting the engine, apply 12 to 14 inches of vacuum to the pressure regulator **(see illustration)**. If the pressure drops, repair the vacuum source to the regulator. If the pressure does not drop, replace the regulator. **Note:** *This test will work only on engines where the fuel pressure regulator is easily accessible (MFI systems). On TBI fuel systems, have the fuel pressure regulator diagnosed by a dealer service department or other qualified repair shop.*

6   If the fuel pressure is not within specifications, check the following: **Note:** *The next two tests will work only on MFI engines where the fuel pressure regulator is easily accessible. On TBI systems, have the fuel pressure regulator diagnosed by a dealer service department.*

3.6 Checking vacuum to the fuel pressure regulator using a vacuum gauge

a) *If the pressure is higher than specified, check for vacuum to the fuel pressure regulator* **(see illustration)**. *Vacuum must fluctuate with the increase or decrease in the engine rpm. If vacuum is present, check for a pinched or clogged fuel return hose or pipe. If the return line is OK, replace the regulator.*

b) *If the pressure is lower than specified, change the fuel filter to rule out the possibility of a clogged filter. If the pressure is still low, install a fuel line shut-off adapter between the pressure regulator and the return line. With the valve open, start the engine (if possible) and slowly close the valve. If the pressure rises above 47 psi, replace the regulator (see Section 13).* **Warning:** *Don't allow the fuel pressure to exceed 60 psi. Also, don't attempt to restrict the return line by pinching it, as the nylon fuel line will be damaged.*

c) *If the pressure is still low with the fuel return line restricted, an injector (or injectors) may be leaking (see Section 13) or the fuel pump may be faulty.*

7   After the testing is done, relieve the fuel

3.11 Check the condition of the fuel pump fuse

pressure (see Section 2) and remove the fuel pressure gauge.

8   If there are no problems with any of the above-listed components, check the fuel pump electrical circuits (see below).

Fuel pump electrical circuit check

Refer to illustrations 3.11, 3.12a and 3.12b

Note: *Refer to Chapter 12 for additional wiring schematics that detail the fuel pump relay and circuit.*

9   If you suspect a problem with the fuel pump, verify the pump actually runs. Have an assistant turn the ignition switch to ON - you should hear a brief whirring noise as the pump comes on and pressurizes the system. Have the assistant start the engine. This time you should hear a constant whirring sound from the pump (but it's more difficult to hear with the engine running).

10   If the pump does not come on (makes no sound), proceed to the next Step.

11   Check the fuel pump fuse **(see illustration)**. If the fuse is blown, replace the fuse and see if the pump works. If the pump still does not work, go to the next Step.

12   Check the fuel pump relay circuit. If the pump does not run, check for an open circuit between the relay and the fuel pump. With the ignition key ON (engine not running), check for battery voltage at the relay connector **(see illustrations)**. The fuel pump relay is located on the fenderwell on TBI models next to the A/C clutch control relay. On MFI models, the fuel pump relay is located in the fuse/relay control box in the engine compartment. **Note:** *If oil pressure drops below the specified pressure level, the oil pressure switch will act as a fuel pressure cut-off device. Be sure to check the oil pressure switch and circuit in the event of a difficult problem diagnosing the fuel pump circuit (refer to the wiring diagrams at the end of Chapter 12).*

13   If battery voltage exists, replace the relay with a known good relay and retest. If necessary, have the relay checked by a qualified automotive electrical specialist.

14   If the fuel pump does not activate, check for power to the fuel pump at the fuel tank. Access to the fuel pump is difficult but it is possible to check for battery voltage at the electrical connector near the tank.

4   Fuel lines and fittings - repair and replacement

Warning: *Gasoline is extremely flammable, so take extra precautions when you work on any part of the fuel system. Don't smoke or allow open flames or bare light bulbs near the work area, and don't work in a garage where a natural gas-type appliance (such as a water heater or a clothes dryer) with a pilot light is present. Since gasoline is carcinogenic, wear latex gloves when there's a possibility of being exposed to fuel, and, if you spill any fuel on your skin, rinse it off immediately with soap and water. Mop up any spills immediately and do not store fuel-soaked rags where they could ignite. The fuel system is under constant pressure, so, if any fuel lines are to be disconnected, the fuel pressure in the system must be relieved first (see Section 2). When you perform any kind of work on the fuel system, wear safety glasses and have a Class B type fire extinguisher on hand.*

1   Always relieve the fuel pressure before servicing fuel lines or fittings on fuel-injected vehicles (see Section 2).

2   The fuel feed, return and vapor lines extend from the fuel tank to the engine compartment. The lines are secured to the underbody with clip and screw assemblies. These lines must be occasionally inspected for leaks, kinks and dents.

3   If evidence of dirt is found in the system or fuel filter during disassembly, the line should be disconnected and blown out. Check the fuel strainer on the fuel level sending unit (see Section 8) for damage and deterioration.

Steel and nylon tubing

4   Because fuel lines used on fuel-injected

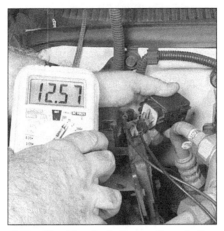

3.12a Checking for battery voltage to the fuel pump relay connector on TBI systems

vehicles are under high pressure, they require special consideration.

5   If replacement of a metal fuel line or emission line is called for, use welded steel tubing meeting GM specification 124-M or its equivalent. Don't use copper or aluminum tubing to replace steel tubing. These materials cannot withstand normal vehicle vibration.

6   If is becomes necessary to replace a section of nylon fuel line, replace it only with the correct part number - don't use any substitutes.

7   Most fuel lines have threaded fittings with O-rings. Any time the fittings are loosened to service or replace components:

a) *Use a backup wrench while loosening and tightening the fittings.*

b) *Check all O-rings for cuts, cracks and deterioration. Replace any that appear worn or damaged.*

c) *If the lines are replaced, always use original equipment parts, or parts that meet the GM standards specified in this Section.*

Rubber hose

Warning: *These models are equipped with electronic fuel injection, use only original equipment replacement hoses or their equivalent. Others may fail from the high pressures of this system.*

8   When rubber hose is used to replace a

3.12b Checking for battery voltage on the fuel pump relay connector (MFI system)

metal line, use reinforced, fuel resistant hose with the word *"Fluoroelastomer"* imprinted on it. Hose(s) not clearly marked like this could fail prematurely and could fail to meet Federal emission standards. Hose inside diameter must match line outside diameter. **Warning:** *Don't substitute rubber hose for metal line on high-pressure systems. Use only genuine factory replacement lines or lines meeting factory specifications.*

9   Don't use rubber hose within four inches of any part of the exhaust system or within ten inches of the catalytic converter. Metal lines and rubber hoses must never be allowed to chafe against the frame. A minimum of 1/4-inch clearance must be maintained around a line or hose to prevent contact with the frame.

Removal and installation

Refer to illustrations 4.10, 4.12a, 4.12b and 4.13

Note: *The following procedure and accompanying illustrations are typical for vehicles covered by this manual. On quick-disconnect (non-threaded) fittings, clean off the fittings before disconnection to prevent dirt from getting in the fittings. After disconnection, clean the fittings with compressed air and apply a few drops of oil.*

10   Relieve the fuel pressure (see Section 2) and disconnect the fuel feed, return or vapor

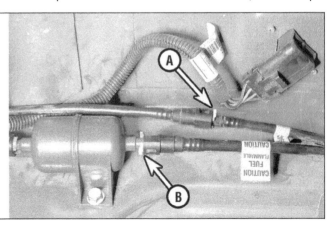

4.10 Some models are equipped with fuel lines that can be disconnected by pinching the tabs and separating each connector

A   *Fuel return line*
B   *Fuel feed line*

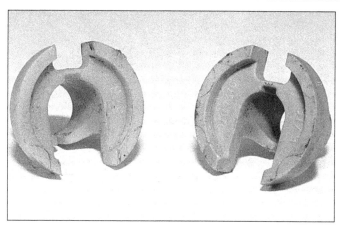

4.12a 3/8 and 5/16-inch fuel line disconnect tools

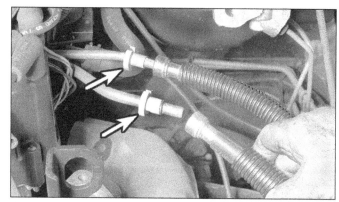

4.12b On quick-connect fuel lines, use a special tool kit, available at most auto parts stores, and follow the kit's instructions to separate the fuel lines

line at the fuel tank **(see illustration)**. **Note:** *Some fuel line connections may be threaded. Be sure to use a back-up wrench when separating the connections.*

11 Remove all fasteners attaching the lines to the vehicle body.

12 Detach the fitting(s) that attach the fuel hoses to the engine compartment metal lines **(see illustrations)**. Twisting them back and forth will allow them to separate more easily.

13 Installation is the reverse of removal. Be sure to use new O-rings at the threaded fittings, if equipped **(see illustration)**.

5 Fuel tank - removal and installation

Refer to illustrations 5.5, 5.6, 5.7, 5.8 and 5.10

Warning: *Gasoline is extremely flammable, so take extra precautions when you work on any part of the fuel system. Don't smoke or allow open flames or bare light bulbs near the work area, and don't work in a garage where a natural gas-type appliance (such as a water heater or a clothes dryer) with a pilot light is present. Since gasoline is carcinogenic, wear latex gloves when there's a possibility*

of being exposed to fuel, and, if you spill any fuel on your skin, rinse it off immediately with soap and water. Mop up any spills immediately and do not store fuel-soaked rags where they could ignite. The fuel system is under constant pressure, so, if any fuel lines are to be disconnected, the fuel pressure in the system must be relieved first (see Section 2). When you perform any kind of work on the fuel system, wear safety glasses and have a Class B type fire extinguisher on hand. **Note:** *Don't begin this procedure until the fuel gauge indicates the tank is empty or nearly empty. If the tank must be removed when it isn't empty (for example, if the fuel pump malfunctions), siphon any remaining fuel from the tank prior to removal.*

1 Unless the vehicle has been driven far enough to completely empty the tank, it's a good idea to siphon the residual fuel out before removing the tank from the vehicle. **Warning:** *DO NOT start the siphoning action by mouth! Use a siphoning kit, available at most auto parts stores.*

2 Relieve the fuel system pressure (see Section 2).

3 Detach the cable from the negative terminal of the battery. **Caution:** *On models equipped with a Delco Loc II or Theftlock audio system, be sure the lockout feature is*

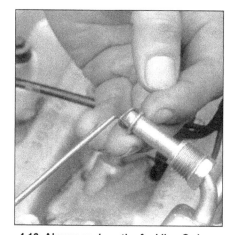

4.13 Always replace the fuel line O-rings (if equipped)

turned off before performing any procedure which requires disconnecting the battery.

4 Raise the vehicle and support it securely on jackstands placed underneath the jacking points.

5 Remove the fuel tank shield from the chassis **(see illustration)**.

6 Disconnect the fuel filler hose and vapor lines from the fuel tank **(see illustration)**.

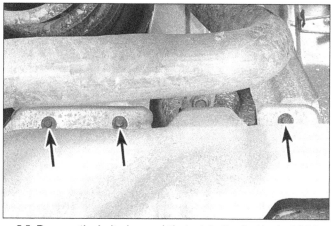

5.5 Remove the bolts (arrows) that retain the fuel tank shield to the frame

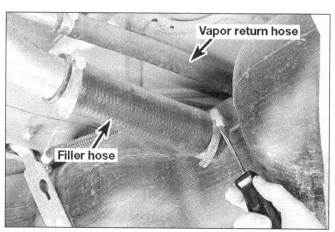

5.6 Loosen the clamps and remove the fuel filler hose and vapor return hose

Vapor return hose

Filler hose

5.7 Use the special fuel line disconnect tools to disengage the fuel lines at the fuel tank

5.8 Remove the bolts (arrows) that retain the fuel filler neck to the body and separate the filler neck from the vehicle

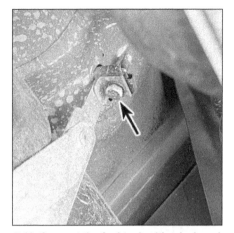

5.10 Support the fuel tank with a jack and remove the fuel tank strap bolts (arrow)

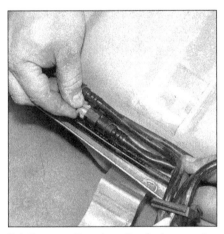

7.4 Squeeze the return line fuel connector and separate the fuel return, feed and vapor line as an assembly

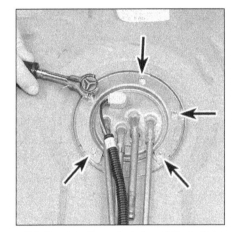

7.5 Remove the nuts (arrows) from the fuel pump/sending unit retaining ring

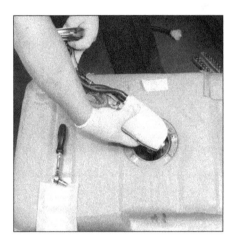

7.6 Lift the fuel pump assembly from the fuel tank

7   Disconnect the fuel feed and return lines and the vapor return line from the fuel pump **(see illustration)**. **Note:** *Disconnect the fuel feed line at the fuel filter* (see Chapter 1).
8   Disconnect the bolts from the fuel filler neck **(see illustration)**. Angle the filler neck down and away from the tank and remove it from the vehicle.
9   Support the fuel tank with a floor jack.
10   Disconnect both fuel tank retaining straps **(see illustration)**.
11   Lower the tank enough to disconnect the wires and ground strap from the fuel pump/fuel level sending unit.
12   Remove the tank from the vehicle.
13   Installation is the reverse of removal.

6   Fuel tank cleaning and repair - general information

1   All repairs to the fuel tank or filler neck should be carried out by a professional who has experience in this critical and potentially dangerous work. Even after cleaning and flushing of the fuel system, explosive fumes can

remain and ignite during repair of the tank.
2   If the fuel tank is removed from the vehicle, it should not be placed in an area where sparks or open flames could ignite the fumes coming out of the tank. Be especially careful inside garages where a natural gas-type appliance is located, because the pilot light could cause an explosion.

7   Fuel pump - removal and installation

Refer to illustrations 7.4, 7.5, 7.6, 7.8, 7.9 and 7.11
Warning: *Gasoline is extremely flammable, so take extra precautions when you work on any part of the fuel system. Don't smoke or allow open flames or bare light bulbs near the work area, and don't work in a garage where a natural gas-type appliance (such as a water heater or a clothes dryer) with a pilot light is present. Since gasoline is carcinogenic, wear latex gloves when there's a possibility of being exposed to fuel, and, if you spill any fuel on your skin, rinse it off immediately with soap and water. Mop up any spills immedi-*

ately and do not store fuel-soaked rags where they could ignite. The fuel system is under constant pressure, so, if any fuel lines are to be disconnected, the fuel pressure in the system must be relieved first (see Section 2). When you perform any kind of work on the fuel system, wear safety glasses and have a Class B type fire extinguisher on hand.
1   Relieve the fuel system pressure (see Section 2).
2   Disconnect the cable from the negative battery terminal. **Caution:** *On models equipped with a Delco Loc II or Theftlock audio system, be sure the lockout feature is turned off before performing any procedure which requires disconnecting the battery.*
3   Remove the fuel tank (see Section 5).
4   Disconnect the fuel feed, return and vapor line as an assembly from the top of the fuel tank **(see illustration)**.
5   The fuel pump/sending unit assembly is located inside the fuel tank. Remove the nuts from the fuel pump/sending unit assembly **(see illustration)**.
6   Lift the fuel pump/sending unit assembly from the fuel tank **(see illustration)**. **Caution:** *The fuel level float and sending unit are delicate. Do not bump them against the tank*

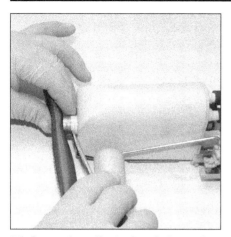

7.8 Pry on the collar to detach the strainer from the fuel pump assembly

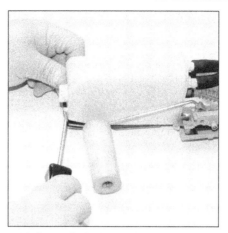

7.9 Press the locking tab to release the fuel pump cover from the frame

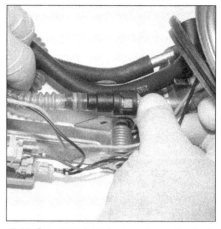

7.11 Squeeze the tabs to release the fuel line from the collar

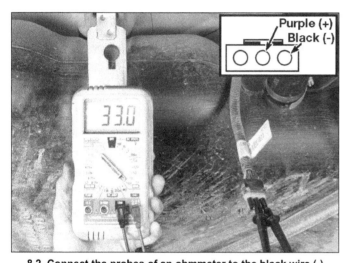

8.2 Connect the probes of an ohmmeter to the black wire (-) and the purple wire (+) and check the resistance of the fuel level sending unit

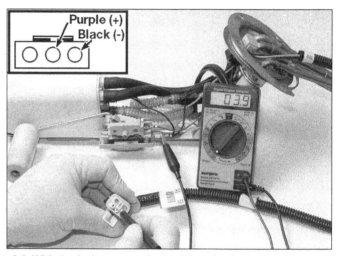

8.5 With the fuel pump on the work bench, check the resistance of the sending unit from empty (lowered float arm) to full (raised float arm)

during removal or the accuracy of the sending unit may be affected.

7   Inspect the condition of the O-ring around the opening of the tank. If it is dried, cracked or deteriorated, replace it.

8   Remove the strainer from the lower end of the fuel pump **(see illustration)**. If it is dirty, clean it with a suitable solvent and blow it out with compressed air. If it is too dirty to be cleaned, replace it.

9   Lift the locking tabs to release the fuel pump cover **(see illustration)**.

10   If it is necessary to separate the fuel pump and sending unit, disconnect the electrical connectors at the pump, noting their position.

11   Disconnect the fuel line from the pump **(see illustration)**.

12   Separate the fuel pump from the assembly.

13   Installation is the reverse of removal. Reassemble the fuel pump to the sending unit bracket assembly and insert the assembly into the fuel tank.

14   Install the fuel tank (see Section 5).

8   Fuel level sending unit - check and replacement

Warning: *Gasoline is extremely flammable, so take extra precautions when you work on any part of the fuel system. Don't smoke or allow open flames or bare light bulbs near the work area, and don't work in a garage where a natural gas-type appliance (such as a water heater or a clothes dryer) with a pilot light is present. Since gasoline is carcinogenic, wear latex gloves when there's a possibility of being exposed to fuel, and, if you spill any fuel on your skin, rinse it off immediately with soap and water. Mop up any spills immediately and do not store fuel-soaked rags where they could ignite. The fuel system is under constant pressure, so, if any fuel lines are to be disconnected, the fuel pressure in the system must be relieved first (see Section 2). When you perform any kind of work on the fuel system, wear safety glasses and have a Class B type fire extinguisher on hand.*

Check

Refer to illustrations 8.2 and 8.5

1   Raise the vehicle and secure it on jackstands.

2   Position the probes of an ohmmeter onto the fuel level sending unit electrical connector terminals **(see illustration)** and check for resistance.

3   First, check the resistance of the sending unit with the fuel tank completely full. The resistance of the sending unit should be about 88 ohms.

4   Wait until the tank is nearly empty and check the resistance of the unit again. The resistance should be 2 ohms.

5   If the readings are incorrect or there is very little change in resistance as the float travels from full to empty, replace the fuel level sending unit assembly. **Note:** *Another check for the fuel level sending unit is to remove the fuel pump from the tank (see Section 7) and check the resistance while moving the float from full (arm at highest point of travel) to empty (arm at lowest point of travel)* **(see illustration)**.

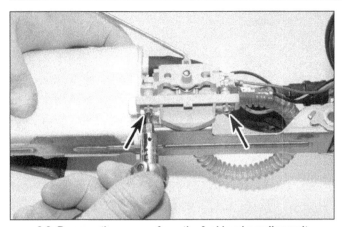

8.8 Remove the screws from the fuel level sending unit

9.3 Remove the air resonator from the engine compartment (MFI system)

9.4 Removing the air intake duct from the air cleaner assembly (MFI system)

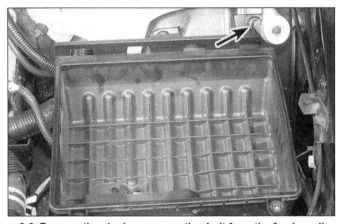

9.6 Remove the air cleaner mounting bolt from the fenderwell

Replacement

Refer to illustration 8.8

6   Remove the fuel tank (see Section 5) and the fuel pump (see Section 7) from the vehicle.

7   Disconnect the sending unit electrical connector from the assembly.

8   Remove the screws from the fuel level sending unit **(see illustration)** and separate it from the assembly.

9   Installation is the reverse of removal.

9   Air cleaner assembly - removal and installation

Refer to illustrations 9.3, 9.4, 9.6 and 9.7

Note: *TBI systems are equipped with the conventional style air cleaner housing mounted directly on top of the throttle body. Refer to Chapter 1 for additional illustrations.*

1   Detach the cable from the negative terminal of the battery. **Caution:** *On models equipped with a Delco Loc II or Theftlock audio system, be sure the lockout feature is turned off before performing any procedure which requires disconnecting the battery.*

2   Disconnect the electrical connector to the intake air temperature (IAT) sensor (see Chapter 6) and the MAF sensor.

3   Remove the air resonator from the engine compartment **(see illustration)**.

4   Remove the clamps that retain the air intake assembly to the air cleaner and remove the assembly from the engine compartment **(see illustration)**.

5   Lift the air filter element from the air cleaner housing (see Chapter 1).

6   Remove the bolts that retain the air cleaner lower assembly to the body **(see illustration)**.

7   Use a punch or a narrow screwdriver and push the locking tabs toward the fender

9.7 Use a punch or a sharp screwdriver to release the locking tabs (MFI system)

to unlock the air cleaner from the bracket **(see illustration)**.

8   Lift the lower air cleaner assembly from the engine compartment. **Note:** *The horns, headlight assembly and the PCM are easily accessed when the lower air cleaner assembly is removed.*

9   Installation is the reverse of removal.

10   Accelerator cable - removal and installation

Refer to illustrations 10.2, 10.4 and 10.5

Removal

1   Detach the screws and the clip retaining the lower instrument panel trim and lower the trim (if necessary) (see Chapter 11).

2   Detach the accelerator cable from the accelerator pedal **(see illustration)**.

3   Squeeze the accelerator cable cover tangs and push the cable through the firewall into the engine compartment.

4   Rotate the throttle lever and detach the accelerator cable from the throttle lever **(see illustration)**.

5   Press the accelerator cable retaining tang and slide the cable grommet out of the accelerator cable bracket **(see illustration)**.

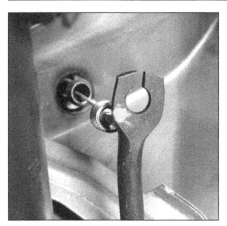

10.2 Slide the accelerator cable through the slot in the pedal arm

10.4 Remove the accelerator cable end through the slot in the throttle lever

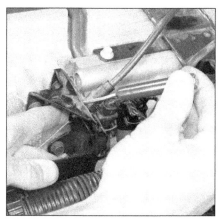

10.5 Press the accelerator cable retaining tabs with a punch or flat-bladed screwdriver and slide the assembly through the bracket housing

Installation

6   Installation is the reverse of removal. **Note:** *To prevent possible interference, flexible components (hoses, wires, etc.) must not be routed within two inches of moving parts, unless routing is controlled.*

7   Operate the accelerator pedal and check for any binding condition by completely opening and closing the throttle.

8   Apply sealant around the accelerator cable at the engine compartment side of the firewall.

11   Fuel injection system - general information

Refer to illustrations 11.3 and 11.6

1   These models are equipped with either the Throttle Body Injection (TBI) system (1993 and earlier models) or the Multiport Fuel Injection (MFI) system (1994 and later models). 1996 models are equipped with the updated OBD II self diagnosis system.

2   The fuel system consists of a fuel tank, an electric fuel pump and fuel pump relay, an air cleaner assembly and an electronic fuel injection system.

Throttle Body Injection (TBI) systems

3   The main component of the TBI system is the Throttle Body Injection (TBI) unit, which is mounted on the intake manifold just like a carburetor **(see illustration)**. The TBI unit is

11.3 Fuel injection component locations on a typical TBI system

1	*Fuel pump relay*	*4*	*Manifold Absolute Pressure sensor*	*8*	*EVAP canister*
2	*Fuel pump/PCM fuse (located behind bracket)*	*5*	*Idle Air Control valve*	*9*	*Engine coolant temperature sensor*
3	*Fuel pump test terminal (red wire)*	*6*	*Fuel injectors*	*10*	*Throttle Position Sensor*
		7	*Air intake duct*		

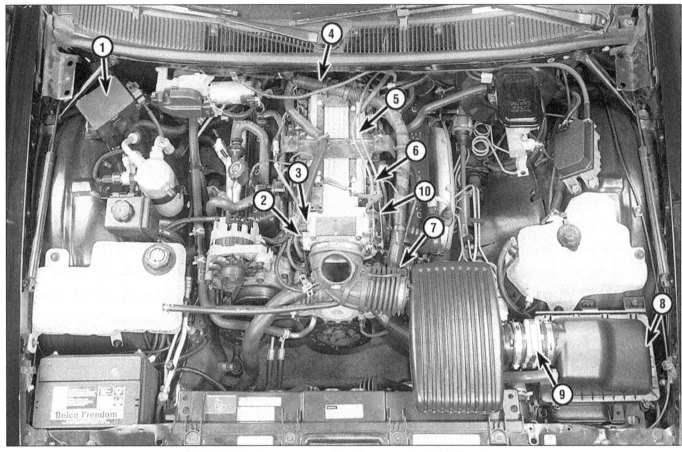

11.6 Fuel injection component locations on a typical MFI system

1	*Fuel pump relay (relay module)*	*5*	*Fuel rail (left bank)*	*8*	*PCM (located under air cleaner)*
2	*Idle Air Control valve*	*6*	*Fuel injector*	*9*	*Mass Air Flow sensor*
3	*Throttle Position Sensor*	*7*	*Secondary Air Injection pump (under*	*10*	*Fuel pressure regulator*
4	*Fuel pressure test port*		*intake duct)*		

made up of two major assemblies: the throttle body and the fuel metering assembly.

4   The throttle body contains a throttle valve, controlled by the accelerator pedal, similar to a carburetor. Attached to the exterior of the body are the Throttle Position Sensor (TPS), which sends throttle position information to the PCM, and the Idle Air Control (IAC) assembly, which is used by the PCM to maintain a constant idle speed during normal engine operation.

5   The fuel metering assembly contains the fuel pressure regulator and the fuel injectors. The regulator dampens the pulsations of the fuel pump and maintains a steady pressure at the injectors. The fuel injectors are controlled by the PCM through an electrically operated solenoid. The amount of fuel injected into the intake manifold is varied by the length of time the injector plunger is held open.

Multiport Fuel Injection (MFI) system

6   Multiport Fuel Injection (MFI) consists of an air intake manifold, the throttle body, the injectors, the fuel rail assembly, an electric fuel pump and associated plumbing **(see illustration)**.

7   Air is drawn through the air cleaner and throttle body. A Mass Airflow Sensor (MAF) informs the PCM of volume and pressure variations.

8   While the engine is running, the fuel constantly circulates through the fuel rail, which removes vapors and keeps the fuel cool while maintaining sufficient pressure to the injectors under all running conditions.

9   As with TBI, the operation of the MFI system is controlled by the PCM so that it works in conjunction with the rest of the vehicle functions to provide optimum driveability and emissions control.

10   Because the MFI system meters fuel and air precisely, it is important to the proper operation of the vehicle that the fuel and air filters be changed at the specified intervals.

12 Fuel injection system - check

Refer to illustrations 12.8, 12.9 and 12.10

Warning: *Gasoline is extremely flammable, so take extra precautions when you work on any part of the fuel system. Don't smoke or allow open flames or bare light bulbs near the work area, and don't work in a garage where a natural gas-type appliance (such as a water heater or a clothes dryer) with a pilot light is present. Since gasoline is carcinogenic, wear latex gloves when there's a possibility of being exposed to fuel, and, if you spill any fuel on your skin, rinse it off immediately with soap and water. Mop up any spills immediately and do not store fuel-soaked rags where they could ignite. The fuel system is under constant pressure, so, if any fuel lines are to be disconnected, the fuel pressure in the system must be relieved first (see Section 2). When you perform any kind of work on the fuel system, wear safety glasses and have a Class B type fire extinguisher on hand.*

Note: *The following procedure is based on the assumption that the fuel pump is working and the fuel pressure is adequate (see Section 3).*

1   Check to see that the battery is fully charged, as the control unit and sensors depend on an accurate supply voltage in order to properly meter the fuel.

2   Check the air filter element - a dirty or partially blocked filter will severely impede performance and economy (see Chapter 1).

3   Check the fuel filter and replace it if necessary (see Chapter 1).

12.8 Use a stethoscope or screwdriver to determine if the injectors are working properly - they should make a steady clicking sound that rises and falls with engine speed changes (MFI system)

12.9 Check the resistance of each injector and compare the readings to the Specifications

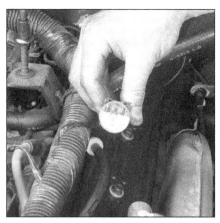

12.10 Install the "noid" light into each injector electrical connector and confirm that it blinks when the engine is cranking or running

5   Check the ground wire connections on the intake manifold for tightness. Check all electrical connectors that are related to the system. Loose connectors and poor grounds can cause many problems that resemble more serious malfunctions.

5   If a blown fuse is found, replace it and see if it blows again. If it does, search for a grounded wire in the harness.

6   Check the air intake duct to the intake manifold for leaks, which will result in an excessively lean mixture. Also check the condition of all vacuum hoses connected to the intake manifold.

7   Remove the air intake duct from the throttle body and check for dirt, carbon or other residue build-up. If it's dirty, clean it with aerosol carburetor cleaner and a rag.

8   With the engine running, place an automotive stethoscope against each injector, one at a time, and listen for a clicking sound, indicating operation **(see illustration)**. If you don't have a stethoscope, place the tip of a screwdriver against the injector and listen through the handle. **Note**: *On the TBI system, you can actually see the fuel being sprayed in the throttle body, it is not necessary to check*

them for a clicking sound using an automotive stethoscope.

9   Unplug the injector electrical connector(s) and test the resistance of each injector. Compare the values to the Specifications listed in this Chapter **(see illustration)**.

10   Install an injector test light ("noid" light) into each injector electrical connector, one at a time **(see illustration)**. Crank the engine over. Confirm that the light flashes evenly on each connector. This will test for voltage to the injectors.

11   The remainder of the system checks can be found in the following Sections.

13   Throttle Body Injection (TBI) - component replacement

Warning: *Gasoline is extremely flammable, so take extra precautions when you work on any part of the fuel system. Don't smoke or allow open flames or bare light bulbs near the work area, and don't work in a garage where a natural gas-type appliance (such as a water heater or a clothes dryer) with a pilot light is present. Since gasoline is carcinogenic, wear latex gloves when there's a possibility*

of being exposed to fuel, and, if you spill any fuel on your skin, rinse it off immediately with soap and water. Mop up any spills immediately and do not store fuel-soaked rags where they could ignite. The fuel system is under constant pressure, so, if any fuel lines are to be disconnected, the fuel pressure in the system must be relieved first (see Section 2). When you perform any kind of work on the fuel system, wear safety glasses and have a Class B type fire extinguisher on hand.

Note: *Because of its relative simplicity, the throttle body assembly does not need to be removed from the intake manifold or disassembled for component replacement. However, for the sake of clarity, the following procedures are shown with the TBI assembly removed from the vehicle.*

1   Relieve system fuel pressure (see Section 2).

2   Detach the cable from the negative terminal of the battery.

3   Remove the air cleaner housing assembly, adapter and gaskets.

Fuel meter cover/fuel pressure regulator assembly

Refer to illustrations 13.6 and 13.7

Note: *The fuel pressure regulator is housed in the fuel meter cover. Whether you are replacing the meter cover or the regulator itself, the entire assembly must be replaced. The regulator must not be removed from the cover.*

4   Unplug the electrical connectors to the fuel injectors.

5   Remove the long and short fuel meter cover screws and remove the fuel meter cover.

6   Remove the fuel meter outlet passage gasket, cover gasket and pressure regulator seal. Carefully remove any old gasket material that is stuck with a razor blade **(see illustration)**. **Caution:** *Do not attempt to re-use either of these gaskets.*

7   Inspect the cover for dirt, foreign material and casting warpage. If it is dirty, clean it with a clean shop rag soaked in solvent. Do

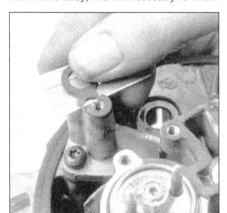

13.6 Carefully peel away the old fuel meter outlet passage gasket and fuel meter cover gasket with a razor blade

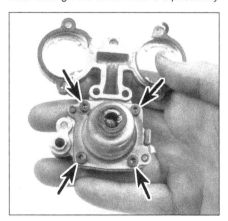

13.7 DO NOT remove the four pressure regulator screws (arrows) from the fuel meter cover

13.14 To remove either injector electrical connector, depress the two tabs on the front and rear of each connector and lift straight up

13.16 To remove an injector, slip the tip of a flat-bladed screwdriver under the lip of the lug on top of the injector and, using another screwdriver as a fulcrum, carefully pry the injector up and out

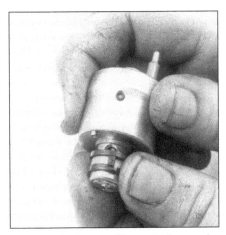

13.21 Slide the new filter onto the nozzle of the fuel injector

13.22 Lubricate the lower O-ring with transmission fluid then place it on the shoulder in the bottom of the injector cavity

13.23 Place the steel back-up washer on the shoulder near the top of the injector cavity

not immerse the fuel meter cover in cleaning solvent - it could damage the pressure regulator diaphragm and gasket. **Warning:** *Do not remove the four screws* **(see illustration)** *securing the pressure regulator to the fuel meter cover. The regulator contains a large spring under compression which, if accidentally released, could cause injury. Disassembly might also result in a fuel leak between the diaphragm and the regulator housing. The new fuel meter cover assembly will include a new pressure regulator.*

8   Install the new pressure regulator seal, fuel meter outlet passage gasket and cover gasket.

9   Install the fuel meter cover using Loctite 262 or equivalent on the screws. **Note:** *The short screws go next to the injectors.*

10   Attach the electrical connectors to both injectors.

11   Attach the cable to the negative terminal of the battery.

12   With the engine off and the ignition on, check for leaks around the gasket and fuel line couplings.

13   Install the air cleaner, adapter and gaskets.

Fuel injector(s)

Refer to illustrations 13.14, 13.16, 13.21, 13.22, 13.23, 13.24 and 13.25

Note: *When replacing a fuel injector, be sure to use one having the identical part number. Injectors from other models are calibrated with different flow rates but can be interchanged with other Model 220 TBI units. Check with a dealer parts department for correct identification.*

14   To unplug the electrical connectors from the fuel injectors, squeeze the plastic tabs and pull straight up **(see illustration)**.

15   Remove the fuel meter cover/pressure regulator assembly. **Note:** *Do not remove the fuel meter cover assembly gasket - leave it in place to protect the casting from damage during injector removal.*

16   Use two screwdrivers **(see illustration)** to pry out the injector(s).

17   Remove the upper (larger) and lower (smaller) O-rings and filter from the injector(s).

18   Remove the steel backup washer from the top of each injector cavity.

19   Inspect the fuel injector filters for evidence of dirt and contamination. If present, check for the presence of dirt in the fuel lines and fuel tank.

20   Be sure to replace the fuel injector with an identical part. Injectors from other models can fit in the Model 220 TBI assembly but are calibrated for different flow rates.

21   Slide the new filter into place on the nozzle of the injector **(see illustration)**.

22   Lubricate the new lower (smaller) O-ring with automatic transmission fluid and place it on the small shoulder at the bottom of the fuel injector cavity in the fuel meter body **(see illustration)**.

23   Install the steel back-up washer in the injector cavity **(see illustration)**.

24   Lubricate the new upper (larger) O-ring with automatic transmission fluid and install it on top of the steel back-up washer **(see illustration)**. **Note:** *The backup washer and the large O-ring must be installed before the injector. If they aren't, improper seating of the large O-ring could cause fuel leakage.*

25   To install an injector, align the raised lug on the injector base with the notch in the fuel meter body cavity **(see illustration)**. Push

13.24 Lubricate the upper O-ring with transmission fluid then install it on top of the steel washer

13.25 Make sure that the lug is aligned with the groove in the bottom of the fuel injector cavity

13.31 The TPS is mounted to the throttle body with two screws

13.37 The IAC valve can be removed with an adjustable wrench (shown) or a 1-1/4 inch wrench

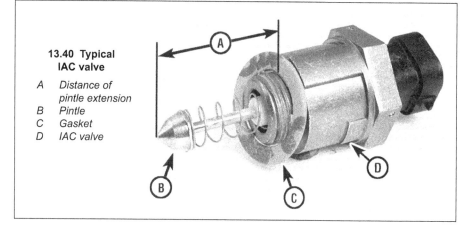

13.40 Typical IAC valve

A   *Distance of pintle extension*
B   *Pintle*
C   *Gasket*
D   *IAC valve*

down on the injector until it is fully seated in the fuel meter body. **Note:** *The electrical terminals should be parallel with the throttle shaft.*

26   Install the fuel meter cover assembly and gasket.

27   Attach the cable to the negative terminal of the battery.

28   With the engine off and the ignition on, check for fuel leaks.

29   Attach the electrical connectors to the fuel injectors.

30   Install the air cleaner housing assembly, adapter and gaskets.

Throttle Position Sensor (TPS)

Refer to illustration 13.31

Note: *For more information on the TPS, refer to Chapter 6.*

31   Remove the two TPS attaching screws and retainers and remove the TPS from the throttle body **(see illustration)**.

32   If you intend to re-use the same TPS, do not attempt to clean it by soaking it in any liquid cleaner or solvent. The TPS is a delicate electrical component and can be damaged by solvents.

33   Install the TPS on the throttle body while lining up the TPS lever with the TPS drive lever.

34   Install the two TPS attaching screws and retainers.

35   Install the air cleaner housing assembly, adapter and gaskets.

36   Attach the cable to the negative terminal of the battery.

Idle Air Control (IAC) valve

Refer to illustrations 13.37, 13.40 and 13.41

Note: *For more information on the IAC, refer to Section 14.*

37   Unplug the electrical connector from the IAC valve and remove the IAC valve **(see illustration)**.

38   Remove and discard the old IAC valve gasket. Clean any old gasket material from the surface of the throttle body assembly to insure proper sealing of the new gasket.

39   All pintles in IAC valves on Model 220 TBI units have the same dual taper. However, the pintles on some units have a 12 mm diameter and the pintles on others have a 10 mm diameter. A replacement IAC valve must have the appropriate pintle taper and diameter for proper seating of the valve in the throttle body.

40   Measure the distance between the tip of the pintle and the gasket mounting surface. If dimension "A" is greater than 1-1/8 inches, it must be reduced to prevent damage to the valve **(see illustration)**.

41   To adjust the pintle of an IAC valve, grasp the valve and exert firm pressure on the pintle with the thumb. Use a slight side-to-side movement on the pintle as you press it in with your thumb **(see illustration)**.

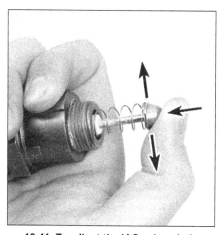

13.41 To adjust the IAC valve pintle distance, gently press the pintle in while rocking it back-and-forth

13.50 Remove the fuel inlet and outlet nuts from the fuel meter body

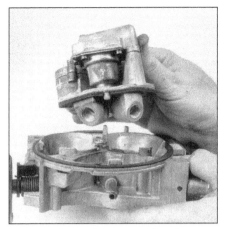

13.52 Once the fuel inlet and outlet nuts are off, pull the fuel meter body straight up to separate it from the throttle body

13.65 When disconnecting the fuel feed and return lines from the fuel inlet and outlet nuts, be sure to use a backup wrench to prevent damage to the lines

42   Install the IAC valve and tighten it to the specified torque. Attach the electrical connector.

43   Install the air cleaner housing assembly, adapter and gaskets.

44   Attach the cable to the negative terminal of the battery.

45   Start the engine and allow it to reach operating temperature, then turn it off. No adjustment of the IAC valve is required after installation. The IAC valve is reset by the PCM when the engine is turned off.

Fuel meter body assembly

Refer to illustrations 13.50 and 13.52

46   Unplug the electrical connectors from the fuel injectors.

47   Remove the fuel meter cover/pressure regulator assembly, fuel meter cover gasket, fuel meter outlet gasket and pressure regulator seal.

48   Remove the fuel injectors.

49   Unscrew the fuel inlet and return line threaded fittings, detach the lines and remove the O-rings.

50   Remove the fuel inlet and outlet nuts and gaskets from the fuel meter body assembly **(see illustration)**. Note the locations of the nuts to ensure proper reassembly. The inlet nut has a larger passage than the outlet nut.

51   Remove the gasket from the inner end of each fuel nut.

52   Remove the fuel meter body-to-throttle body attaching screws and remove the fuel meter body from the throttle body **(see illustration)**.

53   Install the new throttle body-to-fuel meter body gasket. Match the cut-out portions in the gasket with the openings in the throttle body.

54   Install the fuel meter body on the throttle body. Coat the fuel meter body-to-throttle body attaching screws with thread locking compound before installing them.

55   Install the fuel inlet and outlet nuts, with new gaskets, in the fuel meter body and tighten the nuts to the specified torque. Install the fuel inlet and return line threaded fittings

with new O-rings. Use a backup wrench to prevent the nuts from turning.

56   Install the fuel injectors.

57   Install the fuel meter cover/pressure regulator assembly.

58   Attach the cable to the negative terminal of the battery.

59   Attach the electrical connectors to the fuel injectors.

60   With the engine off and the ignition on, check for leaks around the fuel meter body, the gasket and around the fuel line nuts and threaded fittings.

61   Install the air cleaner housing assembly, adapters and gaskets.

Throttle body assembly

Refer to illustrations 13.65 and 13.66

62   Unplug all electrical connectors - the IAC valve, TPS and fuel injectors. Detach the grommet with the wires from the throttle body.

63   Detach the throttle linkage, return spring(s), transmission control cable and, if equipped, cruise control.

64   Clearly label, then detach, all vacuum hoses.

65   Using a backup wrench, detach the inlet and outlet fuel line nuts **(see illustration)**. Remove the fuel line O-rings from the nuts and discard them.

66   Remove the TBI mounting bolts **(see illustration)** and lift the TBI unit from the intake manifold. Remove and discard the TBI manifold gasket.

67   Place the TBI unit on a holding fixture (Kent-Moore J-9789-118 or BT-3553 or equivalent). **Note:** *If you don't have a holding fixture, and decide to place the TBI directly on a work bench surface, be extremely careful when servicing it. The throttle valve can be easily damaged.*

68   Remove the fuel meter body-to-throttle body attaching screws and separate the fuel meter body from the throttle body.

69   Remove the throttle body-to-fuel meter body gasket and discard it.

70   Remove the TPS.

71   Invert the throttle body on a flat surface for greater stability and remove the IAC valve.

72   Clean the throttle body assembly in a cold immersion cleaner. Clean the metal parts thoroughly and blow dry with compressed air. Be sure that all fuel and air passages are free of dirt or burrs. **Caution:** *Do not place the TPS, IAC valve, pressure regulator diaphragm, fuel injectors or other components containing rubber in the solvent or cleaning bath. If the throttle body requires cleaning, soaking time in the cleaner should be kept to a minimum. Some models have throttle shaft dust seals that could lose their effectiveness by extended soaking.*

73   Inspect the mating surfaces for damage that could affect gasket sealing. Inspect the throttle lever and valve for dirt, binds, nicks and other damage.

74   Invert the throttle body on a flat surface for stability and install the IAC valve and the TPS.

75   Install a new throttle body-to-fuel meter body gasket and place the fuel meter body assembly on the throttle body assembly. Coat the fuel meter body-to-throttle body

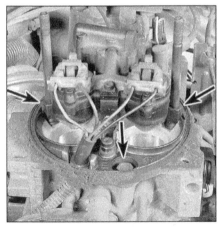

13.66 To remove the Model 220 throttle body from the intake manifold, remove the three bolts (arrows)

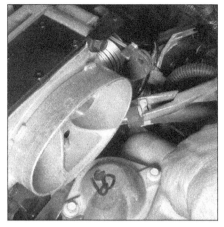

14.2 Clean the throttle body with carburetor cleaner to remove sludge deposits

14.9a Disconnect the coolant lines from the throttle body

14.9b The lower coolant line is very difficult to access so remove the throttle body and angle the assembly to remove the clamp

attaching screws with thread locking compound and tighten them securely.
76   Install the TBI unit and tighten the mounting bolts to the specified torque. Use a new TBI-to-manifold gasket.
77   Install new O-rings on the fuel line nuts. Install the fuel line and outlet nuts by hand to prevent stripping the threads. Using a backup wrench, tighten the nuts to the specified torque once they have been correctly threaded into the TBI unit.
78   Attach the vacuum hoses, throttle linkage, return spring(s), transmission control cable (automatics) and, if equipped, cruise control cable. Attach the grommet, with wire harness, to the throttle body.
79   Plug in all electrical connectors, making sure that the connectors are fully seated and latched.
80   Check to see if the accelerator pedal is free by depressing the pedal to the floor and releasing it with the engine off.
81   Connect the negative battery cable, and, with the engine off and the ignition on, check for leaks around the fuel line nuts.
82   Check the TPS output (see Chapter 6).
83   Install the air cleaner housing assembly, adapter and gaskets.

---

14   Multiport Fuel Injection (MFI) - component check and replacement

---

Warning: *Gasoline is extremely flammable, so take extra precautions when you work on any part of the fuel system. Don't smoke or allow open flames or bare light bulbs near the work area, and don't work in a garage where a natural gas-type appliance (such as a water heater or a clothes dryer) with a pilot light is present. Since gasoline is carcinogenic, wear latex gloves when there's a possibility of being exposed to fuel, and, if you spill any fuel on your skin, rinse it off immediately with soap and water. Mop up any spills immediately and do not store fuel-soaked rags where*

they could ignite. The fuel system is under constant pressure, so, if any fuel lines are to be disconnected, the fuel pressure in the system must be relieved first (see Section 2). When you perform any kind of work on the fuel system, wear safety glasses and have a Class B type fire extinguisher on hand.

Throttle body
Check
Refer to illustration 14.2
1   Detach the air intake duct from the throttle body and move the duct out of the way.
2   Have an assistant depress the throttle pedal while you watch the throttle valve. Check that the throttle valve moves smoothly when the throttle is moved from closed (idle position) to fully open (wide open throttle). **Note:** *Spray carburetor cleaner into the throttle body, especially around the shaft area* **(see illustration)** *to free-up any binding caused by the accumulation of carbon deposits or sludge buildup.*
3   Wiggle the throttle lever while watching the throttle shaft inside the bore. If it appears worn (loose), replace the throttle body unit.

Removal
Refer to illustrations 14.9a, 14.9b and 14.10
Warning: *Wait until the engine is completely cool before beginning this procedure.*
4   Disconnect the cable from the negative terminal of the battery. **Caution:** *On models equipped with a Delco Loc II or Theftlock audio system, be sure the lockout feature is turned off before performing any procedure which requires disconnecting the battery.*
5   Detach the air intake duct.
6   Unplug the Idle Air Control (IAC) valve and the Throttle Position Sensor (TPS) electrical connectors (see Chapter 6).
7   Mark and disconnect any vacuum hoses connected to the throttle body. Also detach the breather hose, if equipped.
8   Disconnect the accelerator cable from the throttle lever, then detach the cable housing from its bracket (see Section 10).

14.10 Remove the throttle body bolts (arrows)

9   Loosen the clamps and disconnect the coolant hoses from the underside of the throttle body **(see illustration)**. Be prepared for some coolant spillage and plug the ends of the hoses. **Note:** *On some models, it will be necessary to remove the throttle body from the plenum to gain access to the coolant hose* **(see illustration)**.
10   Remove the throttle body bolts and detach the throttle body **(see illustration)**.

Installation
11   Clean off all traces of old gasket material from the throttle body and the plenum.
12   Install the throttle body and a new gasket and tighten the bolts to the torque listed in this Chapter's Specifications.
13   The rest of the procedure is the reverse of removal. Be sure to check the coolant level (see Chapter 1) and add, if necessary.

Idle Air Control (IAC) valve
Check
Refer to illustrations 14.15, 14.17a and 14.17b
14   The idle air control valve (IAC) controls the engine idle speed. This output actuator

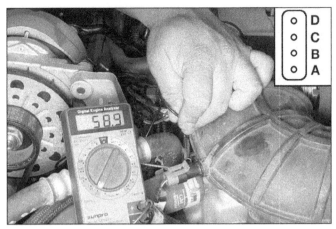

14.15 Measure the resistance across terminals D and C, then across terminals A and B

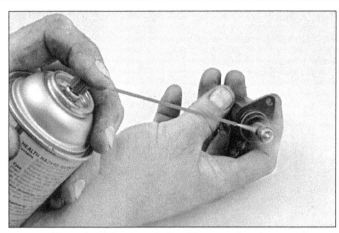

14.17a Clean the IAC valve pintle with aerosol carburetor cleaner to remove carbon deposits

14.17b Spray carburetor cleaner into the IAC valve housing and check for clogged air passages in the air intake plenum

14.19 Remove the IAC valve screws (arrows) and separate the IAC from the throttle body

is mounted on the throttle body and is controlled by voltage pulses sent from the PCM (computer). The IAC valve pintle moves in or out allowing more or less intake air into the system according to the engine conditions. To increase idle speed, the PCM retracts the IAC valve pintle away from the seat and allows more air to bypass the throttle bore. To decrease idle speed, the PCM extends the IAC valve pintle towards the seat, reducing the air flow.

15   To check the IAC valve, unplug the electrical connector and, using an ohmmeter, measure the resistance across terminals A and B, then terminals C and D. Each resistance check should indicate 40 to 80 ohms **(see illustration)**. If not, replace the IAC valve.

16   There is an alternate method for testing the IAC valve. Various SCAN tools are available from auto parts stores and specialty tool companies that can be plugged into the ALDL (diagnostic connector) for the purpose of monitoring the sensors. Connect the SCAN tool and switch to the Idle Air Motor Position mode and monitor the steps (motor winding position). The SCAN tool should indicate between 10 to 200 steps depending upon the rpm range. Allow

the engine to idle for several minutes and while observing the count reading, snap the throttle to achieve high rpm (under 3,500). Repeat the procedure several times and observe the SCAN tool steps (counts) when the engine goes back to idle. The readings should be within 5 to 10 steps each time. If the readings fluctuate greatly, replace the IAC valve. **Note:** *When the IAC valve electrical connector is disconnected for testing, the PCM will have to "relearn" its idle mode. In other words, it will take a certain amount of time before the idle motor resets for the correct idle speed. Make sure the idle is smooth and not misfiring before plugging in the SCAN tool.* **Note:** *Refer to Chapter 6 for additional information and illustrations concerning SCAN tools.*

17   Next, remove the valve (see Step 18) and inspect it:

a) *Check the pintle for excessive carbon deposits. If necessary, clean it with aerosol carburetor cleaner* **(see illustration)**. *Also clean the IAC valve housing to remove any deposits* **(see illustration)**.

b) *Check the IAC valve electrical connections. Make sure the pins are not bent and make good contact with the connector terminals.*

Removal

Refer to illustration 14.19

18   Unplug the electrical connector from the Idle Air Control (IAC) valve.

19   Unscrew the valve or remove the two IAC valve attaching screws and withdraw the valve **(see illustration)**.

20   Check the condition of the rubber O-ring. If it's hardened or deteriorated, replace it. On models equipped with a gasket, remove the gasket.

21   Clean the sealing surface and the bore of the idle air/vacuum signal housing assembly to ensure a good seal. **Caution:** *The IAC valve itself is an electrical component and must not be soaked in any liquid cleaner, as damage may result.*

22   Before installing the IAC valve, the position of the pintle must be checked. If the pintle is extended too far, damage to the assembly may occur.

Installation

23   Measure the distance from the flange or gasket mounting surface of the IAC valve to the tip of the pintle. If the distance is greater than 1-1/8 inch, reduce the distance by applying firm pressure onto the pintle to retract it.

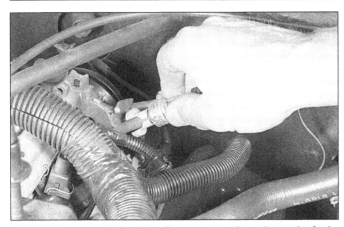

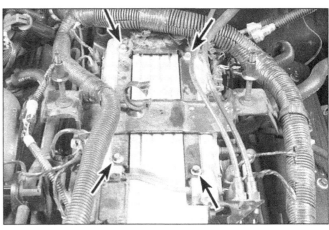

14.29 Use a special fuel line disconnect tool to release the fuel line coupler seal from the fuel rail

14.32 Remove the bolts (arrows) . . .

Try some side-to-side motion in the event the pintle binds (see Section 13).

24   Position the new O-ring or gasket on the IAC valve. Lubricate the O-ring with a light film of engine oil. If the IAC valve is the screw-in type, apply a light film of RTV sealant to the threads of the valve. Install the IAC valve and tighten the valve or the mounting screws securely.

25   Plug in the electrical connector at the IAC valve assembly. **Note:** *No adjustment is made to the IAC assembly after reinstallation. The IAC resetting is controlled by the PCM when the engine is started.*

Throttle Position Sensor (TPS)

Check

26   Check for stored trouble codes in the PCM using the On Board Diagnosis system (see Chapter 6).

27   To check the operation and replacement of the TPS, refer to the *Information Sensors* in Chapter 6.

Fuel rail and injectors

Refer to illustrations 14.29, 14.32, 14.33, 14.34, 14.35a and 14.35b

Warning: *Before any work is performed on*

the fuel lines, fuel rail or injectors, the fuel system pressure must be relieved (see Section 2).

Note: *Refer to Section 12 for the injector checking procedure.*

28   Detach the cable from the negative terminal of the battery. **Caution:** *On models equipped with a Delco Loc II or Theftlock audio system, be sure the lockout feature is turned off before performing any procedure which requires disconnecting the battery.*

29   Using a special fuel line tool to depress the seal inside the fuel line coupler **(see illustration)** and detach the fuel lines from the fuel rail.

30   Detach the vacuum line at the fuel pressure regulator.

31   Label and unplug the injector electrical connectors.

32   Remove the fuel rail retaining bolts **(see illustration)**.

33   Carefully remove the fuel rail with the injectors **(see illustration)**. **Caution:** *Use care when handling the fuel rail assembly to avoid damaging the injectors.* **Note:** *An eight digit identification number is stamped on the side of the fuel rail assembly. Refer to this number if servicing or parts replacement is required.*

14.33 . . . and separate the fuel rail from the intake manifold

34   To remove the fuel injectors, spread the injector retaining clip and pull the injector from the fuel rail **(see illustration)**.

35   Remove the injector O-ring seals **(see illustrations)**. **Note:** *Replace the injector seal with the correct color. Brown seals are used on the engine side of the injector while black seals are used on the fuel rail side of the injector.*

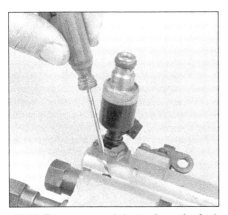

14.34 To remove an injector from the fuel rail, spread the retaining clip with a small screwdriver, then pull the injector from the fuel rail

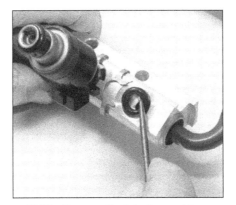

14.35a Remove the injector seal from the fuel rail. The injector seals that are positioned in the fuel rail are black while the seals that go into the engine block are brown

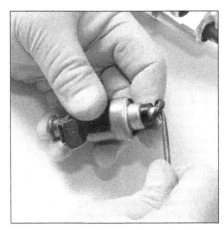

14.35b Carefully pry the seals off the injectors

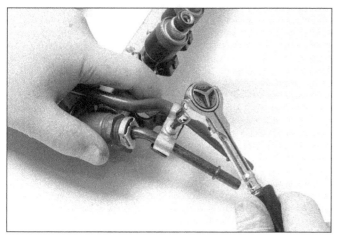

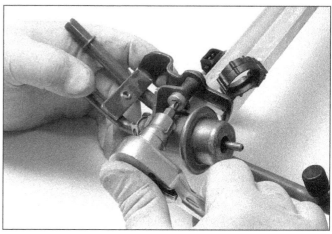

14.44a   Remove the bracket from the fuel pressure regulator fuel lines

14.44b   Remove the fuel pressure regulator mounting screws

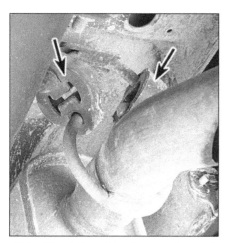

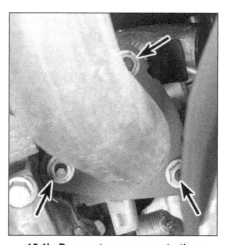

15.1a   Check the rubber hangers (arrows) for deterioration or cracks that may cause the exhaust system to drop

15.1b   Be sure to spray penetrating lubricant onto the flange nuts (arrows) before removing them from the exhaust manifold

15.2   Remove the bracket bolts from the chassis (arrows)

36   Install the new O-ring seal(s), as required, on the injector(s) and lubricate them with a light film of engine oil.

37   Install the injectors on the fuel rail.

38   Secure the injectors with the retainer clips.

39   Installation is the reverse of the removal procedure.

Fuel pressure regulator

Check

40   Refer to Section 3 for the fuel pressure regulator checking procedure.

Replacement

Refer to illustrations 14.44a and 14.44b

41   Relieve the fuel system pressure (see Section 2).

42   Disconnect the cable from the negative terminal of the battery. **Caution:** *On models equipped with a Delco Loc II or Theftlock audio system, be sure the lockout feature is turned off before performing any procedure which requires disconnecting the battery.*

43   Remove the fuel rail following the procedure described earlier in this Section.

44   Remove the fuel line bracket from the fuel pressure regulator assembly **(see illustration)**. Remove the pressure regulator mounting screws **(see illustration)** and separate the two fuel rails from the pressure regulator assembly.

45   Reassembly is the reverse of disassembly. Be sure to replace all gaskets and seals, otherwise a dangerous fuel leak may develop. When installing the seals, lubricate them with a light film of engine oil.

---

15   Exhaust system servicing - general information

Warning: *The vehicle's exhaust system generates very high temperatures and must be allowed to cool down completely before any of the components are touched. Be especially careful around the catalytic converter, where the highest temperatures are generated.*

General Information

Refer to illustrations 15.1a and 15.1b

1   Replacement of exhaust system components is basically a matter of removing the heat shields, disconnecting the component and installing a new one **(see illustration)**. The heat shields and exhaust system hangers must be reinstalled in the original locations or damage could result. Due to the high temperatures and exposed locations of the exhaust system components, rust and corrosion can seize parts together. Penetrating oils are available to help loosen frozen fasteners. However, in some cases it may be necessary to cut the pieces apart with a hacksaw or cutting torch. The latter method should be employed only by persons experienced in this work.

Crossover pipe

Refer to illustration 15.2

2   Remove the bolts or nuts securing the crossover pipe to the exhaust manifolds **(see illustration)**. Remove the crossover pipe.

3   Installation is the reverse of removal. Tighten the fasteners evenly and securely.

Chapter 5
Engine electrical systems

Contents

Specifications

Distributor pick-up coil resistance (1993 and earlier)	500 to 1500 ohms
Ignition coil resistance	
Primary resistance	0.2 to 1.5 ohms
Secondary resistance	5,000 to 25,000 ohms

1 General information

Warning: *Because of the very high voltage generated by the ignition system, extreme care should be taken whenever an operation involving ignition components is performed. This not only includes the distributor, coil(s), module and spark plug wires, but related items that are connected to the systems as well, such as the plug connections, tachometer and testing equipment.*

The models covered by this manual are equipped with two different types of ignition systems. 1993 and earlier models are equipped with a conventional remote mounted coil High Energy Ignition (HEI) system. 1994 and later models with Multi Port Fuel Injection (MFI) are equipped with a unique Distributor Ignition (DI) system (see Section 5).

Although these systems use different ways to generate the ignition signals, they share certain similar components, such as the ignition switch, battery, coil, primary (low tension) and secondary (high tension) wiring circuits and spark plugs.

The charging system consists of a belt-driven alternator with an integral voltage regulator and the battery. These components work together to supply electrical power for the ignition system, the lights and all accessories.

1993 and earlier models are equipped with the CS-130 (100 amp) alternator. Later models are equipped with either a CS-130 or the CS-144 (124 amp) alternator. All types use a conventional pulley and fan. The CS-130 alternators should be considered non-serviceable and, if found to be faulty,

should be exchanged as cores for new or rebuilt units. CS-144 alternators can be rebuilt but it is recommended that the home mechanic exchange the alternator for a rebuilt unit. Because of the expense and the limited availability of parts, no alternator overhaul information is included in this manual.

2 Battery - emergency jump starting

Refer to the *Booster battery (jump) starting* procedure at the front of this manual.

3 Battery - removal and installation

Refer to illustration 3.3

Warning: *Hydrogen gas is produced by the battery, so keep open flames and lighted cigarettes away from it at all times. Always wear eye protection when working around a battery. Rinse off spilled electrolyte immediately with large amounts of water.*

Removal

1   The battery is located at the left front of the engine compartment.
2   Detach the cables from the negative and positive terminals of the battery. **Warning:** *To prevent arcing, disconnect the negative (-) cable first, then remove the positive (+) cable.* **Caution:** *On models equipped with a Delco Loc II or Theftlock audio system, be sure the lockout feature is turned off before performing any procedure which requires disconnecting the battery.*

3   Remove the hold-down clamp bolt and the clamp from the battery carrier **(see illustration)**.
4   Carefully lift the battery from the carrier. **Warning:** *Always keep the battery in an upright position to reduce the likelihood of electrolyte spillage. If you spill electrolyte on your skin, rinse it off immediately with large amounts of water.*

Installation

Note: *The battery carrier and hold-down clamp should be clean and free from corrosion before installing the battery. Make certain that there are no parts in the carrier before installing the battery.*

5   Set the battery in position in its carrier. Don't tilt it.
6   Install the hold-down clamp and bolt. The bolt should be snug, but overtightening it may damage the battery case.

3.3 Remove the battery hold-down bolt (arrow) from the battery carrier

7   Install both battery cables, positive first, then the negative. **Note:** *The battery terminals and cable ends should be cleaned prior to connection* (see Chapter 1).

4   Battery cables - check and replacement

1   Periodically inspect the entire length of each battery cable for damage, cracked or burned insulation and corrosion. Poor battery cable connections can cause starting problems and decreased engine performance.
2   Check the cable-to-terminal connections at the ends of the cables for cracks, loose wire strands and corrosion. The presence of white, fluffy deposits under the insulation at the cable terminal connection is a sign the cable is corroded and should be replaced. Check the terminals for distortion, missing mounting bolts or nuts and corrosion.
3   If only the positive cable is to be replaced, be sure to disconnect the negative cable from the battery first. **Caution:** *On models equipped with a Delco Loc II or Theftlock audio system, be sure the lockout feature is turned off before performing any procedure which requires disconnecting the battery.*
4   Disconnect and remove the cable. Make sure the replacement cable is the same length and diameter.
5   Clean the threads of the starter or ground connection with a wire brush to remove rust and corrosion. Apply a light coat of petroleum jelly to the threads to ease installation and prevent future corrosion.
6   Attach the cable to the starter or ground connection and tighten the mounting nut securely.
7   Before connecting the new cable to the battery, make sure it reaches the terminals without having to be stretched.
8   Connect the positive cable first, followed by the negative cable. Tighten the nuts and apply a thin coat of petroleum jelly to the terminal and cable connection.

5   Ignition system - general information

Warning: *Because of the very high voltage generated by the ignition system, extreme care should be taken whenever an operation involving ignition components is performed. This not only includes the distributor, coil(s), module and spark plug wires, but related items that are connected to the systems as well, such as the plug connections, tachometer and other testing equipment.*

The ignition system consists of the ignition switch, the battery, the coil, the primary (low tension) and secondary (high tension) wiring circuits, the distributor and the spark plugs.

1993 and earlier models

A High Energy Ignition (HEI) distributor with Electronic Spark Timing (EST) is used on 1993 and earlier models. The ignition coil is mounted remotely. The HEI distributor has an internal magnetic pick-up assembly which contains a permanent magnet, a pole piece with internal teeth and a pick-up coil.

All spark timing changes in the HEI/EST distributor are carried out electronically by the Powertrain Control Module (PCM), which monitors data from various engine sensors, computes the desired spark timing and signals the distributor to change the timing accordingly. A back-up spark advance system is incorporated to signal the ignition module in case of PCM failure. No vacuum or mechanical advance is used.

1994 and later models

The 1994 and later models are equipped with a Distributor Ignition (DI) system that incorporates the Opti Spark ignition system. The Opti Spark ignition system consists of the distributor housing, cap and rotor, optical position sensor, sensor disc, pick-up assembly, distributor drive shaft, ignition module, ignition coil, primary and secondary wiring, spark plugs and the necessary control circuits (wiring harness) for the entire system. Although the distributor is composed of components, the entire assembly must be replaced as a single unit.

The Opti Spark distributor is mounted on the front cover and it is driven directly by the camshaft. The cap and rotor directs the spark from the ignition coil to the proper spark plug wire. The distributor cap is marked with the corresponding cylinder numbers.

The distributor contains an optical pick-up system that provides actual crankshaft position (in degrees) to the PCM. The system uses two infrared optical sensors and a flat disc with two rows of notches (slots) cut around the circumference. One row has 360 notches (one-degree widths) and the other row has eight notches. When the optical pick-up turns, it produces two modulated digital signals. The first row (360-degrees) produces the signal for the timing while the second row (8 notches) produces the signal for the RPM reference.

Electronic Spark Control (ESC)

All models are equipped with an Electronic Spark Control (ESC), which uses a knock sensor in connection with the PCM to control spark timing to allow the engine to have maximum spark advance without spark knock. This improves driveability and fuel economy.

Secondary (spark plug) wiring

The secondary (spark plug) wires are a carbon-impregnated cord conductor encased in an 8 mm (5/16-inch) diameter rubber jacket with an outer silicone jacket. This type of wire will withstand very high temperatures and provides an excellent insulator for the high secondary ignition voltage. Silicone spark plug boots form a tight seal on the plug. The boot should be twisted 1/2-turn before removing (for more information on spark plug wiring refer to Chapter 1).

6   Ignition system - check

Warning: *Because of the very high voltage generated by the ignition system, extreme care should be taken whenever an operation is performed involving ignition components. This not only includes the coils, control module and spark plug wires, but related items connected to the system as well, such as the plug connections, tachometer and any test equipment.*

General checks

Refer to illustration 6.3

1   Check all ignition wiring connections for tightness, cuts, corrosion or any other signs of a bad connection. A faulty or poor connection at a spark plug could also result in a misfire. Also check for carbon deposits inside the spark plug boots. Remove the spark plugs, if necessary, and check for fouling.
2   Check for ignition and battery supply to the PCM. Check the ignition fuses (see Chapter 12). **Note:** *The ECM BAT fuse controls the computer and the fuel injection system while the IGN fuse controls power to the ignition system.*
3   Use a calibrated ignition tester to verify adequate available secondary voltage (25,000 volts) at the spark plug **(see illustration)**. Using an ohmmeter, check the resistance of the spark plug wires). Each wire should measure less than 30,000 ohms.
4   Check to see if the fuel pump and relay are operating properly (see Chapter 4). The fuel pump should activate for two seconds when the ignition key is cycled ON. Install an injector test light and monitor the blinks as the injector voltage signal pulses (see Chapter 4, *Fuel injection system - check*).

6.3 To use a calibrated ignition tester (available at most auto parts stores), simply disconnect a spark plug wire, attach the wire to the tester, clip the tester to a convenient ground and operate the starter - if there's enough power to fire the plug, sparks will be visible between the electrode tip and the tester body

1993 and earlier models

Refer to illustration 6.5

5   Refer to the illustration for terminal pin designations for testing the ignition module **(see illustration)**.

6   Check for spark at the spark plugs checking at least two or more spark plug wires **(see illustration 6.3)**. On a NO START condition, check the fuel pump relay and fuel pump systems for fuel delivery problems in the event the ignition spark is available.

7   Unplug the four-wire connector from the ignition module at the distributor and check for spark at the coil wire. If the fuel system is working properly and the ignition system will not start the engine, a spark indicates that the problem must be the distributor cap or rotor. **Note:** *A few sparks followed by no spark is the same condition as no spark at all.*

8   Check the ignition coil (see Section 7). Next, disconnect the two terminal plug on the module and check for voltage on the "C" and the "+" terminals. Normally, there should be battery voltage at the "C" and "+" terminals. Low voltage indicates an open or high resistance circuit from the distributor to the coil or ignition switch. If the "C" terminal voltage is low, but the "+" terminal voltage is 10 volts or more, the circuit from "C" terminal to the ignition coil or ignition coil primary winding is open.

9   Reconnect the "C+" connector onto the ignition module and with the ignition key ON (engine not running), check for voltage at the TACH terminal. The TACH terminal is taped back in the harness near the distributor. If there is less than 10 volts available, check the TACH circuit from the distributor. This test, checks for a shorted module or a grounded circuit from the ignition coil to the module. The distributor module should be turned off, so normal voltage should be about 12 volts. If the module is turned ON, the voltage will be low, but above one volt. This could cause the ignition coil to fail from excessive heat.

10   Check the module itself. Construct a battery pack that is more than 1.5 volts and less than 8 volts (penlight batteries). Disconnect the ignition module four-wire connector from the distributor. Disconnect the pick-up coil two-wire connector and remove the distributor cap and rotor. Connect a voltmeter from TACH to ground. Connect a test light from the battery pack (1.5 to 8.0 volts) and touch the test light probe to terminal P on the ignition module and watch the voltmeter. Voltage should drop. Applying a voltage (1.5V) to module terminal "P" should turn the module on and the TACH terminal voltage should drop to about 7 to 9 volts. **Note:** *Remember, the TACH terminal is usually taped to the harness near the distributor.*

11   This test will determine whether the module or coil is faulty or if the pick-up coil is not generating the proper signal to turn the module on. This test can be performed by using a DC battery with a rating of 1.5 volts. Monitor the change in voltage with a voltmeter connected to the TACH terminal. Some digital multimeters can also be used to trigger the module by selecting ohms, usually the diode position. In this position, the meter may have a voltage across its terminals which can be used to trigger the module. The voltage in the ohms position can be checked by using a second meter or by checking the manufacturer's specifications for the tool being used.

12   Install a spark testing tool **(see illustration 6.3)**. Repeat the previous test but instead observe spark at the spark plug when the test light tip is removed from terminal P. There should be a spark. This test checks the ignition module's ability to cause a spark. If no spark occurs, the fault is most likely in the ignition coil or pick-up coil because most module problems would have been found before this point in the procedure. A GM HEI module tester (Kent Moore J-24642-F or equivalent) can determine which is at fault. If you cannot obtain the module tester, take the vehicle to a dealer service department or other qualified repair shop at this point. Also, check the resistance of the pick-up coil and if the specifications are incorrect, replace it with a new part (see Section 10).

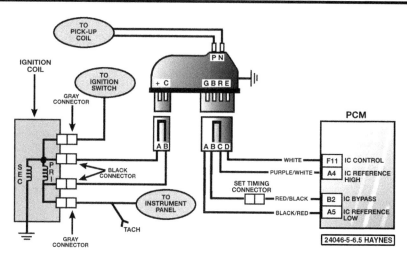

6.5 Ignition schematic of the High Energy Ignition (HEI) system (1993 and earlier models)

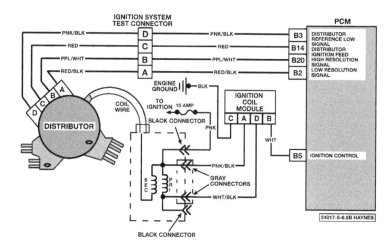

6.13 Ignition schematic of the Distributor Ignition (DI) system (1994 and later models)

1994 and later models

Refer to illustrations 6.13, 6.16, 6.18 and 6.19

Note: *If the PCM sets a code 16, this indicates a low resolution pulse or a defective optical sensor in the distributor. This prevents voltage from reaching the PCM resulting in a no-start condition or a loss of engine performance. Refer to Chapter 6 for additional information on the On-Board Diagnosis (OBD) system and interpreting the coded information.*

Note: *Base timing is preset - there is NO timing adjustment possible. The PCM controls the advance and retard functions of the ignition timing.*

13   Refer to the accompanying schematic for terminal pin designations for testing the Distributor Ignition (DI) ignition system **(see illustration)**.

14   Check for spark at the spark plugs, checking at least two or more spark plug wires **(see illustration 6.3)**. On a NO START condition, check the fuel pump relay and fuel pump systems for fuel delivery problems in the event the ignition spark is available.

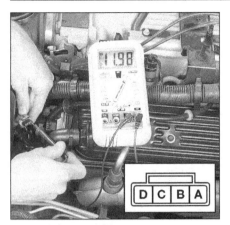

6.16 Disconnect the ignition module harness connector and check for battery voltage to the ignition module on terminals A or D

6.18 Install the LED test light onto the positive (+) battery terminal and connect the probe (tip) onto terminal C and confirm the test light illuminates when the test light is grounded through the circuit

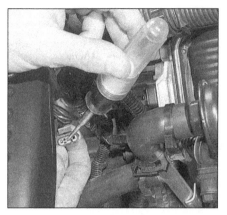

6.19 Clip the lead of the LED test light to the positive (+) battery terminal and connect the probe (tip) to the white/black wire on the ignition coil connector - confirm that the test light flashes when an assistant cranks the engine over

15   Check for spark at the coil wire with a spark tester while an assistant is cranking the engine. A spark indicates that the problem must be the distributor unit. This test separates the ignition wires from the ignition coil. **Note:** *A few sparks followed by no spark is the same condition as no spark at all.*

16   If there is NO spark, check the ignition coil (see Section 7). If the coil checks are correct, disconnect the ignition module connector, turn the ignition key ON (engine not running) and check for battery voltage at the terminals A and D. Normally, there should be battery voltage at the "A" and "D" terminals **(see illustration)**. Low voltage indicates an open or high resistance circuit from the distributor to the coil or ignition switch (primary ignition circuit). Also, check the OBD self diagnosis system for codes that may indicate an ignition system failure (see Chapter 6).

17   Switch the voltmeter to the A/C scale and measure the voltage on terminal B while cranking the engine over. Have an assistant turn the engine over while observing the voltmeter. This voltage signal from the computer acts as reference voltage for ignition control. This test will eliminate the PCM as a source of the problem. If there is a voltage signal present, a no start condition will most likely be caused by a faulty module. Continue testing. If there is no voltage signal from the PCM, replace the computer (see Chapter 6).

18   Turn the ignition key OFF and install a LED type test light onto the battery positive terminal (+). Observe that the test light is ON when terminal C on the ignition module connector (harness side) is probed. This checks the computer ground. This will check for a complete ground wire from the ignition module. The test light should illuminate when touched to the connector terminal C **(see illustration)**. If there is no light, check the wiring harness for an open or shorted circuit.

19   Remove the coil wire and ground it to the engine using a suitable jumper wire. Disconnect the ignition coil harness connector, connect a LED type test light to the positive (+) battery terminal and check for a pulsing

signal voltage at the white/black wire while the engine is cranked over. This test will check for the pulsing signal voltage from the ignition distributor after it has been triggered by the PCM. Have an assistant crank over the engine while observing the LED test light **(see illustration)**. Be sure to remove the coil secondary wire and ground the coil to the engine using a suitable jumper wire. There should be a steady and obvious flashing LED light. Regular 12 volt test lights can be used but the bulb may not respond brightly and clearly as the LED test light.

20   If all the tests results are correct except Step 19 and there still is a NO START condition, it will be very difficult to distinguish a faulty distributor unit from a faulty ignition module. In most cases, the ignition module, when defective, will not produce a pulsing voltage signal with symptoms of a definite start or no-start condition while a faulty DI unit will produce intermittent ignition failures when the engine is running. Take the vehicle to a dealer service department or other qualified repair shop for diagnosis.

21   DI ignition systems are prone to water entering the distributor unit through the venting system. Thoroughly check the vent system for cracks, leaks or damaged components:

a) *Clean the air supply hose from the air intake duct to the distributor*
b) *Check the vacuum supply hose and check valve from the intake manifold*
c) *Check the venting system harness from the lower section of the distributor to the check valve filter*
d) *Make sure the air intake duct is not plugged or damaged*

---

7   Ignition coil - check and replacement

---

1   Disconnect the cable from the negative terminal of the battery. **Caution:** *On models equipped with a Delco Loc II or Theftlock*

audio system, be sure the lockout feature is turned off before performing any procedure which requires disconnecting the battery.

1993 and earlier models
Check
Refer to illustration 7.2

2   Check the coil for opens and grounds by performing the following three tests with an ohmmeter **(see illustration)**.

3   Using the ohmmeter's high scale, hook up the ohmmeter leads as illustrated **(see test 1 in illustration 7.2)**. The ohmmeter should indicate a very high, or infinite, resistance value. If it doesn't, replace the coil.

4   Check the coil primary resistance. Using the low scale, hook up the leads as illustrated **(see test 2 in illustration 7.2)**. The ohmmeter should indicate a very low, or zero, resistance value. If it doesn't, replace the coil. Check the Specifications listed in the beginning of this Chapter.

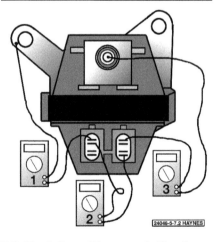

7.2 Check the coil for grounds (1), primary resistance (2) and secondary resistance (3)

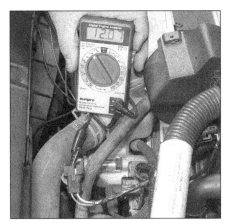

7.9 Unplug the coil electrical connector and check for battery voltage on the pink wire

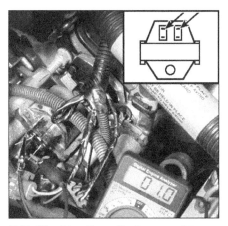

7.11 Checking the coil primary resistance (1994 and later)

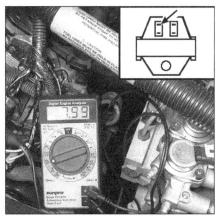

7.12 Checking the coil secondary resistance (1994 and later)

5   Check the coil secondary resistance. Using the high scale, hook up the leads as illustrated **(see test 3 in illustration 7.2)**. The ohmmeter should not indicate an infinite resistance. If it does, replace the coil.

Replacement

6   Unplug the coil high tension wire and both electrical leads from the coil.
7   Remove both mounting nuts and remove the coil from the engine.
8   Installation of the coil is the reverse of the removal procedure.

1994 and later models

Refer to illustrations 7.9, 7.11, 7.12 and 7.14

Check

9   Check to make sure the coil is receiving battery voltage with the ignition key ON (engine not running) **(see illustration)**.
10   Detach the primary electrical connector from the coil.
11   Using the ohmmeter's low scale, hook up the ohmmeter leads to the primary terminals on the ignition coil **(see illustration)**. The ohmmeter should indicate a very low resis-

tance value. If it doesn't, replace the coil.
12   Using the high scale, hook up one lead to the ignition coil primary terminal and the other lead to the secondary terminal **(see illustration)**. The ohmmeter should not indicate an infinite resistance. If it does, replace the coil.

Replacement

13   Unplug the coil high tension wire and both electrical leads from the coil.
14   Remove both mounting nuts and re-move the coil from the engine **(see illustration)**.
15   Installation of the coil is the reverse of the removal procedure.

8   Distributor - removal and installation

1993 and earlier models

Removal

Refer to illustrations 8.5 and 8.6

1   Disconnect the cable from the negative battery terminal. **Caution:** *On models equipped with a Delco Loc II or Theftlock audio system, be sure the lockout feature is*

turned off before performing any procedure which requires disconnecting the battery.
2   Remove the air cleaner assembly (see Chapter 4).
3   Disconnect the electrical connectors from the distributor and the coil.
4   Remove the distributor cap (see Chapter 1).
5   Make a mark on the distributor body directly in-line with the tip of the rotor **(see illustration)**.
6   Mark the position of the distributor body in relation to the engine **(see illustration)**.
7   Remove the distributor hold-down bolt and clamp.
8   Remove the distributor from the engine. Do not rotate the engine with the distributor removed, or the mark you made in Step 5 will be useless.

Installation

Crankshaft not turned after distributor removal

9   Position the rotor in the exact location it was in when the distributor was removed.
10   Lower the distributor into the engine. To mesh the gears at the bottom of the distributor it may be necessary to turn the rotor

7.14 Remove the coil mounting nuts and separate the coil/module assembly from the engine

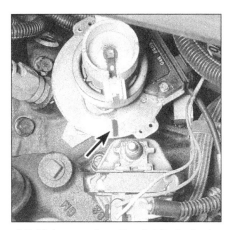

8.5 Make a mark on the distributor body (arrow) to show the direction the rotor is pointing before removing the distributor (1993 and earlier models)

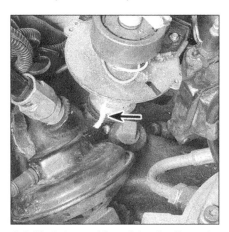

8.6 Mark the position of the distributor in relation to the engine (arrow) before loosening the distributor hold-down clamp

8.33 The distributor is retained by four bolts (1994 and later models)

9.5 The ignition timing indicator is located at the front of the engine

slightly. It is possible that the distributor may not seat down fully against the block because the lower part of the distributor shaft has not properly engaged the oil pump shaft. Make sure the distributor and rotor are properly aligned with the marks made earlier, then use a large socket and breaker bar on the crankshaft bolt to turn the engine in the normal direction of rotation until the two shafts engage and the distributor drops down against the block.

11   With the base of the distributor seated against the engine block turn the distributor housing to align the marks made on the distributor base and the engine block. Make sure the rotor is pointing to its mark.

12   Place the hold-down clamp in position and loosely install the hold-down bolt.

13   Reconnect the distributor wiring harness connectors.

14   Install the distributor cap.

15   Reconnect the coil connector.

16   With the distributor in its original position, tighten the hold-down bolt.

17   Check the ignition timing (Section 9).

Crankshaft turned after distributor removal

18   Remove the number one spark plug.

19   Place your finger over the spark plug hole while turning the crankshaft with a wrench on the pulley bolt at the front of the engine.

20   When you feel compression, continue turning the crankshaft slowly until the timing mark on the vibration damper is aligned with the "0" on the engine timing indicator.

21   Position the rotor between the number one and eight spark plug terminals on the cap.

22   Lower the distributor into the engine. To mesh the gears at the bottom of the distributor, it may be necessary to turn the rotor slightly. If the distributor does not drop down flush against the block it is because the distributor shaft has not mated to the oil pump shaft. Place a large socket and breaker bar on the crankshaft bolt and turn the engine over in the normal direction of rotation until the two shafts engage properly, allowing the

distributor to seat flush against the block.

23   With the base of the distributor properly seated against the engine block, turn the distributor housing to align the marks made on the distributor base and the engine block.

24   Place the hold-down clamp in position and loosely install the hold-down bolt.

25   Reconnect the distributor wiring harness connectors.

26   Install the distributor cap. If the secondary wiring harness was removed from the cap, reinstall it.

27   Reconnect the coil connector.

28   With the distributor in its original position, tighten the hold-down bolt and check the ignition timing.

1994 and later models

Removal

Refer to illustration 8.33

29   Disconnect the cable from the negative battery terminal. **Caution:** *On models equipped with a Delco Loc II or Theftlock audio system, be sure the lockout feature is turned off before performing any procedure which requires disconnecting the battery.*

30   Remove the water pump (see Chapter 3).

31   Remove the drivebelt pulley from the crankshaft (see Chapter 2A).

32   Remove the spark plug wires from the distributor (see Chapter 1).

33   Remove the distributor mounting bolts **(see illustration)**.

34   Remove the distributor from the engine.

Installation

35   Insert the distributor unit onto the splined (or doweled, depending on year) shaft exactly the same relation to the block in which it was removed. To mesh the splines, it may be necessary to turn the distributor unit slightly and wiggle the unit until it slides completely onto the spline and meets the engine block. **Note:** *The splined shaft can be removed for cleaning. In the event the splined shaft gets inadvertently rearranged, it can be inserted into the engine block either direction without assembly trouble.* **Caution:** *Do not*

install the distributor bolts and tighten the unit in such a way as to pull (press) the distributor gears onto the spline. This will damage the distributor assembly.

36   Install the bolts securely.

37   Install the distributor ignition wires and coil wire.

38   Connect the cable to the negative terminal of the battery.

---

9   Ignition timing - check and adjustment (1993 and earlier models)

Refer to illustration 9.5

Note: *If the information in this Section differs from the Vehicle Emission Control Information label in the engine compartment of your vehicle, the label should be considered correct.*

1   Warm the engine to normal operating temperature. Turn the air conditioning and all accessories off.

2   Place the transmission in Park, apply the parking brake and block the wheels to prevent movement of the vehicle.

3   If the SERVICE ENGINE SOON light is On, don't proceed with the ignition timing check until the problem is solved (see Chapter 6).

4   The Electronic Spark Timing (EST) system must be bypassed prior to checking the ignition timing. Locate the EST bypass connector near the air control valve on top of the right cylinder bank valve cover. Disconnect the EST bypass connector.

5   Locate the timing marks at the front of the engine (they should be visible from above) **(see illustration)**. The crankshaft balancer has a groove in it and a small metal plate with notches and numbers is attached to the timing cover. Clean the plate with solvent so the numbers are visible.

6   Use chalk or white paint to mark the groove in the balancer and highlight the notch or point on the timing plate at TDC.

7   With the ignition switch off, connect a timing light according to the manufacturer's instructions. Install the inductive pick-up onto the number one cylinder spark plug wire. Make sure the timing light wires are routed away from the drivebelts and fan.

8   Start the engine and allow it to warm up to normal operating temperature. Aim the timing light at the timing scale on the front engine cover. The mark on the crankshaft pulley should align with the TDC mark on the scale. The timing on these models is 0 degrees TDC. If necessary, loosen the distributor hold-down bolt and slowly rotate the distributor until the timing marks align. Tighten the hold-down bolt and recheck the timing.

9   Turn off the engine, and remove the timing light. Reconnect the EST wire harness connector, then clear any ECM trouble codes set during the ignition timing procedure (see Chapter 6).

10.14 Use a small screwdriver to pry the electrical connectors from the ignition module

10.21 Probe terminal C (red wire) and make sure the distributor is supplied with voltage from the PCM (distributor ignition feed)

10.22 Continuity should exist between terminal D (pink/black wire) and ground

10   Ignition module and pick-up coil - check and replacement

Ignition module

1993 and earlier models

Note: *It is not necessary to remove the distributor to check or replace the module. Refer to illustration 6.5 for the terminal designations on the ignition module.*

Check

1   Disconnect the four-terminal connector from the distributor.
2   Check for a spark at the coil using a spark tester connected to the coil wire (see Section 6). If there is spark, check the cap, rotor and coil wire for opens or other damage.
3   If there is no spark, remove the distributor cap. Reconnect the four-terminal connector to the ignition module (distributor).
4   Unplug the two-wire connector from the distributor. With the ignition switch turned ON, check for voltage at the module positive (+) terminal of the two-wire connector.
5   If the reading is less than ten volts, there is a fault in the wire between the module positive (+) terminal and the ignition coil positive connector or the ignition coil and primary circuit-to-ignition switch.
6   If the reading is ten volts or more, check the "C" terminal on the module (two wire connector).
7   If the reading is less than one volt, there is an open or grounded lead in the distributor-to-coil "C" terminal connection or ignition coil or an open primary circuit in the coil itself.
8   If the reading is one to ten volts, replace the module with a new one and check for a spark (see Section 6). If there is a spark the module was faulty and the system is now operating properly. If there is no spark, there is a fault in the ignition coil.
9   If the reading in Step 4 is 10 volts or more, unplug the pick-up coil connector (two wire) from the module (terminals P and N). Locate the TACH harness terminal (taped to the harness near the coil) and install a volt-

meter positive probe (+). With the ignition key ON (engine not running), use a 1.5 volt power source (flashlight battery) and jumper wires and carefully touch the positive probe to terminal P on the module **(see illustration 6.5)**. Voltage at TACH should momentarily drop then return to normal.
10   If there is no drop in voltage, check the module ground and, if it is good, replace the module with a new one.
11   If the voltage drops, check for spark at the coil wire (spark tester hooked to the coil wire) as the test light is removed from the module terminal. If there is no spark, the module is faulty and should be replaced with a new one. If there is a spark, the pick-up coil or connections are faulty or not grounded.

Replacement

Refer to illustration 10.14

12   Detach the cable from the negative terminal of the battery. **Caution:** *If the vehicle is equipped with a Delco Loc II or Theftlock audio system, make sure you have the correct activation code before disconnecting the battery. See the information at the front of this manual for the radio re-activation procedure.*
13   Remove the distributor cap and rotor (see Chapter 1).
14   Disconnect the electrical connectors from the module **(see illustration)**. Note that the connectors cannot be interchanged.
15   Remove both module attaching screws and lift the module up and away from the distributor.
16   Do not wipe the grease from the module or the distributor base if the same module is to be reinstalled. If a new module is to be installed, a package of silicone grease will be included with it. Wipe the distributor base and the new module clean, then apply the silicone grease on the face of the module and on the distributor base where the module seats. This grease is necessary for heat dissipation.
17   Install the module and attach both electrical leads.
18   Install the distributor rotor and cap (see Chapter 1).

19   Attach the cable to the negative terminal of the battery.

1994 and later models

Refer to illustrations 10.21, 10.22, 10.23 and 10.29

Check

20   Refer to Steps 13 through 20 of Section 6, *Ignition system - check*, and follow the procedures before performing the module checks.
21   With the ignition key OFF, disconnect the distributor electrical connector near the right bank fuel rail **(see illustration)**. Turn the ignition key ON (engine not running) and probe Terminal C (red wire) with a test light. The light should illuminate to indicate the ignition feed circuit in the PCM is activated **(see illustration 6.13)**.
22   Next, use an ohmmeter and check for continuity on Terminal D (pink/black wire) and ground. Continuity should exist **(see illustration)**.
23   Change the voltmeter to DC scale and measure the voltage signal on Terminal A and Terminal B. It should be approximately 5.0 volts **(see illustration)**.
24   If the readings of any of the tests are incorrect, have the PCM diagnosed by a dealer service department.

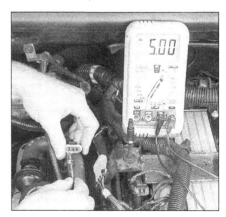

10.23 Probe terminal A (red/black wire) and check for a 5.0 volt signal

10.29 Remove the ignition module mounting bolts (arrows)

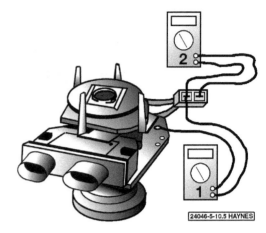

10.37 Pick-up coil test connections on the HEI distributor

25   If the readings of all the tests are correct, the PCM is signaling the module correctly.
26   If all the tests' results are correct and there is still a NO START condition, it will be very difficult to distinguish a faulty distributor unit from a faulty ignition module. In most cases, the ignition module when defective, will NOT produce a pulsing voltage signal with symptoms of a definite start or no-start condition while a faulty DI unit will produce intermittent ignition failures when the engine is running. Take the vehicle to a dealer service department or other qualified repair shop for diagnosis.

Replacement
27   Detach the cable from the negative terminal of the battery. **Caution:** *On models equipped with a Delco Loc II or Theftlock audio system, be sure the lockout feature is turned off before performing any procedure which requires disconnecting the battery.*
28   Disconnect the ignition module electrical connector.
29   Remove both module attaching screws and lift the module up and away from its mount **(see illustration)**.
30   Install the module and tighten the screws securely.
31   Plug in the electrical connector.

32   Attach the cable to the negative terminal of the battery.

Pick-up coil (1993 and earlier models)
33   Detach the cable from the negative terminal of the battery. **Caution:** *On models equipped with a Delco Loc II or Theftlock audio system, be sure the lockout feature is turned off before performing any procedure which requires disconnecting the battery.*
34   Remove the distributor cap and rotor.
35   Remove the distributor from the engine (only do this if you have difficulty in accessing the terminals of the pick-up coil) (see Section 8).
36   Detach the pick-up coil leads from the module.

Check
Refer to illustration 10.37
37   Connect one lead of an ohmmeter to the terminal of the pick-up coil lead and the other to the distributor body as shown in test 1 **(see illustration)**. Flex the leads by hand to check for intermittent opens. The ohmmeter should indicate infinite resistance at all times. If it doesn't, the pick-up

coil is defective and must be replaced.
38   Connect the ohmmeter leads to both terminals of the pick-up coil lead (see test 2 in **illustration 10.38**). Flex the leads by hand to check for intermittent opens. The ohmmeter should read one steady value between 500 and 1,500 ohms as the leads are flexed by hand. If it doesn't, the pick-up coil is defective and must be replaced.

Replacement
Refer to illustrations 10.39, 10.41a, 10.41b and 10.42
39   Remove the distributor, if not already done. Remove the spring clip from the distributor shaft **(see illustration)**.
40   Mark the distributor tang drive and shaft so they can be reassembled in the same position.
41   Carefully set the distributor in a vise with a shop rag to protect the shaft. Using a hammer and punch, remove the roll pin from the distributor shaft and gear **(see illustrations)**.
42   Remove the distributor shaft **(see illustration)**.
43   Lift the pick-up coil assembly straight up and remove it from the distributor. Note the order in which you remove the pieces.
44   Reassembly is the reverse of disassembly.

10.39 Remove the spring clip from the distributor shaft

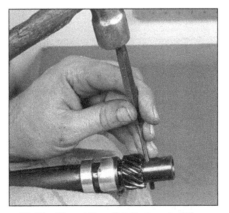

10.41a Mount the distributor shaft in a soft-jawed vise and using a drift punch and hammer, knock out the roll pin

10.41b Remove the driven gear and spacer washers from the end of the shaft, making sure to note the order in which you remove any spacers

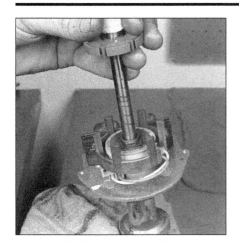

10.42 Remove the shaft from the distributor

11   Charging system - general information and precautions

Caution: *On models equipped with a Delco Loc II or Theftlock audio system, be sure the lockout feature is turned off before performing any procedure which requires disconnecting the battery.*

The charging system consists of a belt-driven alternator with an integral voltage regulator and the battery. These components work together to supply electrical power for the ignition system, the lights and all accessories.

1993 and earlier models are equipped with the CS-130 (100 amp) alternator. Later models are equipped with either the CS-130 or the CS-144 (124 amp) alternator. All types use a conventional pulley and fan. The CS-130 alternators should be considered non-serviceable and, if found to be faulty, should be exchanged as cores for new or rebuilt units. CS-144 alternators can be rebuilt but it is recommended that the home mechanic exchange the alternator for a rebuilt unit. Because of the expense and the limited availability of parts, no alternator overhaul information is included in this manual.

The purpose of the voltage regulator is to limit the alternator's voltage to a preset value. This prevents power surges, circuit overloads, etc., during peak voltage output. On all models with which this manual is concerned, the voltage regulator is contained within the alternator housing.

The charging system does not ordinarily require periodic maintenance. The drivebelt, electrical wiring and connections should, however, be inspected at the intervals suggested in Chapter 1.

Take extreme care when making circuit connections to a vehicle equipped with an alternator and note the following. When making connections to the alternator from a battery, always match correct polarity. Before using arc welding equipment to repair any part of the vehicle, disconnect the wires from the alternator and the battery terminals. Never start the engine with a battery charger connected. Always disconnect both battery leads before using a battery charger.

The charging indicator light on the dash lights up when the ignition switch is turned on and goes out when the engine starts. If the light stays on or comes on once the engine is running, a charging system problem has occurred. See Section 12 for the proper diagnosis procedure for each type of alternator.

12   Charging system - check

1   If a malfunction occurs in the charging circuit, do not immediately assume that the alternator is causing the problem.

2   First, check the following items:

  a) *Make sure the battery cable connections at the battery are clean and tight.*
  b) *The battery electrolyte specific gravity (if possible). If it is low, charge the battery.*
  c) *Check the external alternator wiring and connections. They must be in good condition.*
  d) *Check the drivebelt condition and tension (Chapter 1).*
  e) *Make sure the alternator mounting bolts are tight.*
  f) *Run the engine and check the alternator for abnormal noise (may be caused by a loose drive pulley, loose mounting bolts, worn or dirty bearings, defective diode or defective stator).*

3   Check the charge light bulb and circuit. With the ignition key ON and the engine not running, the lamp should be ON. If not detach the wiring harness at the alternator.

  a) *Install a fused jumper wire (5 amp) to ground and connect the other end to the lead that was removed from the L terminal on the alternator.*
  b) *If the charging lamp on the dash comes ON, the alternator is defective. Replace the alternator.*
  c) *If the charging lamp on the dash remains OFF, locate the open circuit between the alternator and the bulb on the dash. First check the bulb to make sure it is not blown.*
  d) *With the ignition key ON and the engine running, the lamp should be OFF. If it remains ON while running, stop the engine and remove the lead from the L terminal on the alternator.*
  e) *If the lamp on the dash goes OFF, the alternator is defective.*
  f) *If the lamp on the dash remains ON, there is a grounded L terminal in the wiring harness.*

4   Using a voltmeter, check the battery voltage with the engine off. It should be approximately 12 volts.

5   Start the engine and check the battery voltage again. It should now be approximately 14 to 15 volts.

13   Alternator - removal and installation

Refer to illustration 13.4

1   Detach the cable from the negative terminal of the battery. **Caution:** *On models equipped with a Delco Loc II or Theftlock audio system, be sure the lockout feature is turned off before performing any procedure which requires disconnecting the battery.*

2   Clearly label, if necessary, then remove the electrical connectors from the alternator.

3   Remove the serpentine drivebelt (see Chapter 1).

4   Remove the alternator mounting bolts **(see illustration)** and remove the alternator.

5   Installation is the reverse of removal.

14   Starting system - general information

The function of the starting system is to crank the engine quickly enough for it to start. The starting system is composed of a starter motor, solenoid, ignition switch and battery. The battery supplies the electrical energy to the solenoid, which then completes the circuit to the starting motor, which does the actual work of cranking the engine.

The solenoid and starter motor are mounted together at the lower front side of the engine. No periodic lubrication or maintenance is required.

The electrical circuitry of the vehicle is arranged so that the starter motor can only be operated when the transmission selector lever is in Park or Neutral.

Never operate the starter motor for more than 15 seconds at a time without pausing to allow it to cool for at least two minutes.

Excessive cranking can cause overheating, which can seriously damage the starter.

There are two types of starters used in these models. 1993 and earlier models are equipped with either the SD-260 and SD-300 plunger and lever type starter assemblies. 1994 and later models are equipped with either the SD-260 (4.3L engines) or the

13.4 Remove the bolts (arrows) retaining the alternator to its bracket

PG-250/260 (5.7L engines) direct drive type starters. SD-260/300 starters use a solenoid mounted on the starter motor while PG-250/260 is a direct drive gear reduction system with the pinion (gear and shaft) driven by the armature shaft. The PG-250/260 starters found on 1994 and later 5.7L engines are not serviceable and in the event of failure, they must be replaced as a single unit.

Although the PG-250 and PG-260 starters look similar, the PG-260 is lighter and uses a different solenoid which requires a different battery positive cable due to the terminal locator. Both PG starter motor assemblies have an anodized metallic color (gold) finish while the SD-260 is painted black.

The starter attachment and pinion-to-flywheel clearance is the same for all starters.

15 Starter motor - testing in vehicle

1 If the starter motor does not turn at all when the switch is operated, make sure that the shift lever is in Neutral or Park.
2 Make sure that the battery is charged and that all cables, both at the battery and starter solenoid terminals, are secure.
3 If the starter motor spins but the engine is not cranking, the overrunning clutch in the starter motor is slipping and the motor must be removed from the engine for replacement.
4 If, when the switch is actuated, the starter motor does not operate at all but the solenoid clicks, then the problem lies with either the battery, the main solenoid contacts or the starter motor itself. **Note:** *Before diagnosing starter problems, make sure the battery is fully charged.*
5 If the solenoid plunger cannot be heard when the switch is actuated, the solenoid itself is defective or the solenoid circuit is open.
6 To check the solenoid, connect a jumper lead between the battery (+) and the "S" terminal on the solenoid. If the starter motor now operates, the solenoid is OK and the problem is in the ignition switch, neutral start switch or in the wiring.
7 If the starter motor still does not operate, remove the starter/solenoid assembly for disassembly, testing and repair.
8 If the starter motor cranks the engine at an abnormally slow speed, first make sure that the battery is charged and that all terminal connections are clean and tight. If the engine is partially seized, or has the wrong viscosity oil in it, it will crank slowly.
9 Run the engine until normal operating temperature is reached, then stop the engine, disconnect the coil wire from the distributor cap and ground it on the engine.

16.3 Remove the starter electrical connections on the solenoid (arrow)

10 Connect a voltmeter positive lead to the starter motor terminal of the solenoid and then connect the negative lead to ground.
11 Crank the engine and take the voltmeter readings as soon as a steady figure is indicated. Do not allow the starter motor to turn for more than 15 seconds at a time. A reading of 9 volts or more, with the starter motor turning at normal cranking speed, is normal. If the reading is 9 volts or more but the cranking speed is slow, the motor is faulty. If the reading is less than 9 volts and the cranking speed is slow, the solenoid contacts are probably burned.

16 Starter motor - removal and installation

Refer to illustrations 16.3 and 16.4
Note: *These vehicles were designed for starter mounting without using shims. In the event the starter assembly is equipped with shims, it will be necessary to replace them in their original location to ensure proper pinion-to-flywheel engagement.*

Removal

1 Disconnect the negative battery cable. **Caution:** *On models equipped with a Delco Loc II or Theftlock audio system, be sure the lockout feature is turned off before performing any procedure which requires disconnecting the battery.*
2 Raise the front of the vehicle and support it securely on jackstands.
3 From under the vehicle, disconnect the solenoid wire and battery cable from the terminals on the solenoid **(see illustration)**.
4 Remove the starter motor bolts **(see illustration)**.
5 Remove the starter motor. Note the location of the spacer shim(s), if equipped.

16.4 Remove the starter bolts (arrows)

Installation

6 Installation is the reverse of removal. Be sure to install the spacer shim(s) in exactly the same location, if equipped (see Steps 7 through 12 for the correct shim adjustment procedure).

Shim adjustment procedure

7 Remove the flywheel housing cover (see Chapter 7).
8 Inspect the flywheel or driveplate for signs of unusual wear such as chipped or missing gear teeth.
9 Using a wire type feeler gauge (round diameter), measure the clearance between the top of the flywheel ring gear tooth and the bottom of the pinion tooth on the starter. Normal clearance should be 0.01 to 0.06 inches.
10 If the starter clearance is less than 0.02 inches and the starter whined after engagement, the starter is too close to the flywheel. Add a 0.04 shim to gain clearance. Do not use more than two shims.
11 If the starter clearance is more than 0.06 inches and the starter whines during engagement then the starter is far from the flywheel. Remove a 0.04 shim to gain clearance. Do not remove more than two shims.
12 To install a shim, loosen the inside bolt, remove the outer bolt and slide the shim between the engine and the starter without removing the starter from the engine block.

17 Starter solenoid - replacement

Note: *The starter solenoid is not easily removed from the main starter body without complete disassembly. It is recommended that the starter/solenoid assembly be exchanged as a complete unit in the event of failure.*

Chapter 6
Emissions and engine control systems

Contents

Specifications

Torque specifications

Crankshaft sensor bolts .... 60 in-lbs

1   General information

Refer to illustrations 1.5a and 1.5b

To prevent pollution of the atmosphere from burned and evaporating gases, a number of emissions control systems are incorporated on the vehicles covered by this manual. The combination of systems used depends on the year in which the vehicle was manufactured, the locality to which it was originally delivered and the engine type. The major systems incorporated on the vehicles with which this manual is concerned include the:

Fuel Control System (TBI and MFI systems)
Exhaust Gas Recirculation (EGR) system
Evaporative Emissions Control (EVAP) system
Transmission Converter Clutch (TCC) system
Positive Crankcase Ventilation (PCV) system
Secondary Air Injection (AIR) system
Catalytic converter

All of these systems are linked, directly or indirectly, to the On Board Diagnostic (OBD) system. The Sections in this Chapter include general descriptions, checking procedures (where possible) and component replacement procedures (where applicable) for each of the systems listed above.

Before assuming that an emissions control system is malfunctioning, check the fuel and ignition systems carefully. In some cases special tools and equipment, as well as specialized training, are required to accurately diagnose the causes of a rough running or difficult to start engine. If checking and servicing become too difficult, or if a procedure is beyond the scope of the home mechanic, consult your dealer service department. This does not necessarily mean, however, that the emissions control systems are particularly difficult to maintain and repair. You can quickly and easily perform many checks and do most (if not all) of the regular maintenance at home with common tune-up and hand tools. **Note:** *The most frequent cause of emissions system problems is simply a loose or broken vacuum hose or wiring connection. Therefore, always check the hose and wiring connections first.*

Pay close attention to any special precautions outlined in this Chapter. It should be noted that the illustrations of the various systems may not exactly match the system installed on your particular vehicle due to changes made by the manufacturer during production or from year to year.

A Vehicle Emissions Control Information (VECI) label is located in the engine compartment of all vehicles with which this manual is concerned **(see illustrations)**. This label contains important emissions specifications and

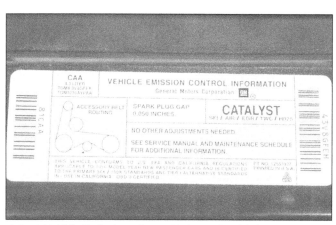

1.5a A Vehicle Emissions Control Information (VECI) label will be found in the engine compartment of all vehicles - if it's missing, obtain a new one from a dealer parts department

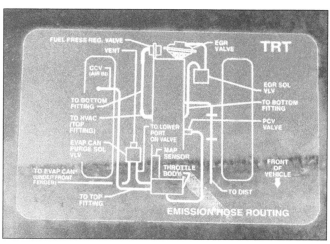

1.5b Typical vacuum hose schematic

2.1 Digital multimeters can be used for testing all types of circuits; because of their high impedance, they are much more accurate than analog meters for measuring millivolts in low-voltage computer circuits

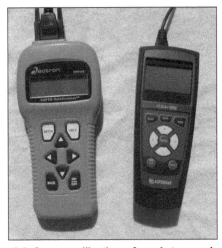

2.2 Scanners like these from Actron and AutoXray are powerful diagnostic aids - they can tell you just about anything that you want to know about your engine management system

specially designed scanner must also be developed. At this time, several manufacturers plan to release OBD II scan tools for the home mechanic. Ask the parts salesman at a local auto parts store for additional information concerning dates and costs. **Note:** *Although OBD II codes cannot be accessed on 1996 models without a Scan tool, follow the simple component checks in Section 4.*

4   Another type of code reader and less expensive is available at parts stores **(see illustration)**. These tools simplify the procedure for extracting codes from the engine management computer by simply plugging in to the diagnostic connector on the vehicle wiring harness. **Note:** *Some diagnostic connectors are located under the dash, kick panel or glovebox while others are located in the engine compartment.*

General description

5   The electronically controlled fuel and emissions system is linked with many other related engine management systems. It consists mainly of sensors, output actuators and a Powertrain Control Module (PCM). Completing the system are various other components which respond to commands from the PCM.

6   In many ways, this system can be compared to the central nervous system in the human body. The sensors (nerve endings) constantly gather information and send this data to the PCM (brain), which processes the data and, if necessary, sends out a command for some type of vehicle change (limbs).

7   Here's a specific example of how one portion of this system operates: An oxygen sensor, mounted in the exhaust manifold and protruding into the exhaust gas stream, constantly monitors the oxygen content of the exhaust gas as it travels through the exhaust pipe. If the percentage of oxygen in the exhaust gas is incorrect, an electrical signal is sent to the PCM. The PCM takes this information, processes it and then sends a command to the fuel injector(s) on the fuel injection system, telling it to change the fuel/air mixture. To be effective, all this happens in a fraction of a second, and it goes on continuously while the engine is running. The end result is a fuel/air mixture which is constantly kept at a predetermined ratio, regardless of driving conditions.

setting procedures, as well as a vacuum hose schematic with emissions components identified. When servicing the engine or emissions systems, the VECI label in your particular vehicle should always be checked for up-to-date information. **Note:** *Because of a federally mandated extended warranty which covers the emission control system components (and any components which have a primary purpose other than emission control but have significant effects on emissions), check with your dealer about warranty coverage before working on any emission related systems.*

The number of emissions control system components on later model fuel-injected vehicles has actually decreased due to the high efficiency of the new fuel injection and ignition systems. These models are equipped with a three way catalytic converter containing beads which are coated with a catalyst material containing platinum, palladium and rhodium to reduce the level of nitrogen oxides.

---

2   On Board Diagnostic (OBD) system and trouble codes

Diagnostic tool information

Refer to illustrations 2.1, 2.2 and 2.4

1   A digital multimeter is a necessary tool for checking fuel injection and emission related components **(see illustration)**. A digital volt-ohmmeter is preferred over the older style analog multimeter for several reasons. The analog multimeter cannot display the volts-ohms or amps measurement in hundredths and thousandths increments. When working with electronic circuits which are often very low voltage, this accurate reading is most important. Another good reason for the digital multimeter is the high impedance circuit. The digital multi-

meter is equipped with a high resistance internal circuitry (10 million ohms). Because a voltmeter is hooked up in parallel with the circuit when testing, it is vital that none of the voltage being measured should be allowed to travel the parallel path set up by the meter itself. This dilemma does not show itself when measuring larger amounts of voltage (9 to 12 volt circuits) but if you are measuring a low voltage circuit such as the oxygen sensor signal voltage, a fraction of a volt may be a significant amount when diagnosing a problem.

2   Hand-held scanners are the most powerful and versatile tools for analyzing engine management systems used on later model vehicles **(see illustration)**. Early model scanners handle codes and some diagnostics for many OBD I systems. Each brand scan tool must be examined carefully to match the year, make and model of the vehicle you are working on. Often interchangeable cartridges are available to access the particular manufacturer; Ford, GM, Chrysler, etc.). Some manufacturers will specify by continent; Asia, Europe, USA, etc. Seek the advice of your local auto parts retailer.

3   With the arrival of the Federally mandated emission control system (OBD II), a

2.4 Trouble code tools simplify the task of extracting the trouble codes

2.11a The 12 pin Assembly Line Data Link (ALDL) terminal identification

A   Ground
B   Diagnostic TEST terminal

2.11b The 16 pin ALDL terminal on a 1996 MFI engines with OBD II

Testing

Refer to illustrations 2.11a and 2.11b

8   One might think that a system which uses exotic electrical sensors and is controlled by an on-board computer would be difficult to diagnose. This is not necessarily the case.

9   The On Board Diagnostic (OBD) system has a built-in self-diagnostic system, which indicates a problem by turning on a SERVICE ENGINE SOON light on the instrument panel when a fault has been detected. **Note:** *Since some of the trouble codes do not set the SERVICE ENGINE SOON light, it is a good idea to access the OBD system and look for any trouble codes that may have been recorded and need tending.*

10   Perhaps more importantly, the PCM will recognize this fault, in a particular system monitored by one of the various information sensors, and store it in its memory in the form of a trouble code. Although the trouble code cannot reveal the exact cause of the malfunction, it greatly facilitates diagnosis as you or a dealer mechanic can tap into the PCMs memory and be directed to the problem area.

11   To retrieve this information from the PCM on 1991 through 1995 models (which are OBD I models) you must use a short jumper wire to ground a diagnostic terminal. The terminal is part of an electrical connector called the Assembly Line Data Link (ALDL) **(see illustrations)**. The ALDL is located under the dashboard, just below the instrument panel and to the left of the center console. To use the ALDL, remove the plastic cover (if equipped). With the electrical connector exposed to view, push one end of the jumper wire into the diagnostic TEST terminal and the other end into the GROUND terminal.

12   Turn the ignition to the ON position. **Caution:** *Do not start the engine with the TEST*

terminal grounded. The SERVICE ENGINE SOON light should flash Trouble Code 12, indicating that the diagnostic system is working. Code 12 will consist of one flash, followed by a short pause, then two more flashes in quick succession. After a longer pause, the code will repeat itself two more times. If no other codes have been stored, Code 12 will continue to repeat itself until the jumper wire is disconnected. If additional Trouble Codes have been stored, they will follow Code 12. Again, each Trouble Code will flash three times before moving on.

13   Once the code(s) have been noted, use the Trouble Code Identification information which follows to locate the source of the fault. **Note:** *Whenever the battery cable is disconnected, all stored Trouble Codes in the PCM are erased. Be aware of this before you disconnect the battery.*

14   To retrieve this information from 1996 models (OBD II), a SCAN tool must be connected to the Assembly Line Diagnostic Link (ALDL). The SCAN tool is a hand-held digital computer scanner that interfaces with the on-board computer. The SCAN tool is a very powerful tool; it not only reads the trouble codes but also displays the actual operating conditions of the sensors and actuators. SCAN tools are expensive, but they are necessary to accurately diagnose a modern computerized fuel injected engine. SCAN tools are available from automotive parts stores and specialty tool companies.

15   It should be noted that the self-diagnosis feature built into this system does not detect all possible faults. If you suspect a problem with the On Board Diagnostic (OBD) system,

but the SERVICE ENGINE SOON light has not come on and no trouble codes have been stored, take the vehicle to a dealer service department or other qualified repair shop for diagnosis.

16   Furthermore, when diagnosing an engine performance, fuel economy or exhaust emissions problem (which is not accompanied by a SERVICE ENGINE SOON light) do not automatically assume the fault lies in this system. Perform all standard troubleshooting procedures, as indicated elsewhere in this manual, before turning to the On Board Diagnostic (OBD) system.

17   Finally, since this is an electronic system, you should have a basic knowledge of automotive electronics before attempting any diagnosis. Damage to the PCM, Programmable Read Only Memory (PROM) calibration unit or related components can easily occur if care is not exercised.

Trouble Code Identification

18   Following is a list of the typical Trouble Codes which may be encountered while diagnosing the On Board Diagnostic (OBD I) system. Also included are simplified troubleshooting procedures. If the problem persists after these checks have been made, the vehicle must be diagnosed by a professional mechanic who can use specialized diagnostic tools and advanced troubleshooting methods to check the system. Procedures marked with an asterisk (*) indicate component replacements which may not cure the problem in all cases. For this reason, you may want to seek professional advice before purchasing replacement parts.

19   To clear the Trouble Code(s) from the PCM memory, momentarily remove the PCM/IGN fuse from the fuse box for 10 seconds. Clearing codes may also be accomplished by removing the fusible link (main power fuse) located near the battery positive terminal (see Chapter 12) or by disconnecting the cable from the positive terminal (+) of the battery. **Caution:** *On models equipped with a Delco-Loc II or Theftlock audio system, be sure the lockout feature is turned off before disconnecting the battery cable.* Disconnecting the power to the PCM to clear the memory can be an important diagnostic tool, especially on intermittent problems. **Caution:** *To prevent damage to the PCM, the ignition switch must be OFF when disconnecting or connecting power to the PCM.* **Note:** *Disconnecting the negative battery terminal will erase any radio preset codes that have been stored.*

OBD I Trouble Codes

Code	Circuit or system	Probable cause
Note: *Not all codes apply to all models.*		
13	Oxygen sensor circuit	Check the wiring and connectors from the oxygen sensor. Replace oxygen sensor (see Section 4).*
14	Coolant sensor circuit	If the engine is experiencing overheating problems, the problem must be rectified (high temperature indicated) before continuing (see Chapters 1 and 3). Check all wiring and connectors associated with the sensor. Replace the coolant sensor (see Section 4).*

Code	Circuit or system	Probable cause
Note: *Not all codes apply to all models.*		
15	Coolant sensor circuit	See above. Also, check the thermostat for proper operation (low temperature indicated).
16	Distributor Ignition (DI) system	The low resolution signal (4 pulses/revolution) is not being detected by the PCM. Have the DI unit checked by a dealer service department.
18	Injector circuits	Check the fuel injectors and circuits as described in Chapter 4.
21	TPS circuit (signal voltage high)	Check for sticking or misadjusted TPS. Check all wiring and connections at the TPS and at the PCM. Adjust or replace TPS* (see Section 4).
22	TPS circuit (signal voltage low)	See above.
23	Intake Air Temperature (IAT) sensor	Low temperature indicated (see Section 4).
24	Vehicle Speed Sensor (VSS)	A fault in this circuit should be indicated only while the vehicle is in motion. Disregard code 24 if set when drive wheels are not turning. Check connections at the PCM. Check the TPS setting (see Section 4).
25	Intake Air Temperature (IAT) sensor	Check the resistance of the IAT sensor. Check the wiring and connections to the sensor. Replace the (IAT) sensor.*
26	EVAP purge solenoid	Check the EVAP purge solenoid circuit for an open or short. Check the purge solenoid (see Section 7).
27	EGR vacuum solenoid	Check the EGR vacuum solenoid circuit for an open or short. Check the vacuum solenoid. (see Section 5).
28	Transmission Range (TR) pressure switch	TR pressure switch in the valve body indicates a fault in the shift detection system between the 5 pressure switches and the TFT sensor. Have the vehicle diagnosed by a dealer service department.
29	Secondary Air Injection Pump circuit failure	PCM detects high voltage potential in the Secondary AIR pump relay circuit. Check the relay, pump and circuit (see Section 6).
32	Exhaust Gas Recirculation (EGR) failure	Check the vacuum source and all vacuum lines. Check the system electrical connectors at the PCM and EGR valve. Replace the EGR valve or PCM as necessary (see Section 5).*
33	Manifold Absolute Pressure (MAP) signal voltage high	Check vacuum hose(s) from MAP sensor. Check electrical sensor or circuit connections at the PCM. Replace MAP sensor (see Section 4).*
34	Manifold Absolute Pressure (MAP) signal voltage low	Check vacuum hose(s) from MAP sensor. Check electrical sensor or circuit connections at the PCM. Replace MAP sensor (see Section 4).*
36	Distributor Ignition (DI) system	The high resolution signal (180 pulses/revolution) is not being detected by the PCM. Have the DI unit checked by a dealer service department.
37	Brake switch stuck "ON"	TCC brake switch indicates an open circuit. Have the TCC brake switch checked by a dealer service department.
38	Brake switch stuck "OFF"	TCC brake switch indicates a short circuit. Have the TCC brake switch checked by a dealer service department.
41	Ignition control circuit (open circuit)	Check the wiring and connectors between the ignition module and the PCM. Check the ignition module (see Chapter 5). Replace the PCM.*
42 (1991 thru 1993)	Electronic Spark Timing (EST) system	Open or short to ground in the EST or Bypass circuits. See Chapter 5 for HEI ignition system diagnostics.*
42 (1994 and 1995)	Ignition Control Circuit (shorted circuit)	Check the wiring and connectors between the ignition module and the PCM. Check the ignition module (see Chapter 5). Replace the PCM.*
43 (1991 thru 1993)	Electronic Spark Control (ESC) system	ESC module is not receiving a knock signal from the knock sensor. Check the knock sensor (see Section 4) or the ESC module. Have the module checked by a dealer service department.

43	Knock Sensor (KS) circuit (1994 and 1995)	Check the PCM for an open or short to ground; if necessary, reroute the harness away from other wires such as spark plugs, etc. Replace the knock sensor (see Section 4).*
44	Lean exhaust	Check the wiring and connectors from the oxygen sensor to the PCM. Check the PCM ground terminal. Check the fuel pressure (Chapter 4). Replace the oxygen sensor (see Section 4).*
45	Rich exhaust	Check the evaporative charcoal canister and its components for the presence of fuel. Check for fuel or contaminated oil. Check the fuel pressure regulator. Check for a leaking fuel injector. Check for a sticking EGR valve. Replace the oxygen sensor (see Section 4).*
46	PASS-Key circuit	If engine will not start, have theft deterrent system diagnosed by a dealership service department.
47	Knock Sensor (KS) circuit	Check the PCM for a missing knock sensor module. Replace the knock sensor module (see Section 4).*
48	Mass Airflow (MAF) sensor	MAF sensor reference voltage is detected but no signal pulses (frequency) from MAF sensor. Replace MAF sensor (see Section 4).*
50	System voltage low	Code 50 will set if the voltage at the PCM is less than 8 volts Check the charging system (see Chapter 5).
51 (1991 thru 1993)	PROM (MEMCAL)	Faulty or incorrect PROM (MEMCAL). Diagnosis should be performed by a dealer service department or other qualified repair shop
51 (1994 and 1995)	EEPROM	Have EEPROM programmed by dealer service department.
53	System voltage high	Code 53 will set if the voltage at the PCM is greater than 17.1-volts. Check the charging system (see Chapter 5).
54	Fuel pump relay low voltage	Check the fuel pump relay and circuit for shorts or damage (see Chapter 4).
55 (1991 thru 1993)	PCM failure	Check the PCM power and ground circuits. If OK, replace the PCM.*
55 (1994 and 1995)	Fuel system lean	Fuel system lean under heavy load or acceleration. Check the fuel filter and the fuel pressure regulator (see Chapter 4).
58	Transmission Fluid Temp (TFT) sensor	The TFT sensor located in the valve body indicates a high fluid temperature. Have the transmission diagnosed by a dealer service department.
59	Transmission Fluid Temp (TFT) sensor	The TFT sensor located in the valve body indicates a low fluid temperature. Have the transmission diagnosed by a dealer service department.
63	Heated oxygen sensor circuit (right side)	Heater circuit is open or no available voltage to the oxygen heaters from fused 10 amp circuit. Check circuit and oxygen sensor (see Section 4).
64	Heated oxygen sensor circuit (right side)	Signal voltage remains below 200 millivolts indicating lean running condition. Check circuit and oxygen sensor (see Section 4).
65	Heated oxygen sensor circuit (right side)	Signal voltage remains above 700 millivolts indicating rich running condition. Check circuit and oxygen sensor (see Section 4).
66	A/C refrigerant pressure sensor circuit	Short in circuit indicates the A/C pressure is below 8 psi or above 448 psi. See Chapter 3.
67	A/C refrigerant pressure sensor circuit	A/C pressure switch does not indicate any change in psi (voltage values) when clutch is cycled off. See Chapter 3 and Section 4 in this Chapter.
68	A/C relay circuit	PCM detects voltage at the clutch from the relay after request signal. Check the clutch and relay circuit. See Section 4 and the wiring diagrams in Chapter 12.
69	A/C clutch circuit	PCM detects no voltage at the clutch from the relay after request signal. Check the clutch and relay circuit. See Section 4 and the wiring diagrams in Chapter 12.
70 (1994 and 1995)	A/C clutch relay driver circuit	PCM detects incorrect voltage potential to relay from driver within PCM. Have the PCM diagnosed by a dealer service department.

Code	Circuit or system	Probable cause
Note: *The following codes apply to 1994 and 1995 models only.*		
72	Vehicle speed sensor loss	Output speed remains undetected or inconsistent. Check the VSS (See Section 4).
73	Pressure control solenoid circuit	PCM detects variable amperage ratings at the solenoid. Have the pressure control solenoid diagnosed by a dealer service department.
75	Transmission system voltage	PCM detects incorrect shift patterns. Have the transmission control system diagnosed by a dealer service department.
77	Primary cooling fan relay control circuit	PCM detects variable amperage ratings at the cooling fan relay. Have the cooling fan system diagnosed by a dealer service department.
78	Secondary cooling fan relay control circuit	PCM detects variable amperage ratings at the cooling fan relay. Have the cooling fan system diagnosed by a dealer service department.
79	Transmission fluid temperature sensor circuit	PCM detects high transmission temperature. Have the transmission fluid temp sensor and circuit diagnosed by a dealer service department.
81	2-3 shift solenoid circuit	PCM detects incorrect voltage values at the shift solenoid. Have the 2-3 shift solenoid circuit diagnosed by a dealer service department.
82	1-2 shift solenoid circuit	PCM detects incorrect voltage values at the shift solenoid. Have the 1-2 shift solenoid circuit diagnosed by a dealer service department.
83	TCC PWM solenoid circuit	PCM detects high voltage at the TCC PWM solenoid during activation. Have the TCC system diagnosed by a dealer service department.
84	3-2 shift solenoid circuit	PCM detects incorrect voltage values at the shift solenoid. Have the 3-2 shift solenoid circuit diagnosed by a dealer service department.
85	TCC solenoid circuit	PCM detects stuck solenoid in the TCC solenoid system during activation. Have the TCC system diagnosed by a dealer service department.
90	TCC solenoid circuit	PCM detects incorrect voltage values at the TCC solenoid. Have the TCC system diagnosed by a dealer service department.
95	CHANGE OIL lamp circuit	PCM detects incorrect voltage values for the CHANGE OIL lamp circuit. Have the circuit diagnosed by a dealer service department.
96	LOW OIL lamp circuit	PCM detects incorrect voltage values for the LOW OIL lamp circuit. Have the LOW OIL lamp circuit diagnosed by a dealer service department.
97	VSS output circuit	PCM detects incorrect voltage values at the VSS. Check the VSS and circuit (see Section 4).

*Component replacement may not cure the problem in all cases. For this reason, you may want to seek professional advice before purchasing replacement parts.

OBD II Trouble Codes

Code	Code Definition	Location
P0101	Mass Air Flow (MAF) sensor error	See Section 4
P0107	Manifold Absolute Pressure (MAP) sensor circuit low input	See Section 4
P0108	Manifold Absolute Pressure (MAP) sensor circuit high input	See Section 4
P0112	Intake Air Temperature (IAT) sensor circuit low input	See Section 4
P0113	Intake Air Temperature (IAT) sensor circuit high input	See Section 4
P0117	Electronic Coolant Temperature (ECT) sensor circuit low input	See Section 4
P0118	Electronic Coolant Temperature (ECT) sensor circuit high input	See Section 4
P0121	Throttle Position Sensor (TPS) range/performance fault	See Section 4
P0122	Throttle Position Sensor (TPS) circuit low input	See Section 4
P0123	Throttle Position Sensor (TPS) circuit high input	See Section 4
P0131	Upstream heated O2 sensor circuit low voltage (Bank 1, Sensor 1)	See Section 4

Code	Code Definition	Location
P0133	Upstream heated O2 sensor circuit high voltage (Bank 1, Sensor 1)	See Section 4
P0137	Downstream heated O2 sensor circuit low voltage (Bank 1, Sensor 2)	See Section 4
P0138	Downstream heated O2 sensor circuit high voltage (Bank 1, Sensor 2)	See Section 4
P0141	O2 sensor heater circuit fault (Bank 1, Sensor 2)	See Section 4
P0171	System Adaptive fuel too lean	See Chapter 4
P0172	System Adaptive fuel too rich	See Chapter 4
P0191	Injector Pressure sensor system performance	See Chapter 4
P0192	Injector Pressure sensor circuit low input	See Chapter 4
P0193	Injector Pressure sensor circuit high input	See Chapter 4
P0301	Cylinder number 1 misfire detected	See Chapter 5
P0302	Cylinder number 2 misfire detected	See Chapter 5
P0303	Cylinder number 3 misfire detected	See Chapter 5
P0304	Cylinder number 4 misfire detected	See Chapter 5
P0325	Knock sensor circuit 1 fault	See Section 4
P0326	Knock sensor circuit performance	See Section 4
P0351	COP ignition coil 1 primary circuit fault	See Chapter 5
P0352	COP ignition coil 2 primary circuit fault	See Chapter 5
P0353	COP ignition coil 3 primary circuit fault	See Chapter 5
P0354	COP ignition coil 4 primary circuit fault	See Chapter 5
P0400	EGR flow fault	See Section 5
P0401	EGR insufficient flow detected	See Section 5
P0402	EGR excessive flow detected	See Section 5
P0420	Catalyst system efficiency below threshold (Bank 1)	See Section 8
P0421	Catalyst system efficiency below threshold (Bank 1)	See Section 8
P0430	Catalyst system efficiency below threshold (Bank 2)	See Section 8
P0431	Catalyst system efficiency below threshold (Bank 2)	See Section 8
P0441	EVAP incorrect purge flow	See Section 6
P0443	EVAP VMV circuit fault	See Section 6
P0452	EVAP fuel tank pressure sensor low input	See Section 4
P0453	EVAP fuel tank pressure sensor high input	See Section 4
P0502	VSS circuit low input	See Section 4
P0503	VSS circuit range performance	See Section 4
P0506	IAC system rpm lower than expected	See Chapter 4
P0507	IAC system rpm higher than expected	See Chapter 4
P0602	PCM control module programming error	See Section 4
P0705	Transaxle Range sensor circuit malfunction	See Section 4

Component replacement may not cure the problem in all cases. For this reason, you may want to seek professional advice before purchasing replacement parts.

3   Powertrain Control Module (PCM) - check and replacement

Note: *1993 and earlier models are equipped with a replaceable PROM in the PCM that must be installed into the new PCM when exchanged. 1994 and later models are equipped with an (non-replaceable) EEPROM that must be recalibrated with a special factory SCAN tool (TECH 1) after replacement of the PCM.*

Check

1   The PCM on 1993 and earlier models is located in the passenger compartment under the glovebox and behind the right side kick panel. The PCM on 1994 and later models is located in the left corner of the engine compartment under the air cleaner assembly.
2   Using the tips of the fingers, tap vigorously on the side of the computer while the engine is running. If the computer is not functioning properly, the engine may stumble or stall and display glitches on the engine data stream obtained using a SCAN tool or other diagnostic equipment.
3   If the PCM fails this test, check the electrical connectors. Each connector is color coded to fit the respective slot in the computer body. If there are no obvious signs of damage, have the unit checked at a dealer service department.

Replacement

Refer to illustration 3.7

Caution: *To prevent damage to the PCM, the ignition switch must be turned Off when disconnecting or connecting the PCM connectors.*

4   The PCM on 1993 and earlier models is located in the passenger compartment under the glovebox and behind the right side kick panel. The PCM on 1994 and later models is located in the left corner of the engine compartment under the air cleaner assembly.
5   Disconnect the cable from the negative battery terminal (see Chapter 5)
6   On 1993 and earlier models, remove the lower glovebox panel screws and right side kick panel (see Chapter 11). Remove the mounting nuts and lower the PCM.
7   On 1994 and later models, remove the air cleaner assembly (see Chapter 4) and lift the PCM from its mounting location in the engine compartment **(see illustration)**.
8   Carefully lift the PCM from the engine compartment without damaging the electrical connectors and wiring harness to the computer.
9   Unplug the electrical connectors from the PCM. Each connector is color coded to fit its respective receptacle in the PCM.
10   Installation is the reverse of removal.

PROM replacement (1993 and earlier models)

Refer to illustrations 3.13 and 3.14

11   1993 and earlier models are equipped with a memory calibration chip called the PROM. New PCMs are not supplied with a PROM, therefore it is necessary to remove the

3.7 Remove the locking tabs from the PCM electrical connectors

original and install it into the new PCM. Use care not to damage the PROM when working with these delicate electrical components.
12   Remove the PCM from the area under the glovebox (see Step 6).
13   Remove the screws and detach the PROM cover from the PCM **(see illustration)**.
14   Using two fingers, carefully push both retaining clips back away from the PROM and lift the unit straight out of the PCM **(see illustration)**. **Note:** *There are two types of clips used on the PROM sockets; hollow type tabs or solid type tabs.*
15   Installation is the reverse of removal. Be sure to use the alignment notches in the seat area of the PCM when sliding the PROM back in place. Press only on the ends of the PROM assembly until it snaps back into place.

EEPROM reprogramming (1994 and later models)

16   The 1994 and later models are equipped with Electrical Erasable Programmable Read Only Memory (EEPROM) chip that is permanently soldered to the PCM circuit board. The EEPROM can be reprogrammed using the GM dealer's TECH 1 SCAN tool. Do not attempt to remove this component from the PCM. Have the EEPROM reprogrammed at a dealer service department.

3.14 Press the retaining tabs out to release the PROM

3.13 Remove the PROM cover screws from the PCM

4   Information sensors - check and replacement

Caution: *When performing the following tests, use only a high-impedance (10 meg-ohm) digital multimeter to prevent damage to the PCM.*
Caution: *On models equipped with a Delco-Loc II or Theftlock audio system, be sure the lockout feature is turned off before disconnecting the battery cable.*
Note: *After performing any checking procedure to any of the information sensors, be sure to clear the PCM of all trouble codes by removing the PCM/IGN fuse from the fuse box or disconnecting the cable from the negative terminal of the battery for at least ten seconds.*

Engine Coolant Temperature (ECT) sensor

General description

Refer to illustration 4.1

1   The coolant temperature sensor is a thermistor (a resistor which varies the value of its resistance in accordance with temperature changes). The change in the resistance value will directly affect the voltage signal from the coolant sensor. As the sensor temperature DECREASES, the resistance will INCREASE. As the sensor temperature INCREASES, the resistance will DECREASE. A failure in the coolant sensor circuit should set either a Code 14 or a Code 15. These codes indicate a failure in the coolant temperature circuit, so the appropriate solution to the problem will be either repair of a wire or replacement of the sensor. The sensor can be checked with an ohmmeter. **Note:** *The coolant temperature sensor on 1993 and later models is located on the thermostat housing and on 1994 and later models, it is located on the front of the water pump **(see illustration)**. If access to the coolant temperature sensor makes it difficult to position probes of the meter onto the terminals. It may be necessary to remove the sensor and perform the tests in a pan of heated water to simulate the conditions.*

4.1 Location of the engine coolant temperature (ECT) sensor (arrow) (1994 and later models)

4.2a Check the ECT resistance at cold and warm temperatures

Temperature (degrees-F)	Resistance (ohms)
212	176
194	240
176	332
158	458
140	668
122	972
112	1182
104	1458
95	1800
86	2238
76	2795
68	3520
58	4450
50	5670
40	7280
32	9420

4.2b Coolant temperature and intake air temperature sensors approximate temperature vs. resistance relationships

Check

Refer to illustrations 4.2a, 4.2b and 4.3

2   To check the sensor, check the resistance of the coolant temperature sensor while it is completely cold **(see illustration)**. Next, start the engine and warm it up until it reaches operating temperature and check it again. Compare your resistance values with the accompanying chart **(see illustration)**. If the sensor resistance does not correspond to the specified temperature, replace the sensor.

3   Check the reference voltage from the PCM to the ECT with the ignition key ON (engine not running). It should be approximately 5.0 volts **(see illustration)**.

Replacement

Refer to illustration 4.4

Warning: *Wait until the engine is completely cool before beginning this procedure.*

4   Before installing the new sensor, wrap the threads with Teflon sealing tape to prevent leakage and thread corrosion **(see illustration)**.

5   To remove the sensor, release the locking tab, unplug the electrical connector, then carefully unscrew the sensor. **Caution:** *Handle the coolant sensor with care. Damage to this sensor will affect the operation of the entire fuel injection system.*

6   Installation is the reverse of removal. Check the coolant level and add some, if necessary (see Chapter 1).

Manifold Absolute Pressure (MAP) sensor

General description

Refer to illustration 4.8

7   The Manifold Absolute Pressure (MAP) sensor monitors the intake manifold pressure changes resulting from changes in engine load and speed and converts the information into a voltage output. The PCM uses the MAP sensor to control fuel delivery and ignition timing. A failure in the MAP sensor circuit should set a Code 33 or 34.

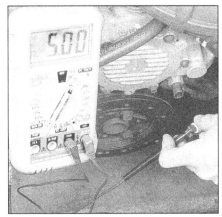

4.3 Check the REFERENCE voltage from the computer to the ECT sensor

8   The MAP sensor SIGNAL voltage to the PCM varies from below 2 volts at idle (high vacuum) to above 4 volts with the ignition key ON (engine not running) or wide open throttle (WOT) (low vacuum). These values correspond with the altitude and pressure changes the vehicle experiences while driving **(see illustration)**. **Note:** *The MAP sensor is located on the engine compartment fenderwell on 1993 and earlier models or on the intake manifold on 1994 and later models.*

VOLTAGE RANGE	ALTITUDE
3.8 - 5.5V	Below 1,000
3.6 - 5.3V	1,000 - 2,000
3.5 - 5.1V	2.000 - 3,000
3.3 - 5.0V	3,000 - 4,000
3.2 - 4.8V	4,000 - 5,000
3.0 - 4.6V	5,000 - 6,000
2.9 - 4.5V	6,000 - 7,000
2.8 - 4.3V	7,000 - 8,000
2.6 - 4.2V	8,000 - 9,000
2.5 - 4.0V	9,000 - 10,000 FEET

LOW ALTITUDE = HIGH PRESSURE = HIGH VOLTAGE

4.8 Typical Manifold Absolute Pressure (MAP) sensor altitude (pressure) vs. voltage values

4.4 To prevent coolant leakage, be sure to wrap the temperature sensor threads with Teflon tape before installation

Check

Refer to illustrations 4.9, 4.10a and 4.10b

9   Check the REFERENCE voltage from the computer to the MAP sensor. Unplug the sensor, turn the ignition On and probe terminal A (gray [+]) and terminal C (purple [-]) - the two outside terminals. The voltage should be approximately 5.0 volts **(see illustration)**.

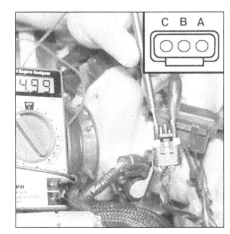

4.9 Check the MAP sensor REFERENCE voltage on the purple wire (-) and the gray wire (+). It should be approximately 5.0 volts

4.10a Check the MAP sensor SIGNAL voltage on the light green wire (+) and the purple wire (-) with the ignition key ON (engine not running)

4.10b Start the engine and observe the MAP sensor voltage at idle. With the engine idling, vacuum in the intake manifold is high, therefore the voltage value will be low

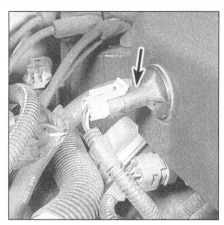

4.12 Location of the Intake Air Temperature (IAT) sensor (arrow) on a 1993 and earlier model

10   Turn the ignition off, plug in the electrical connector, backprobe the MAP sensor using pins and check the SIGNAL voltage. Connect the positive lead of the voltmeter to terminal B (light green [+]) of the sensor electrical connector and the negative lead to terminal C (purple) [-]). With the ignition ON (engine not running) the voltage reading should be about 4.0 to 4.5 volts **(see illustration)**. **Note:** *It will probably be necessary to use pins or straightened-out paper clips to backprobe the terminals in the connector.* Start the engine and let it warm up. The voltage should be approximately 1.5 to 2.0 volts at idle **(see illustration)**. Raise the engine rpm and observe that as vacuum decreases, voltage increases. At wide open throttle the MAP sensor SIGNAL should produce approximately 4.5 volts. **Note:** *It is easy to simulate engine vacuum conditions on the TBI systems by installing a handheld vacuum pump to the MAP sensor. On MFI systems, it is difficult to simulate driving conditions while observing the voltmeter with the engine rpm raised with the vehicle sitting in the driveway. The voltage fluctuations will be slight. It is recommended to use a SCAN tool plugged into the ALDL while the vehicle is driving to observe "real time" data.*

Replacement

11   To replace the sensor, detach the vacuum hose, unplug the electrical connector and remove the mounting screws. Installation is the reverse of removal.

Intake Air Temperature (IAT) sensor

General description

Refer to illustration 4.12

12   The Intake Air Temperature (IAT) sensor is located inside the air duct **(see illustration)** directly downstream of the air filter housing. This sensor acts as a thermistor (a resistor which changes the value of its resistance as the temperature changes). When the intake air is cold, the sensor resistance is high, therefore the PCM will read a high signal voltage. If the intake air is warm, the resistance is low giving the PCM a low voltage reading. The sensor values range from 2,700 ohms at 70-degrees F to 240 ohms at 190-degrees F **(see illustration 4.2b)**. The PCM supplies approximately 5.0 volts (REFERENCE voltage) to the IAT sensor. A failure in the IAT sensor circuit should set either a Code 23 or Code 25.

Check

Refer to illustrations 4.13 and 4.14

13   Measure the REFERENCE voltage. Disconnect the IAT electrical sensor and with the ignition key ON (engine not running), probe terminal A ground (black) and terminal B REF (tan) **(see illustration)**. The meter should read approximately 5.0 volts.

14   Next, measure the resistance across the sensor terminals **(see illustration)**. By measuring its resistance when cold, then warming it up (a hair dryer can be used for this) and taking another measurement. The resistance of the IAT sensor should be HIGH when the temperature is low. Next, start the engine and let it idle. Wait awhile and let the engine reach operating temperature. Turn the ignition key OFF, disconnect the IAT temperature sensor and check the resistance across the terminals. The resistance should be LOW when the air temperature is high. If the sensor does not exhibit this change in resistance, replace the sensor with a new part.

Replacement

15   To remove an IAT sensor, unplug the electrical connector and remove the sensor from the air intake duct. Carefully twist the sensor to release it from the rubber boot.

16   Installation is the reverse of removal.

Oxygen sensor

General description

Refer to illustration 4.17

17   The oxygen sensor, which is located in the exhaust manifold or exhaust pipe **(see illustration)**, monitors the oxygen content of the exhaust gas stream. The oxygen content in the exhaust reacts with the oxygen sensor to produce a voltage output which varies from 0.1-volt (high oxygen, lean mixture) to 0.9-volts (low oxygen, rich mixture). The PCM constantly monitors this variable voltage output to determine the ratio of oxygen to fuel in the mixture. The PCM alters the air/fuel mixture ratio by controlling the pulse width (open time) of the fuel injectors. A mixture ratio of 14.7 parts

4.13 Check the REFERENCE voltage of the IAT sensor. It should be approximately 5.0 volts

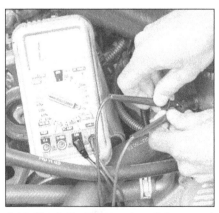

4.14 Check the IAT sensor resistance and compare the value at the existing temperature

4.17 Location of the left bank oxygen sensor on a 1996 and later model

4.23 Check the oxygen sensor millivolt signal with the engine idling

air to 1 part fuel is the ideal mixture ratio for minimizing exhaust emissions, thus allowing the catalytic converter to operate at maximum efficiency. It is this ratio of 14.7 to 1 which the PCM and the oxygen sensor attempt to maintain at all times. 1993 and earlier models are equipped with a single wire oxygen sensor (no heater) while 1994 and 1995 models are equipped with a single heated oxygen sensor. 1996 models are equipped with a left bank oxygen sensor (left exhaust manifold) and right bank oxygen sensor (right exhaust manifold) and two post-catalytic converter oxygen sensors (total of 4 oxygen sensors). When checking the oxygen sensor system, it will be necessary to test all oxygen sensors.

18   The oxygen sensor produces no voltage when it is below its normal operating temperature of about 600-degrees F. During this initial period before warm-up, the PCM operates in OPEN LOOP mode.

19   If the engine reaches normal operating temperature and/or has been running for two or more minutes, and if the computer detects a malfunction, the PCM will set a code 13, 44 or 45.

20   When there is a problem with the oxygen sensor or its circuit, the PCM operates in the open loop mode - that is, it controls fuel delivery in accordance with a programmed default value instead of feedback information from the oxygen sensor.

21   The proper operation of the oxygen sensor depends on four conditions:

a) *Electrical* - *The low voltages generated by the sensor depend upon good, clean connections which should be checked whenever a malfunction of the sensor is suspected or indicated.*

b) *Outside air supply* - *The sensor is designed to allow air circulation to the internal portion of the sensor. Whenever the sensor is removed and installed or replaced, make sure the air passages are not restricted.*

c) *Proper operating temperature* - *The PCM will not react to the sensor signal until the sensor reaches approximately*

600 degrees-F. This factor must be taken into consideration when evaluating the performance of the sensor.

d) *Unleaded fuel* - *The use of unleaded fuel is essential for proper operation of the sensor. Make sure the fuel you are using is of this type.*

22   In addition to observing the above conditions, special care must be taken whenever the sensor is serviced.

a) *The oxygen sensor has a permanently attached pigtail and electrical connector which should not be removed from the sensor. Damage or removal of the pigtail or electrical connector can adversely affect operation of the sensor.*

b) *Grease, dirt and other contaminants should be kept away from the electrical connector and the louvered end of the sensor.*

c) *Do not use cleaning solvents of any kind on the oxygen sensor.*

d) *Do not drop or roughly handle the sensor.*

e) *The silicone boot must be installed in the correct position to prevent the boot from being melted and to allow the sensor to operate properly.*

Check

Refer to illustration 4.23

Note: *Access to the oxygen sensor(s) makes it very difficult but not impossible to backprobe the harness electrical connectors for testing purposes. The exhaust manifolds and pipes are extremely hot and will melt stray electrical probes and leads that touch the surface during testing. If possible, use a SCAN tool that plugs into the ALDL (diagnostic link). This tool will access the PCM data stream and indicates the millivolt changes for each individual oxygen sensor.*

23   Check the oxygen sensor millivolt signal. Locate the oxygen sensor electrical connector and carefully backprobe it using a long pin(s) into the appropriate wire terminals. In most models, connect the positive probe of a voltmeter onto the signal wire (purple) and the negative probe to the ground wire. **Note:**

Consult the wiring diagrams at the end of Chapter 12 for additional information on the oxygen sensor electrical connector wire color designations. Monitor the SIGNAL voltage (millivolts) as the engine goes from cold to warm **(see illustration)**.

24   The oxygen sensor will produce a steady voltage signal of approximately 0.1 to 0.2 volts with the engine cold (open loop). After a period of approximately two minutes, the engine will reach operating temperature and the oxygen sensor will start to fluctuate between 0.1 to 0.9 volts (closed loop). If the oxygen sensor fails to reach the closed loop mode or there is a very long period of time until it does switch into closed loop mode, replace the oxygen sensor with a new part. **Note:** *Post-catalytic converter oxygen sensors will not change voltage values as quickly as pre-catalytic converter oxygen sensors. Because the post-catalytic converter oxygen sensors detect oxygen content after the exhaust has been catalyzed, voltage values should fluctuate much slower and deliberate.*

25   Also inspect the oxygen sensor heater. Disconnect the oxygen sensor electrical connector and working on the oxygen sensor side, connect an ohmmeter between the black wire (-) and pink wire (+). It should measure approximately 5 to 7 ohms. **Note 1:** *Early engines are not equipped with heated oxygen sensors. This can be determined by the single wire.* **Note 2:** *On models equipped with oxygen sensor heaters, the wire colors often change from the harness to the actual oxygen sensor wires according to manufacturer's specifications. Follow the wire colors to the oxygen sensor electrical connector and determine the matching wires and their colors before testing the heater resistance.*

26   Check for proper supply voltage to the heater. Disconnect the oxygen sensor electrical connector and working on the engine side of the harness, measure the voltage between the black wire (-) and pink wire (+) on the oxygen sensor electrical connector. There should be battery voltage with the ignition key ON (engine not running). If there is no voltage, check the circuit between the main relay, the fuse and the sensor. **Note:** *It is important to remember that supply voltage will only last approximately 2 seconds because the system uses a relay to divert the voltage, therefore it will be necessary to have an assistant turn the ignition key ON.*

27   If the oxygen sensor fails any of these tests, replace it with a new part.

28   Access to the oxygen sensors can be difficult making it impossible to monitor the SIGNAL voltage changes without removing several components to gain access to the oxygen sensor electrical connectors. Install the SCAN tool and switch to the Oxygen Sensor mode and monitor the oxygen sensor crosscounts (varying millivolt signals). The SCAN tool should indicate approximately 100 to 200 millivolts when cold (LEAN condition) and then fluctuate from 300 to 800 millivolts warm (CLOSED LOOP).

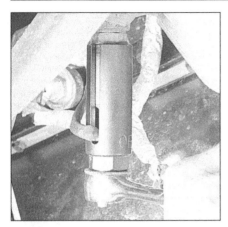

4.32 Use a special slotted socket to remove the oxygen sensor from the exhaust pipe

Replacement

Refer to illustration 4.32

Note: *Because it is installed in the exhaust manifold or pipe, which contracts when cool, the oxygen sensor may be very difficult to loosen when the engine is cold. Rather than risk damage to the sensor (assuming you are planning to reuse it in another manifold or pipe) or the threads which it screws into, start and run the engine for a minute or two, then shut it off. Be careful not to burn yourself during the following procedure.*

29   Disconnect the cable from the negative terminal of the battery (see Chapter 5).

30   Raise the vehicle and place it securely on jackstands.

31   Disconnect the electrical connector from the sensor.

32   Carefully unscrew the sensor from the exhaust manifold **(see illustration)**.

33   Anti-seize compound must be used on the threads of the sensor to facilitate future removal. The threads of new sensors will already be coated with this compound, but if an old sensor is removed and reinstalled, recoat the threads.

34   Install the sensor and tighten it securely.

35   Reconnect the electrical connector of the pigtail lead to the main engine wiring harness.

4.40a Check the SIGNAL voltage on the dark blue (+) wire and the black (-) wire at closed throttle . . .

4.39 Check the TPS REFERENCE voltage on the gray (+) wire and the black (-) wire. It should be approximately 5.0 volts

36   Lower the vehicle, take it on a test drive and check to see that no trouble codes set.

Throttle Position Sensor (TPS)

General description

37   The TPS is located on the end of the throttle shaft on the throttle body. By monitoring the output voltage from the TPS, the PCM can determine fuel delivery based on throttle valve angle (driver demand). A broken or loose TPS can cause intermittent bursts of fuel from the injector and an unstable idle because the PCM thinks the throttle is moving.

38   A problem in any of the TPS circuits will set a Code 21 or 22. Once a trouble code is set, the PCM will use an artificial default value for TPS and some vehicle performance will return.

Check

Refer to illustrations 4.39, 4.40a and 4.40b

39   Locate the Throttle Position Sensor (TPS) on the throttle body. Using a voltmeter, check the REFERENCE voltage from the PCM. Install the positive probe (+) onto the gray wire (reference) and negative probe (-) onto the black wire (ground) **(see illustration)**. The meter should read approximately 5.0 volts.

40   Next, check the TPS signal voltage. With the throttle fully closed, install the posi-

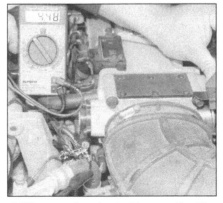

4.40b . . . and again at wide open throttle. The voltage should increase smoothly to approximately 4.5 volts

tive probe (+) of the voltmeter onto the dark blue wire and the negative probe (-) onto the black wire **(see illustration)**. Gradually open the throttle valve and observe the TPS voltage. The voltage should increase from approximately 0.2 to 1.0 volt at closed throttle to approximately 4.0 to 4.5 volts at wide open throttle **(see illustration)**. Confirm a smooth change in the voltage values as the sensor travels from closed to open throttle. If the readings are incorrect, replace the TPS.

Replacement

Refer to illustration 4.42

41   Disconnect the electrical connector from the TPS.

42   Remove the mounting screws from the TPS **(see illustration)** and remove the TPS from the throttle body.

43   When installing the TPS, be sure to align the socket locating tangs on the TPS with the throttle shaft in the throttle body.

44   Installation is the reverse of removal.

Neutral Start switch

45   The Neutral Start switch, located on the rear upper part of the automatic transmission, indicates to the PCM when the transmission is in Park or Neutral. This information is used for Transmission Converter Clutch (TCC), Exhaust Gas Recirculation (EGR) and Idle Air Control (IAC) valve operation. **Caution:** *The vehicle should not be driven with the Neutral Start switch disconnected because idle quality will be adversely affected.*

46   For more information regarding the Neutral Start switch, which is part of the Neutral start and back-up light switch assembly, see Chapter 7.

Air conditioning control

Air conditioning clutch control

47   During air conditioning operation, the PCM controls the application of the air conditioning compressor clutch. The PCM controls the air conditioning clutch control relay to delay clutch engagement after the air conditioning is turned ON to allow the IAC valve to adjust the idle speed of the engine to compensate for the additional load. The PCM also controls the relay to disengage the clutch on WOT (wide

4.42 Remove the TPS mounting screws (arrows)

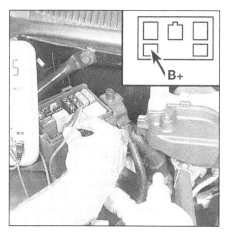

4.49 Check for battery voltage on the air conditioning clutch relay connector

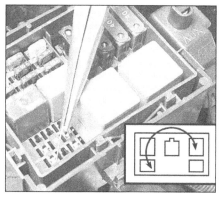

4.50 Apply battery voltage to the air conditioning clutch terminal using a jumper wire and make sure the clutch activates

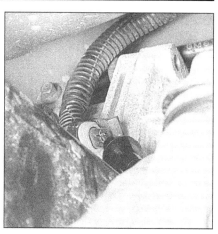

4.52 Location of the Vehicle Speed Sensor (VSS)

open throttle) to prevent excessively high rpm on the compressor. Be sure to check the air conditioning system as detailed in Chapter 3 before attempting to diagnose the air conditioning clutch or electrical system.

Air conditioning "On" signal

Refer to illustrations 4.49 and 4.50

48   Turning on the air conditioning supplies battery voltage to the air conditioning compressor clutch of the PCM electrical connector to increase idle air rate and maintain idle speed. In most cases, if the air conditioning does not function, the problem is probably related to the air conditioning system relays and switches and not the PCM.

49   Remove the air conditioning relay from the relay center and check for battery voltage to the relay **(see illustration)**. Battery voltage should exist with the ignition key ON (engine not running).

50   If battery voltage exists, install a jumper wire into the relay connector and observe that the air conditioning clutch activates **(see illustration)**. If the air conditioning clutch and relay system are working properly, check the air conditioning system pressures (see Chapter 3).

Vehicle Speed Sensor (VSS)

General description

51   The VSS is located in the transmission housing at the rear section near the output shaft. This permanent magnet generator sends a pulsing voltage signal to the PCM, which the PCM converts to miles per hour. The VSS is part of the Transmission Converter Clutch (TCC) system. A problem in the VSS control circuits will set a Code 24.

Check

Refer to illustration 4.52

52   To check the VSS, disconnect the electrical connector in the wiring harness near the sensor. Check the REFERENCE voltage from the PCM by probing the yellow (+) and the purple (-) with a voltmeter **(see illustration)**. There should be approximately 5.0 volts present. If reference voltage is present, have the VSS diagnosed by a dealer service depart-

ment. **Note:** *Refer to Chapter 7 for additional information on the VSS.*

Replacement

53   To replace the VSS, detach the sensor retaining screw and bracket, unplug the sensor and remove it from the transmission.

54   Installation is the reverse of removal.

Crankshaft sensor (1996 only)

Note: *1996 models are equipped with a crankshaft sensor and a crankshaft target wheel located on the front cover. Because this system requires a special TECH 1 SCAN tool to access the self diagnosis system, have the vehicle codes extracted by a dealer service department or other qualified repair facility.*

General description

55   The crankshaft sensor is mounted on the engine front cover and perpendicular to the crankshaft target wheel. This sensor reads positions on the crankshaft target wheel from 4 keyed slots positioned 60-degrees apart. The air gap between the target wheel and the sensor is preset at the factory and cannot be adjusted. The sensor detects pulses as the target wheel rotates and creates changes in the magnetic field between the slots and the crankshaft position sensor. These pulses relay information to the computer which allows precise timing controls for the fuel injection and ignition systems.

Check

Refer to illustration 4.57

56   To check the crankshaft sensor, it is necessary to use a special GM SCAN tool (Tech 1). The crankshaft sensor and circuit must be monitored while the engine is in motion. However, there is a quick voltage check that can be performed on the crankshaft sensor circuit to indicate a possible defective voltage supply. It is recommended the sensor be diagnosed by a dealer service department or other qualified repair shop.

57   Locate the crankshaft sensor, disconnect the electrical connector. Check for battery voltage from the PCM to the sensor **(see illustration)**.

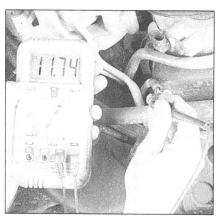

4.57 Check for battery voltage to the crankshaft sensor on the red wire (+)

Replacement

Refer to illustration 4.60

58   Disconnect the negative terminal from the battery.

59   Disconnect the electrical connector from the crankshaft sensor.

60   Remove bolts from the crankshaft sensor and remove the sensor **(see illustration)**.

61   Installation is the reverse of removal. Tighten the bolts to the torque listed in this Chapter's Specifications.

4.60 Remove the crankshaft sensor mounting bolt

Knock sensors

General description

Refer to illustration 4.62

62   The knock sensors are located in the engine block. 1994 and later models use two knock sensors; one under the left bank cylinder head and the other under the right bank cylinder head **(see illustration)**. 1993 and earlier models are equipped with an Electronic Spark Control module mounted on the right inner fender near the hood hinge. 1994 and later models are equipped with a knock sensor module in the PCM. Octane ratings vary the performance of engines and often detonation leads to spark knock. To control spark knock, the knock control system detects abnormal vibration in the engine. This system is designed to reduce spark knock up to 20-degrees during periods of heavy detonation. This allows the engine to use maximum spark advance to improve driveability. The knock sensors produce an AC output voltage which increases with the severity of the knock. The signal is fed into the PCM and the timing is retarded up to 20-degrees to compensate for the severe detonation. Any problems with the knock sensor circuit will set a code 43.

Check

1993 and earlier models

63   Connect a timing light to the number one spark plug wire. Start the engine and lightly rap on the engine block near the knock sensor with a hammer while watching the timing indicator with the timing light. The timing should momentarily retard with each rap, indicating the system is operating. If the timing does not retard, connect a 12-volt test light to the battery, disconnect the connector from the knock sensor and momentarily touch the connector terminal with the test light. The timing should retard each time you touch the terminal. If it doesn't, there's a problem with the ESC module, PCM or related wiring. If timing retards while touching the knock sensor connector with a test light connected to 12-volts, but not when connected to the knock sensor, the knock sensor is probably faulty. Be sure to check for power to the ESC module on the pink/black wire with the ignition key On and check for continuity to ground on the brown wire.

1994 and later models

64   Disconnect the electrical connector for the knock sensor and with the ignition key ON (engine not running), check the REFERENCE voltage from the computer using a voltmeter. It should be approximately 5.0 volts. Using an ohmmeter, check the resistance of the sensor. It should be approximately 3300 to 4500 ohms. If the knock sensor resistance is correct but there is no REFERENCE voltage available, have the vehicle checked at a dealer service department or other repair shop equipped with the necessary tools.

Component replacement

Refer to illustrations 6.66, 4.67a, 4.67b and 4.67c

Knock sensor

Warning: *Wait for the engine to cool completely before performing this procedure.*

4.62 Location of the knock sensor (on the side of the engine block above the oil pan rail)

65   The knock sensor is threaded into the engine block coolant passage, when it is removed the coolant will drain from the engine block. Drain the cooling system (see Chapter 3). Place a drain pan under the sensor, disconnect the electrical connector and remove the knock sensor. A new sensor is pre-coated with thread sealant, do not apply any additional sealant or the operation of the sensor may be effected. Install the knock sensor and tighten it securely (approximately 14 ft-lbs). Don't overtighten the sensor or damage may occur. Plug in the electrical connector, refill the cooling system and check for leaks.

ESC module (1993 and earlier models)

66   Disconnect the electrical connector from the ESC module and remove the mounting screws **(see illustration)**.

Knock sensor module (1994 and later models)

67   Remove the PCM (see Section 3) and remove the knock sensor module **(see illustrations)**.

Mass Airflow (MAF) sensor (1994 and later models)

General description

68   The Mass Airflow Sensor (MAF) is located on the air intake duct. This sensor uses a hot wire sensing element to measure the amount of air entering the engine. The air passing over the hot wire causes it to cool. Consequently, this change in temperature can be converted into an analog voltage signal to the PCM which in turn calculates the required fuel injector pulse width. A problem with the MAF sensor or circuit will set a code 48.

Check

Refer to illustrations 4.69 and 4.70

69   Check for power to the MAF sensor. Disconnect the MAF sensor electrical connector and check the pink wire of the harness connector for battery voltage **(see illustration)**. Check the black/white wire for continuity to ground.

70   Reconnect the electrical connector and backprobe the MAF signal wire (yellow) and ground (black/white) with the voltmeter and

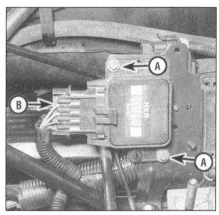

4.66 To remove the ESC module, disconnect the electrical connector (B) and remove the mounting screws (A)

check for the voltage **(see illustration)**. The voltage should be 0.2 to 1.5 volts at idle.

71   Raise the engine rpm. The signal voltage from the MAF sensor should increase to about 4.0 volts at 60 mph. It is impossible to simulate these conditions in the driveway at home but it is necessary to observe the voltmeter for a fluctuation in voltage as the engine speed is raised. The vehicle will not be under load conditions but it should manage to vary slightly. This parameter is easily monitored on a SCAN tool. Look for a steady increase in the voltage signal.

72   If the voltage readings are correct, check the wiring harness for open circuits or a damaged harness (see Chapter 12).

Replacement

73   Disconnect the electrical connector from the MAF sensor.

74   Remove the upper section of the air cleaner assembly (see Chapter 4).

75   Remove the bolts and lift the MAF sensor from the engine compartment.

76   Installation is the reverse of removal.

---

5   Exhaust Gas Recirculation (EGR) system

General description

1   The EGR system meters exhaust gases into the engine induction system through passages cast into the intake manifold. From there the exhaust gases pass into the fuel/air mixture for the purpose of lowering combustion temperatures, thereby reducing the amount of oxides of nitrogen (NOx) formed.

2   The amount of exhaust gas admitted is regulated by a negative or positive backpressure controlled (EGR) valve in response to engine operating conditions. The EGR valve is under the control of the EGR vacuum control solenoid valve, which, in turn, is under the control of the PCM. The vacuum signal to the EGR valve is controlled by varying the duty cycle (on-time) of the solenoid valve. The duty cycle is calculated by the computer using information from the ECT, VSS and IAT sen-

4.67a Remove the bolts from the knock sensor cover

4.67b Remove the bolts and lift the cover to expose the knock sensor module

4.67c Pinch the tabs and lift the knock sensor module from the PCM

sors. Problems with the EGR control circuit or the EGR valve will set a Code 32.

3   Common engine problems associated with the EGR system are rough idling or stalling at idle, rough engine performance during light throttle application and stalling during deceleration.

Check

Refer to illustrations 5.6 and 5.7

4   Start the engine, warm it up to normal operating temperature and allow the engine to idle. Manually lift the EGR diaphragm. The engine should stall, or at least the idle should drop considerably.

5   If the EGR valve appears to be in proper operating condition, carefully check all hoses connected to the valve for breaks, leaks or kinks. Replace or repair the valve/hoses as necessary.

6   With the engine idling at normal operating temperature, disconnect the vacuum hose from the EGR valve and connect a vacuum pump **(see illustration)**. When vacuum is applied, the engine should stumble or die, indicating the vacuum diaphragm is operating properly. **Note:** *These models use a backpressure-type EGR valve. Backpressure must be created in the exhaust system before the vacuum pump will actuate the valve. To create backpressure, install a large socket into the tailpipe and clamp it to the pipe to prevent it from dropping out during the test. The 1/2 inch hole in the socket will allow the engine to idle but still create backpressure on the EGR valve. Don't restrict the exhaust system any longer than necessary to perform this test.* Replace the EGR valve with a new one if the test does not affect the idle.

7   Check the operation of the EGR control solenoid. Check for vacuum to the solenoid **(see illustration)**. If vacuum exists, check for battery voltage to the solenoid.

8   Further testing of the EGR system, vacuum solenoid and the PCM will require a SCAN tool to access computer information that directly controls the EGR system. Have the vehicle checked by a dealer service department or other qualified repair facility.

4.69 Check for battery voltage to the MAF sensor

Component replacement

EGR valve

9   Disconnect the vacuum hose at the EGR valve.

10   Remove the nuts or bolts which secure the valve to the intake manifold or adapter.

11   Lift the EGR valve from the engine.

12   Clean the mounting surfaces of the EGR

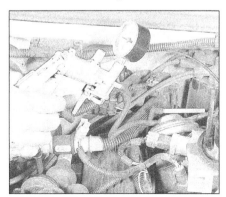

5.6 Install a hand-held vacuum pump to the EGR valve while the engine is idling and observe that the engine stumbles and dies when the EGR valve is opened

4.70 Backprobe the MAF sensor yellow (+) and black/white (-) wires - slowly raise the engine rpm and observe that voltage increases

valve. Remove all traces of gasket material.

13   Place the new EGR valve, with a new gasket, on the intake tube or adapter and tighten the attaching nuts or bolts.

14   Connect the vacuum signal hose.

EGR vacuum solenoid

15   Remove the hoses from the EGR vac-

5.7 Remove the vacuum line and check for manifold vacuum to the EGR control solenoid

6.2a Location of the air pump (arrow) - 1993 and earlier

uum control solenoid, labeling them to ensure proper installation.

16 Remove the solenoid and replace it with the new one.

EGR valve cleaning

17 With the EGR valve removed, inspect the passages for excessive deposits.

18 It is a good idea to place a rag securely in the passage opening to keep debris from entering. Clean the passages by hand, using a drill bit.

---

6   Air Injection (AIR) system

General Information

Refer to illustrations 6.2a and 6.2b

1 The AIR system controls emissions during the early stages (rich conditions) of engine operation through its warm-up cycle. Early models are equipped with a mechanical air pump while later models are equipped with an electronically controlled air pump. Both pumps force air down into the exhaust manifolds to oxidize the hydrocarbons (HC) and carbon monoxide (CO) created by the rich running conditions.

2 1993 and earlier systems consist of an air pump, air control valve, diverter valve, check

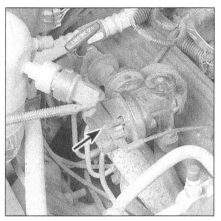

6.2b Location of the diverter valve (arrow) - 1993 and earlier

valves, injection manifolds and hose routing, upstream and downstream check valves and three-way catalytic converters **(see illustrations)**. Later systems are equipped with an electric air pump, check valves, an AIR cross under pipe, a relay, PCM and connecting harness and vacuum lines.

3 1994 and later systems operate electronically. The PCM receives ECT, IAT and engine rpm information from the information sensors to initiate Secondary AIR activation. The relay provides the start-up signal to the air pump. The electric pump provides the necessary air to oxidize the emission gasses created by the rich running start-up condition.

Check

Air pump

Mechanical air pump

4 Check and adjust the drivebelt tension (see Chapter 1).

5 Disconnect the air supply hose at the air bypass valve inlet.

6 The pump is operating satisfactorily if airflow is felt at the pump outlet with the engine running at idle, increasing as the engine speed is increased.

7 If the air pump doesn't pass the above tests, replace it with a new or rebuilt unit.

Electric air pump

Refer to illustrations 6.9a, 6.9b and 6.10

8 With the engine cold, disconnect the AIR line at the check valve, start the engine and observe that fresh air is being pumped to the exhaust system. If no air is felt, continue checking the system.

9 With the engine completely cooled down, turn the ignition key ON (engine not running) and check for battery voltage to the air pump **(see illustration)**. Check for battery voltage to the relay. Power to the AIR system relay is supplied by fuse number 4 (10 amp) and fuse number 3 (25 amp) (see Chapter 12) **(see illustration)**. If no battery voltage is present, have the PCM and the AIR circuit diagnosed by a dealer service department.

10 Check the operation of the electric air pump. Disconnect the air pump electrical connector and using jumper wires, apply battery voltage to the relay terminal **(see illustration)**. The pump should activate and there should be a noticeable sound as the pump produces air volume. If there is no sound, replace the air pump.

11 Physically inspect the air pump and lines for damage, broken connectors or water that might have been ingested into the impeller housing from the motor cup (bowl) attached to the body.

Check valve

12 Disconnect the hoses from both ends of the check valve, carefully noting the installed position of the valve and the hoses.

13 Blow through both ends of the check valve, verifying that air flows in one direction only.

14 If air flows in both directions or not at all, replace the check valve with a new one.

15 When reconnecting the valve, make sure it is installed in the proper direction.

Mechanical air pump noise test

16 Check for unusual noises from the air pump. Under normal conditions, noise rises in pitch as the engine speed increases. To determine if noise is the fault of the air injection system, detach the drivebelt (after verifying that the belt tension is correct) and operate

6.9a Check for battery voltage to the AIR pump - 1994 and later

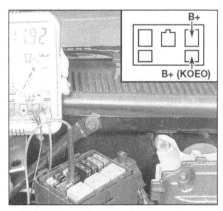

6.9b Check for battery voltage on the air pump relay connector with the ignition key OFF and with the ignition key ON

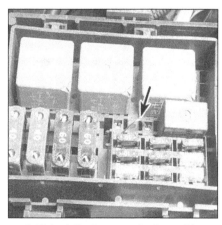

6.10 Install a jumper wire (arrow) to activate the AIR pump

the engine. If the noise disappears, proceed with the following diagnosis. **Caution:** *The pump must accumulate 500 miles (vehicle miles) before the following check is valid.*

17   If the belt noise is excessive:

a) *Check for a loose belt and tighten as necessary (refer to Chapter 1).*

b) *Check for a seized pump and replace it if necessary.*

c) *Check for a loose pulley. Tighten the mounting bolts as required.*

d) *Check for loose, broken or missing mounting brackets or bolts. Tighten or replace as necessary.*

18   If there is excessive mechanical noise:

a) *Check for an overtightened mounting bolt.*

b) *Check for an overtightened drivebelt (see Chapter 1).*

c) *Check for excessive flash on the air pump adjusting arm boss and remove as necessary.*

d) *Check for a distorted adjusting arm and, if necessary, replace the arm.*

19   If there is excessive thermactor system noise (whirring or hissing sounds):

a) *Check for a leak in the hoses (use a soap and water solution to find the leaks) and replace the hose(s) as necessary.*

b) *Check for a loose, pinched or kinked hose and reassemble, straighten or replace the hose and/or clamps as required.*

c) *Check for a hose touching other engine parts and adjust or reroute the hose to prevent further contact.*

d) *Check for an inoperative bypass valve (refer to Step 12 through 19) and replace if necessary.*

e) *Check for an inoperative check valve (refer to Step 22 through 25) and replace if necessary.*

f) *Check for loose pump or pulley mounting fasteners and tighten as necessary.*

g) *Check for a restricted or bent pump outlet fitting. Inspect the fitting and remove any casting flash blocking the air passageway. Replace bent fittings.*

h) *Check for dumping past the air control valve (only at idle). On many vehicles, the AIR system has been designed to dump air at idle to prevent overheating the catalytic converter. This condition is normal. Determine that the noise persists at higher speeds before proceeding.*

l) *Check for air dumping through the diverter valve (the decel and idle dump). Make sure that the hoses are connected properly and not cracked.*

20   If there is excessive pump noise, make sure the pump has had sufficient break-in time (at least 500 miles). Check for a worn or damaged pump and replace as necessary.

Component replacement

21   To replace the air control valve, air diverter valve, check valves, cross-under pipe or other components of the AIR system,

clearly label, then disconnect, the hoses leading to them, replace the faulty component and reattach the hoses to the proper ports. Make sure the hoses are in good condition. If not, replace them with new ones.

22   To replace the air supply pump, first loosen the engine drivebelt (see Chapter 1), then remove the pump mounting bolts from the mounting bracket.

23   On mechanical air pumps, after the new pump is installed, adjust the drivebelts (see Chapter 1).

7   Evaporative Emissions Control (EVAP) system

1   The fuel evaporative emissions control (EVAP) system absorbs fuel vapors from the fuel tank and during engine operation releases them into the engine intake system where they mix with the incoming air/fuel mixture. The main components of the evaporative emissions system are the canister (filled with activated charcoal to absorb fuel vapors), the purge valve/purge control solenoid, the vent valve, the fuel tank pressure control valve, the pressure/vacuum relief fuel filler cap, the fuel tank and the vapor and purge lines. The EVAP system components vary slightly with different years and applications.

2   After passing through a fuel tank pressure control valve, fuel tank vapor is carried through the vapor hose to the charcoal canister. The activated charcoal in the canister absorbs and stores the vapors. When a programmed set of conditions are met (engine running, warmed to a pre-set temperature, etc.), the purge valve is opened. Fuel vapors from the canister are then drawn through the purge hose by intake manifold vacuum into the intake manifold and combustion chamber where they are consumed during normal engine operation. The vacuum hose from the engine to the EVAP canister is designed in three different ways depending upon the engine and the fuel system:

a) **Ported vacuum source** - this type uses ported vacuum from the throttle body directly to the EVAP control valve.

b) **Manifold vacuum with EVAP purge control valve** - this type uses a manifold vacuum signal to a separate purge control valve that is mounted on the engine or on the canister itself.

c) **EVAP purge control solenoid** - this type uses a computer controlled signal to regulate the purge control solenoid and engine vacuum.

3   The ECM/PCM regulates the rate of vapor flow from the canister to the intake manifold by controlling the EVAP purge control solenoid. During cold running conditions and hot start time delay, the ECM/PCM does not energize the solenoid. After the engine has warmed up to the correct operating temperature, the ECM/PCM purges the vapors into the intake manifold according to the running conditions of the engine. The ECM/PCM

will cycle (ON then OFF) the purge control solenoid about 5 to 10 times per second. The flow rate will be controlled by the pulse width, or length of time, the solenoid is allowed to be energized. **Note:** *Early systems are equipped with a purge control valve mounted on the top of the charcoal canister that is controlled by a purge control solenoid.*

4   Models equipped with the OBD-II system (1995 and 1996) can perform a self-diagnostic check to monitor the EVAP system. The PCM monitors the flow of fuel vapor using the purge vacuum switch. The switch detects the flow of fuel vapor and stops the 12 volt signal signaling a purge event. A scan tool can be used to read these actions by the computer. The PCM then monitors the system and sets a diagnostic code if a leak or flow problem is detected.

5   On OBD-II models, it is possible to detect a clogged vent on the charcoal canister as the fuel vapor backs up and the purge vacuum switch detects the excess fuel vapor indicating a stuck open condition and setting a diagnostic trouble code. Check the integrity of the charcoal canister, the canister vent tube and the hoses and connections.

Check

Note: *The evaporative control system, like all emission control systems, is protected by a Federally-mandated warranty (5 years or 50,000 miles at the time this manual was written). The EVAP system probably won't fail during the service life of the vehicle; however, if it does, the hoses or charcoal canister are usually to blame.*

6   Always check the hoses first. A disconnected, damaged or missing hose is the most likely cause of a malfunctioning EVAP system. Refer to the Vacuum Hose Routing Diagram (attached to the radiator support) to determine whether the hoses are correctly routed and attached. Repair any damaged hoses or replace any missing hoses as necessary.

7   Inspect the charcoal canister, purge control valve and the vacuum hoses for cracks, damage, misrouted hoses or component failures.

8   On models equipped with electronic purge controls, check the related fuses and wiring to the purge and vent valves. The purge and vent valves are normally closed - no vapors will pass through the ports. When the PCM energizes the solenoid (by completing the circuit to ground), the valve opens and vapors flow through. A scan tool is required to thoroughly check the system. If the above checks fail to identify the problem area, have the system diagnosed by a dealer service department or other qualified repair shop.

Component replacement

9   Be sure to check all EVAP system hoses for damage and leaks, all hose connections for proper sealing, the charcoal canister for leaks and damage and the electrical connectors for any damaged harness connections before replacing electrical components on this system.

EVAP canister

Refer to illustration 7.10

10   The EVAP canister is located in different locations according to the year and application **(see illustration)**.

11   On late models, the canister is located on the frame rail toward the front of the engine compartment on the passenger side. Raise the vehicle and support it securely on jackstands.

12   Disconnect the hoses from the canister. Remove the bracket mounting bolt and remove the canister.

13   Installation is the reverse of removal.

Purge valve/purge control solenoid

14   The purge valve is mounted on the charcoal canister on early models or on the intake manifold on late models.

15   On late models, disconnect the electrical connector from the purge control solenoid. Depress the locking tab and remove the hose from the purge control solenoid.

16   Remove the mounting nuts/bolt. If equipped, remove the wiring harness retainer from the mounting stud. Remove the purge control solenoid.

17   Installation is the reverse of removal.

Fuel tank pressure control valve

18   The fuel tank pressure control valve is mounted on a bracket near the fuel tank.

19   Raise the vehicle and support it securely on jackstands.

20   Disconnect the electrical connector. Remove the hose from the valve.

21   Release the retainers and remove the valve from the bracket.

22   Installation is the reverse of removal.

Purge vacuum switch

Refer to illustration 7.23

23   The purge vacuum switch monitors the operation of the purge control solenoid and the flow of the fuel vapors to the charcoal canister **(see illustration)**.

24   Disconnect the electrical connector to the purge vacuum switch. Depress the locking tab and remove the hose from the purge vacuum switch.

25   Remove the mounting nuts/bolt. If equipped, remove the wiring harness retainer from the mounting stud. Remove the purge vacuum switch.

26   Installation is the reverse of removal.

8   Positive Crankcase Ventilation (PCV) system

General description

1   The positive crankcase ventilation system reduces hydrocarbon emissions by circulating fresh air through the crankcase to pick-up blow-by gases, which are then rerouted through the throttle body and burned in the engine.

7.10 The charcoal canister is located in the corner of the engine compartment near the windshield washer tank on early models

2   The main components of this system are vacuum hoses and a PCV valve, which regulates the flow of gases according to engine speed and manifold vacuum.

Check and component replacement

3   Checking the system and PCV valve replacement are covered in Chapter 1.

9   Transmission Converter Clutch (TCC) system

General information

1   The purpose of the Torque Converter Clutch (TCC) system, equipped in automatic transmissions, is to eliminate the power loss of the torque converter stage when the vehicle is in the cruising mode (usually above 35 mph). This economizes the automatic transmission to the fuel economy of the manual transmission. The lock-up mode is controlled by the PCM through the activation of the TCC apply solenoid which is built into the automatic transmission. When the vehicle reaches a specified speed, the PCM energizes the solenoid and allows the torque converter to lock-up and mechanically couple the engine to the transmission, under which conditions emissions are at their minimum. However, because of other operating condition demands (deceleration, passing, idle, etc.), the transmission must also function in its normal, fluid-coupled mode. When such latter conditions exist, the solenoid de-energizes, returning the torque converter to normal operation. The converter also returns to normal operation whenever the brake pedal is depressed.

Check

2   Due to the requirement of special diagnostic equipment for the testing of this system, and the possible requirement for dismantling of the automatic transmission to replace com-

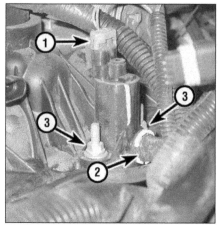

7.23 Details of the EVAP canister purge vacuum switch

1   Electrical connector
2   EVAP purge hose
3   Mounting nuts

ponents of this system, checking and replacing of the components should be handled by a dealer service department or other qualified repair facility.

10   Catalytic converter

General description

1   The catalytic converter is an emission control device added to the exhaust system to reduce pollutants from the exhaust gas stream. These systems are equipped with a single bed monolith catalytic converter. This monolithic converter contains a honeycomb mesh which is also coated with two types of catalysts. One type is the oxidation catalyst while the other type is a three-way catalyst that contains platinum and palladium. The three-way catalyst lowers the levels of oxides of nitrogen (NOx) as well as hydrocarbons (HC) and carbon monoxide (CO) emissions. The oxidation catalyst lowers the levels of hydrocarbons and carbon monoxide.

Check

2   The test equipment for a catalytic converter is expensive and highly sophisticated. If you suspect the converter is malfunctioning, take it to a dealer service department or authorized emissions inspection facility for diagnosis and repair.

3   Whenever the vehicle is raised for service of underbody components, check the converter for leaks, corrosion and other damage. If damage is discovered, the converter should be replaced.

4   Because the converter is welded to the exhaust system, converter replacement requires removal of the exhaust pipe assembly (see Chapter 4). Take the vehicle, or the exhaust system, to a dealer service department or a muffler shop.

Chapter 7
Automatic transmission

Contents

Specifications

Torque specifications — Ft-lbs

Transmission-to-engine bolts	35
Torque converter-to-driveplate bolts	46

1 General information

All vehicles covered in this manual come equipped with an automatic transmission. 1993 and earlier models use a Hydra-Matic 4L60 (formerly the THM 700-R4); 1994 and later models use a Hydra-matic 4L60-E electronic four-speed unit. These transmissions are equipped with a lock-up torque converter that engages in high gear. The lock-up torque converter provides a direct connection between the engine and the drive wheels for improved efficiency and economy. The lock-up converter consists of a solenoid-controlled clutch on the torque converter that engages to lock up the converter in high gear.

Due to the complexity of the automatic transmissions covered in this manual and the need for specialized equipment to perform most service operations, this Chapter contains only general diagnosis, routine maintenance, adjustment and removal and installation procedures.

If the transmission requires major repair work, it should be left to a dealer service department or an automotive or transmission repair shop. You can, however, remove and install the transmission yourself and save the expense, even if the repair work is done by a transmission shop.

2 Diagnosis - general

Note: *Automatic transmission malfunctions may be caused by five general conditions:*

poor engine performance, improper adjustments, hydraulic malfunctions, mechanical malfunctions or malfunctions in the computer or its signal network. Diagnosis of these problems should always begin with a check of the easily repaired items: fluid level and condition (see Chapter 1) and shift linkage adjustment. Next, perform a road test to determine if the problem has been corrected or if more diagnosis is necessary. If the problem persists after the preliminary tests and corrections are completed, additional diagnosis should be done by a dealer service department or transmission repair shop. Refer to the Troubleshooting *section at the front of this manual for information on symptoms of transmission problems.*

Preliminary checks

1   Drive the vehicle to warm the transmission to normal operating temperature.
2   Check the fluid level as described in Chapter 1:
   a) *If the fluid level is unusually low, add enough fluid to bring the level within the designated area of the dipstick, then check for external leaks (see below).*
   b) *If the fluid level is abnormally high, drain off the excess, then check the drained fluid for contamination by coolant. The presence of engine coolant in the automatic transmission fluid indicates that a failure has occurred in the internal radiator walls that separate the coolant from the transmission fluid (see Chapter 3).*
   c) *If the fluid is foaming, drain it and refill the transmission, then check for coolant in the fluid or a high fluid level.*

3   Check the engine idle speed. **Note:** *If the engine is malfunctioning, do not proceed with the preliminary checks until it has been repaired and runs normally.*
4   Check the shift linkage or cable (see Section 4 or 5). Make sure it's properly adjusted and operates smoothly.
5   On models equipped with a 4L60 (1993 and earlier models), check the throttle valve (TV) cable (see Section 6). Make sure it's properly adjusted and operates smoothly.

Fluid leak diagnosis

6   Most fluid leaks are easy to locate visually. Repair usually consists of replacing a seal or gasket. If a leak is difficult to find, the following procedure may help.
7   Identify the fluid. Make sure it's transmission fluid and not engine oil or brake fluid (automatic transmission fluid is a deep red color).
8   Try to pinpoint the source of the leak. Drive the vehicle several miles, then park it over a large sheet of cardboard. After a minute or two, you should be able to locate the leak by determining the source of the fluid dripping onto the cardboard.
9   Make a careful visual inspection of the suspected component and the area immediately around it. Pay particular attention to gasket mating surfaces. A mirror is often helpful for finding leaks in areas that are hard to see.
10   If the leak still cannot be found, clean the suspected area thoroughly with a degreaser or solvent, then dry it.
11   Drive the vehicle for several miles at normal operating temperature and varying

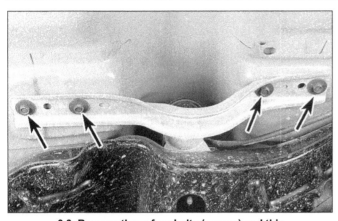

3.3 Remove these four bolts (arrows) and this small crossmember

3.4 Pry off the extension housing seal with a large screwdriver as shown, or use a chisel and hammer to drive it out

speeds. After driving the vehicle, visually inspect the suspected component again.

12   Once the leak has been located, the cause must be determined before it can be properly repaired. If a gasket is replaced but the sealing flange is bent, the new gasket will not stop the leak. The bent flange must be straightened.

13   Before attempting to repair a leak, check to make sure that the following conditions are corrected or they may cause another leak. **Note:** *Some of the following conditions cannot be fixed without highly specialized tools and expertise. Such problems must be referred to a transmission repair shop or a dealer service department.*

Gasket leaks

14   Check the pan periodically. Make sure the bolts are tight, no bolts are missing, the gasket is in good condition and the pan is flat (dents in the pan may indicate damage to the valve body inside).

15   If the pan gasket is leaking, the fluid level or the fluid pressure may be too high, the vent may be plugged, the pan bolts may be too tight, the pan sealing flange may be warped, the sealing surface of the transmission housing may be damaged, the gasket may be damaged or the transmission casting may be cracked or porous. If sealant instead of gasket material has been used to form a seal between the pan and the transmission housing, it may be the wrong type of sealant.

Seal leaks

16   If a transmission seal is leaking, the fluid level or pressure may be too high, the vent may be plugged, the seal bore may be damaged, the seal itself may be damaged or improperly installed, the surface of the shaft protruding through the seal may be damaged or a loose bearing may be causing excessive shaft movement.

17   Make sure the dipstick tube seal is in good condition and the tube is properly seated. Periodically check the area around the speedometer gear or sensor for leakage. If transmission fluid is evident, check the O-ring for damage.

3.5 If you can't remove the extension housing seal with a screwdriver or chisel, pry it out with a seal removal tool (available at most auto parts stores)

Case leaks

18   If the case itself appears to be leaking, the casting is porous and will have to be repaired or replaced.

19   Make sure the oil cooler hose fittings are tight and in good condition.

Fluid comes out vent pipe or fill tube

20   If this condition occurs, the transmission is overfilled, there is coolant in the fluid, the case is porous, the dipstick is incorrect, the vent is plugged or the drain-back holes are plugged.

---

3   Oil seal replacement

---

1   Oil leaks frequently occur due to wear of the extension housing oil seal, and/or the speedometer drive gear oil seal and O-ring. Replacement of these seals is relatively easy, since the repairs can usually be performed without removing the transmission from the vehicle.

Extension housing oil seal

Refer to illustrations 3.3, 3.4, 3.5 and 3.6

2   The extension housing oil seal is located

3.6 Drive the new seal into the extension housing with a large socket

at the extreme rear of the transmission, where the driveshaft is attached. If leakage at the seal is suspected, raise the vehicle and support it securely on jackstands. If the seal is leaking, transmission fluid will be built up on the front of the driveshaft and may be dripping from the rear of the transmission.

3   Remove the driveshaft (see Chapter 8). Remove the small crossmember located right behind the extension housing seal **(see illustration)**.

4   Using a screwdriver, or chisel and hammer, carefully pry the oil seal out of the rear of the transmission **(see illustration)**. DO NOT damage the splines on the transmission output shaft!

5   If the oil seal cannot be removed with a screwdriver or chisel, obtain a special oil seal removal tool **(see illustration)**. Seal removal tools are available at most auto parts stores.

6   Using a large section of pipe or a very large deep socket as a drift, install the new oil seal **(see illustration)**. Drive it into the bore squarely and make sure it's completely seated.

7   Lubricate the splines of the transmission output shaft and the outside of the driveshaft sleeve yoke with lightweight grease, then install the driveshaft. Be careful not to damage the lip of the new seal.

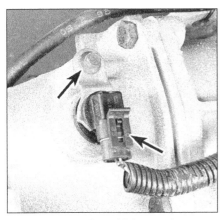

3.8 To remove the vehicle speed sensor, unplug the electrical connector and remove the hold-down bolt (arrows)

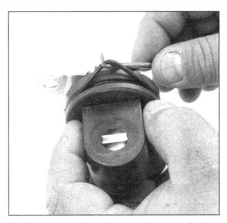

3.11 Remove the old speed sensor O-ring and install a new one

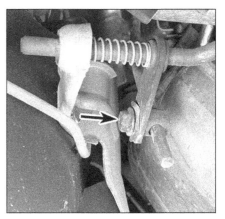

4.3 Loosen the bolt (arrow) on the trunnion block to allow the range selector rod to slide freely in the trunnion

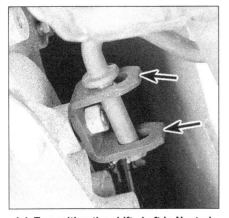

4.4 To position the shift shaft in Neutral, rotate the lever (arrows) to the Park position (counterclockwise, as far as it will go), then click it back two detent positions to Neutral

4.5 Before tightening the trunnion bolt, make sure the shift lever (arrow) on the steering column is also in the Neutral position (which it should be if the gear indicator needle inside the car is pointing at the Neutral position)

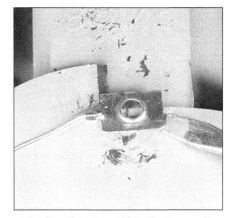

4.8a To adjust the shift indicator needle, pry the guide clip loose from the shift bowl, place the shift lever in the Neutral position . . .

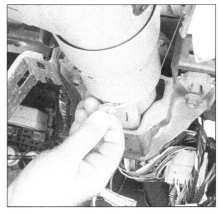

4.8b . . . move the clip until the indicator needle is pointing at the middle of the "N" on the shift indicator panel, make sure the wire between the guide clip and the indicator is taut and push the clip back onto the shift bowl

Speed sensor O-ring

Refer to illustrations 3.8 and 3.11

Note: *Any time the speed sensor is removed, you MUST install a new O-ring.*

8   The speed sensor **(see illustration)** is located on the left (driver's) side of the extension housing. To determine if the O-ring is leaking, look for transmission fluid around the sensor.

9   Unplug the sensor electrical connector.

10   Remove the sensor hold-down bolt and remove the sensor.

11   Remove the old O-ring **(see illustration)** and install a new O-ring on the sensor.

12   Installation is the reverse of removal. Tighten the sensor hold-down bolt securely.

4   Shift linkage - adjustment

Refer to illustrations 4.3, 4.4, 4.5, 4.8a and 4.8b

1   Put the shift lever in the Neutral position.

2   Raise the vehicle and place it securely on jackstands.

3   Loosen the trunnion screw **(see illustra-**

tion) enough to allow the range selector rod to slide freely in the trunnion.

4   Rotate the transmission shift shaft **(see illustration)** to the Park position (click it counterclockwise, as far as it will go), then click it back two detents to the Neutral position.

5   With the steering column selector lever in the Neutral position **(see illustration)**, hold the shift linkage rod to prevent it from sliding in the trunnion and tighten the trunnion bolt securely.

6   Check the shift linkage adjustment: the shift lever must go into all positions and the engine must start only in the P or N positions. If necessary, readjust the shift linkage until your adjustment meets both these criteria. If the shift linkage can't be adjusted so that the engine starts only in the P and N positions, have the mechanical Neutral start system checked out by a dealer service department.

7   Once the shift linkage is properly adjusted, adjust the back-up light switch (see Section 8).

8   If, even with the shift linkage properly adjusted, the gear position indicator needle

doesn't line up with the numbers/letters on the shift indicator panel, alter the position of the needle by changing the position of the indicator cable clip on the steering column shift bowl **(see illustrations)**.

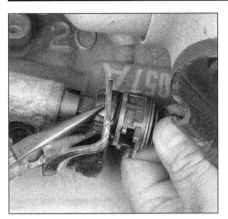

5.3 To detach the shift cable from the transmission bracket, pinch the plastic tangs together and disengage the cable from the bracket

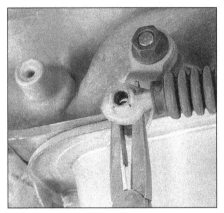

5.4 To disconnect the shift cable from the shift lever at the transmission, pry it off the pin on the shift lever with a screwdriver or a pair of pliers

5.7 To disconnect the shift cable from the shift lever inside the vehicle, pry it off the pin on the shift lever with a screwdriver or a pair of pliers

5 Shift cable (1996 Impala SS) - removal and installation

Refer to illustrations 5.3, 5.4, 5.7 and 5.8

Warning: *The models covered by this manual are equipped with airbags. Always disable the airbag system before working in the vicinity of the impact sensors, steering column or instrument panel to avoid the possibility of accidental deployment of the airbag(s), which could cause personal injury (see Chapter 12). The yellow wires and connectors routed through the console are for this system. Do not use electrical test equipment on these yellow wires or tamper with them in any way while working around the console.*

1   Disconnect the cable from the negative battery terminal. **Caution:** *On models equipped with a Delco Loc II or Theftlock audio system, be sure the lockout feature is turned off before performing any procedure which requires disconnecting the battery.*

2   Raise the vehicle and place it securely on jackstands.

3   To detach the shift cable from the transmission bracket, remove the U-clip retainer, then squeeze the tangs on the cable housing

and pull the cable through the bracket **(see illustration)**.

4   Disconnect the shift cable from the shift lever on the transmission **(see illustration)**.

5   Lower the vehicle.

6   Remove the center console (see Chapter 11).

7   Disconnect the shift cable from the shift lever **(see illustration)**.

8   Pry out the grommet **(see illustration)** from the base of the shift lever assembly and pull out the shift cable through the hole in the floor.

9   Guide the new cable through the hole in the floor and install the grommet.

10   Connect the shift cable to the shift lever.

11   Install the center console trim panel (see Chapter 11).

12   Place the shift lever (inside the vehicle) in Neutral. Make sure it remains in the Neutral position until the shift cable is installed.

13   Raise the vehicle and place it securely on jackstands.

14   Attach the shift cable to the transmission bracket with the adjustment button unlocked.

15   Place the shift lever on the transmission in the Neutral position by rotating it counter-

clockwise from Park through Reverse into Neutral.

16   Connect the shift cable to the shift lever on the transmission.

17   Lock the adjustment button on the shift cable to the transmission bracket by turning it counterclockwise.

18   Lower the vehicle.

19   Reconnect the cable to the negative battery terminal.

6 Throttle valve (TV) cable (1993 and earlier models) - replacement and adjustment

1   The throttle valve (TV) cable controls line pressure, shift feel, shift points, detent downshifts and part throttle downshifts. The function of the cable is similar to the combined functions of a detent cable and a vacuum modulator.

Replacement

Refer to illustrations 6.3, 6.4 and 6.7

2   Remove the air cleaner (see Chapter 4).

3   Detach the cable terminal (end) from the

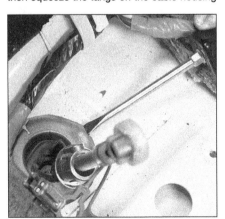

5.8 Pry the shift cable grommet out of its hole in the floor, then pull the shift cable assembly through the hole

6.3 Detach the throttle valve (TV) cable from the throttle lever by pulling it forward, off the mounting lug

6.4 Pinch the locking tangs of the TV cable housing with a pair of pliers and push the housing through the bracket

6.7 Remove the screw and washer holding the cable to the transmission, then pull up the cable housing and disconnect the cable from the link

7.2 Locate the Brake Transmission Shift Interlock (BTSI) solenoid (arrow) on the steering column (column-shift models)

7.9 To lower the steering column for access to the BTSI solenoid or the back-up light switch, remove these nuts (arrows)

throttle lever **(see illustration)**.

4   Compress the locking tangs on the cable housing and detach the housing from the bracket **(see illustration)**.

5   Remove the routing clips or straps.

6   Trace the cable to the transmission. Remove the bolt and the flat washer.

7   Pull up on the cable housing at the transmission until the cable is visible, then disconnect the cable from the transmission link **(see illustration)**.

8   Remove the seal.

9   Installation is the reverse of removal. Be sure to tighten the bolt securely. When you're finished reattaching the TV cable assembly, adjust it (see below).

Adjustment

10   With the engine off, depress and hold down the metal re-adjust tab at the engine end of the TV cable.

11   Move the slider until it stops against the fitting.

12   Release the re-adjust tab.

13   Rotate the throttle lever to its full travel position.

14   The slider must move (ratchet) toward the lever when the lever is rotated to its full travel position.

15   Make sure the cable moves freely. The cable may appear to function properly with the engine stopped and cold. Recheck it after the engine is hot.

16   Road test the vehicle.

---

7   Brake/transmission shift interlock system - description, check and component replacement

---

Warning: *The models covered by this manual are equipped with airbags. Always disable the airbag system before working in the vicinity of the impact sensors, steering column or instrument panel to avoid the possibility of acciden-*

tal deployment of the airbag(s), which could cause personal injury (see Chapter 12). The yellow wires and connectors routed through the console are for this system. Do not use electrical test equipment on these yellow wires or tamper with them in any way while working around the console.

Description

1   The brake transmission interlock system prevents the (column or floor-mounted) shift lever from being moved out of Park unless the brake pedal is depressed simultaneously. When the car is started, a solenoid is energized, locking the shift lever in Park; when the brake pedal is depressed, the solenoid is deenergized, unlocking the shift lever so that it can be moved out of Park.

Check

Refer to illustration 7.2

2   On column-shift models, remove the knee bolster (the trim panel under the dash, in front of the steering column) and the metal plate above it (see Chapter 11). Using a flashlight, locate the brake/transmission shift interlock solenoid **(see illustration)**.

3   On the 1996 Impala SS, remove the center console (see Chapter 11) and locate the brake/transmission shift interlock solenoid, in front of the shift lever.

4   Verify that the brake/transmission shift interlock solenoid operates as follows:

a) *When the ignition key is in the Lock position, the solenoid plunger should be in and you should not be able to move the shift lever, even with the brake pedal applied.*

b) *With the ignition key is turned to the Off position, the solenoid plunger should be pop out and you should be able to move the shift lever to any gear position without applying the brake pedal.*

c) *When the ignition key is turned to the Run position, the solenoid should go back into the solenoid and you should*

not be able to move the shift lever; except with the brake pedal applied, the solenoid plunger should be released (pop out) and you should be able to move the shift lever out of Park into any gear.

d) *Place the key in the Lock position, then turn the key to the Acc (accessory) position. The solenoid plunger should go in a little further than it does when the key is turned to Lock, and you should not be able to move the shift lever (even with the brake applied).*

5   If the solenoid doesn't operate as described above, unplug the electrical connector from the solenoid, apply battery voltage to the solenoid and verify that it "clicks" on, then open the circuit and verify that the solenoid clicks off.

a) *If the solenoid doesn't operate as described, replace it.*

b) *If the solenoid is operating properly, troubleshoot the circuit between the battery and the solenoid, and between the solenoid and ground.*

c) *If the solenoid circuit is okay, take the vehicle to a dealer service department.*

Component replacement

Solenoid

6   Disconnect the cable from the negative battery terminal. **Caution:** *On models equipped with a Delco Loc II or Theftlock audio system, be sure the lockout feature is turned off before performing any procedure which requires disconnecting the battery.*

Column-shift models

Refer to illustrations 7.9, 7.10a, 7.10b, 7.11, 7.12, 7.14a, 7.14b and 7.15

7   Disable the airbag system (see Chapter 12).

8   Remove the trim panel under the left side of the dash (see Chapter 11) and the metal plate above it.

9   Lower the steering column **(see illustration)**.

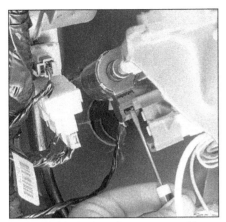

7.10a Pry the locking tab loose with a small screwdriver . . .

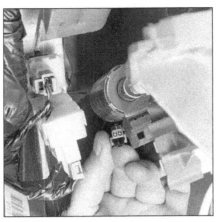

7.10b . . . and unplug the electrical connector from the BTSI solenoid

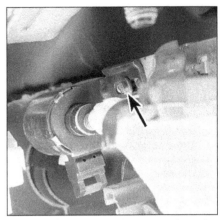

7.11 To detach the BTSI solenoid from the steering column, remove this screw (arrow)

10   Unplug the electrical connector from the solenoid **(see illustrations)**.
11   Remove the solenoid mounting screw **(see illustration)**.
12   Disengage the solenoid from the actuator rod **(see illustration)**.

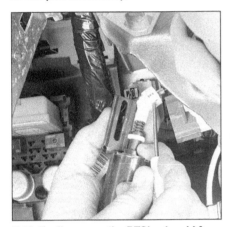

7.12 To disengage the BTSI solenoid from the actuator rod, pry the locking fingers loose and pull the solenoid-half of the actuator rod connector from the other half of the connector

13   Remove the solenoid assembly, spring and solenoid-half of the actuator rod connector.
14   Remove the solenoid-half of the actuator rod connector and spring **(see illustrations)** and install them on the new solenoid.
15   Connect the new solenoid assembly to the actuator rod **(see illustration)**. Installation is otherwise the reverse of removal.

Floor-shift models

16   Remove the center console (see Chapter 11).
17   Unplug the electrical connector from the solenoid.
18   Pry the end of the solenoid rod from its pin on the shift lever assembly and remove the solenoid.
19   Detach the solenoid from the shift lever base.
20   Installation is the reverse of removal.

Park/lock cable (floor-shift models only)

21   Disable the airbag module (see Chapter 12).
22   Pull out the retainer from the back of the shift lever handle and remove the handle by pulling it straight up.

23   Remove the center console, the left lower dash trim panel, the knee bolster and the knee bolster deflector (see Chapter 11).
24   Detach the Park/lock cable from the shift lever.
25   Detach the Park/lock cable from the bracket on the shift lever assembly.
26   Remove the brake light switch (see Chapter 9).
27   Unbolt and lower the steering column assembly.
28   Disconnect the Park/lock cable from the ignition switch.
29   Remove the Park/lock cable.
30   Installation is the reverse of removal.

---

8   Back-up light switch/ transmission position sensor - description, check and replacement

---

Description

Refer to illustration 8.1

Warning: *The models covered by this manual are equipped with airbags. Always disable the airbag system before working*

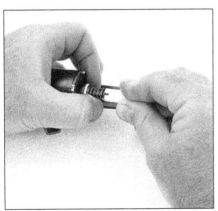

7.14a Remove the solenoid-half of the actuator rod connector from the solenoid . . .

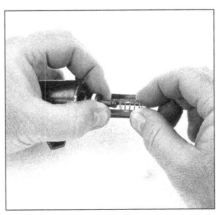

7.14b . . . then remove the spring; install these two parts on the new solenoid

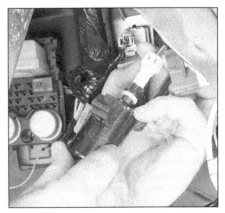

7.15 To reconnect the BTSI solenoid to the actuator rod, push the solenoid-half of the actuator rod connector into the other half until it snaps into place

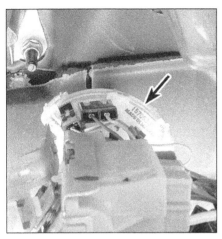

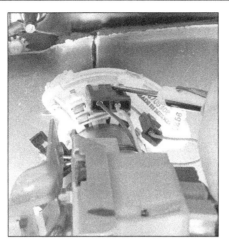

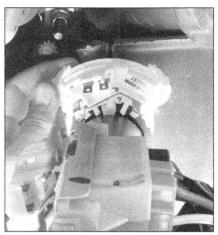

8.1 On column-shift models, the back-up light switch/transmission position sensor (arrow) is located on top of the steering column, near the firewall

8.4a To unplug the electrical connectors from the switch, unlock the locking tabs on top of the connector with a small screwdriver

8.4b To disengage the back-up light switch/transmission position sensor from the steering column, tilt the top of the switch to the rear (toward you), then slide the switch forward slightly and lift up; to install the switch, position the switch over the hole, tilt the top of the switch slightly to the rear, insert the switch into the hole, slide it back until it snaps into place (the locking tabs along the rear edge of the switch must grip the steering column)

in the vicinity of the impact sensors, steering column or instrument panel to avoid the possibility of accidental deployment of the airbag(s), which could cause personal injury (see Chapter 12). The yellow wires and connectors routed through the console are for this system. Do not use electrical test equipment on these yellow wires or tamper with them in any way while working around the console.

Note: *The models covered by this manual do not use an electrical neutral start switch. A mechanical block in the steering column prevents the engine from cranking in positions other than Park or Neutral. However, these models use a back-up light switch/transmission position sensor that's physically similar in appearance to the earlier neutral start/ backup light switch and is adjusted basically the same way. Diagnosis and disassembly of the mechanical-type neutral start system is not recommended.*

1   On column-shift models, the back-up light switch/transmission position sensor **(see illustration)** is located on top of the steering column, near the firewall. On the 1996 Impala SS, it's located on the shift lever base. When the switch is operating properly, the back-up lights should come on when the shift lever is moved to the Reverse. If the back-up lights don't come on, either the shift linkage is out of adjustment, a fuse is blown, the bulbs are burned out, there's an open in the circuit, or the back-up light switch is bad. The transmission position sensor portion of the switch is used to inform the PCM when the transmission is in gear.

Check

Refer to illustrations 8.4a, 8.4b and 8.4c

2   Make sure the shift linkage is properly adjusted (see Section 4).

3   Verify that there's voltage available at the switch on the pink wire with the ignition key On. If there isn't, check the fuse (see Chapter 12). If the fuse is okay, check the cir-

cuit to the switch (see Wiring Diagrams at the end of Chapter 12).

4   If the shift linkage is adjusted, the fuse is okay and the circuit to the switch is good, unplug the electrical connectors and remove the switch **(see illustrations)**. Test the switch on the bench with an ohmmeter as follows **(see illustration)**:

   a) *To test the back-up light switch, verify that there's continuity between the two terminals of the center connector when the switch is in Reverse, and no continuity when it's in any other gear position.*

   b) *To test the transmission position sensor, verify that there's continuity between the two terminals of the left connector when the switch is in Park and Neutral, and no continuity when it's in any other gear position.*

   c) *To test the brake/transmission interlock switch, verify that there's continuity between the terminal corresponding to the pink wire of the center connector and the single (light green) wire connector when the switch is in Park, and no continuity when it's in any other gear position.*

5   If the switch doesn't operate as described, replace it.

Replacement

Column-shift models

6   Unplug the electrical connector from the switch.

7   To disengage the switch from the steering column, slide it forward and lift it up **(see illustration 8.4b)**.

8   If you're installing the old switch, put the shift lever in Neutral, align the actuator on the underside of the switch with the hole in the shift tube, position the connector side of the switch assembly to fit into the cutout on the steering column jacket, push down on the switch assembly to lock the tangs on the switch into place on the steering column jacket, move the switch assembly to the Low

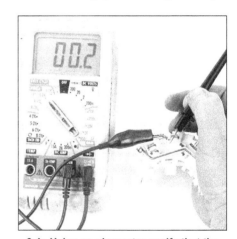

8.4c Using an ohmmeter, verify that the back-up light switch terminals have no continuity when the switch is in Park, or in other gear except Reverse, then verify that the terminals have continuity when the switch is in Reverse

gear position, and place the transmission in Park. The switch assembly will ratchet as it adjusts itself. If you're installing a new switch, use the same procedure, but skip the second-to-last step, *i.e.* don't put the switch assembly into the Low gear position.

9   Verify that the switch has continuity with the carrier tang in Reverse.

10   Installation is otherwise the reverse of removal.

Floor-shift models

11   Disconnect the cable from the negative battery terminal. **Caution:** *On models equipped with a Delco Loc II or Theftlock*

9.2 To check the transmission mount, insert a prybar as shown, between the mount and the crossmember, and note whether the mount is split or cracked when you pry it up and down

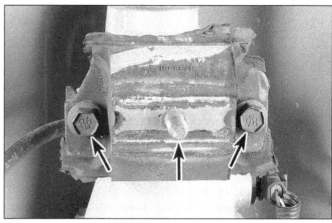

9.3 To remove the transmission mount, remove the nut from the stud (middle arrow) protruding through the crossmember, then remove the mount bolts (outer arrows) (crossmember removed for clarity)

audio system, be sure the lockout feature is turned off before performing any procedure which requires disconnecting the battery. Disable the airbag system (see Chapter 12).
12   Remove the center console trim panel (see Chapter 11).
13   Unplug the electrical connector from the back-up light switch/transmission position sensor.
14   Remove the switch retaining screws and remove the switch.
15   Installation is the reverse of removal.
16   Enable the airbag system (see Chapter 12).

9   Transmission mount - check and replacement

Refer to illustrations 9.2 and 9.3
1   Raise the vehicle and place it securely on jackstands.
2   Place a prybar or large screwdriver between the crossmember and the transmission **(see illustration)** and try to lever the

transmission up and down. If the mount is cracked or torn, replace it.
3   Remove the nut that secures the mount to the crossmember and the two bolts that attach the mount to the transmission **(see illustration)**.
4   Support the transmission with a floor jack, raise the transmission until the mount stud is clear of the crossmember, then remove the mount.
5   Installation is the reverse of removal.

10   Transmission - removal and installation

Warning: *The models covered by this manual are equipped with airbags. Always disable the airbag system before working in the vicinity of the impact sensors, steering column or instrument panel to avoid the possibility of accidental deployment of the airbag(s), which could cause personal injury (see Chapter 12). The yellow wires and connectors routed through the console are for this system. Do not use electrical test equip-*

ment on these yellow wires or tamper with them in any way while working around the console.

Removal

Refer to illustrations 10.8a, 10.8b, 10.8c, 10.8d, 10.10, 10.12, 10.13, 10.15 and 10.17
1   Disconnect the cable from the negative battery terminal. **Caution:** *On models equipped with a Delco Loc II or Theftlock audio system, be sure the lockout feature is turned off before performing any procedure which requires disconnecting the battery.* Disable the airbag system (see Chapter 12).
2   Remove the transmission fluid level dipstick.
3   Raise the vehicle and place it securely on jackstands.
4   Drain the transmission fluid (see Chapter 1).
5   Disconnect the driveshaft from the transmission (see Chapter 8).
6   Disconnect the shift linkage (see Section 4) or the shift cable (see Section 5).
7   Clearly label, then unplug, all electrical connectors.

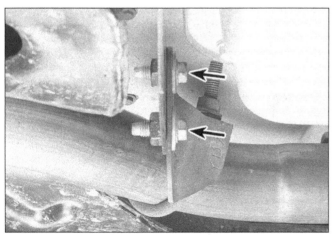

10.8a To detach the exhaust pipe crossmember, remove these two bolts (arrows) from the left end . . .

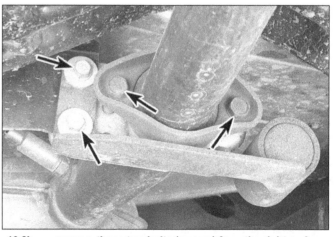

10.8b . . . remove these two bolts (arrows) from the right end . . .

10.8c . . . and remove these two bolts from the middle
of the crossmember

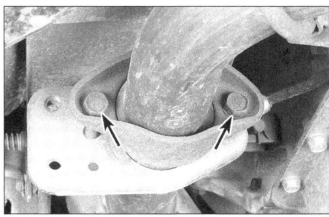

10.8d To detach the forward exhaust pipe sections from the rear
part of the exhaust system, remove the right exhaust flange
bolts (arrows) shown in illustration 10.8b and these
two (arrows) at the left flange

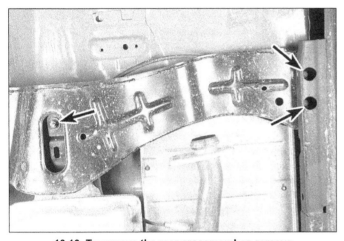

10.10 To remove the rear crossmember, remove
these bolts (arrows)

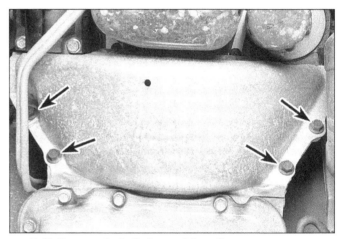

10.12 Remove the bolts (arrows) from the torque converter
access cover and remove the cover

8   On 1993 and earlier models, remove the catalytic converter support bracket. On 1994 and later models, remove the exhaust pipe crossmember **(see illustrations)** and unbolt the two forward sections of the exhaust system from the exhaust manifolds (see Chap-ter 4).

9   Support the engine assembly with a floor jack (place a block of wood between the oil pan and the jack head). Support the transmission with another jack, preferably one made for this purpose (available at most tool rental yards). Safety chains will help steady the transmission on the jack.

10   Remove the transmission crossmember **(see illustration)**.

11   Remove the transmission mount (see Section 9).

12   Remove the bolts from the torque converter access cover **(see illustration)**. Remove the cover.

13   Mark the relationship of the torque converter and driveplate with white paint so they can be installed in the same position, then remove the flywheel-to-torque converter bolts **(see illustration)**. Turn the crankshaft for access to each bolt. Turn the crankshaft in a clockwise direction only (as viewed from the front).

10.13 Mark the relationship of the
torque converter to the driveplate, then
rotate the driveplate to reach each
flywheel-to-torque converter bolt

14   Lower the transmission and engine slightly to access the dipstick tube and the oil cooler lines.

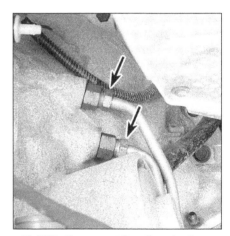

10.15 To disconnect the oil cooler lines,
unscrew these two fittings (arrows)

15   Disconnect the oil cooler lines **(see illustration)**. Use a flare-nut wrench and a back-up wrench to unscrew the cooler line fittings. Cap the open lines to prevent transmission fluid from leaking out.

16   Remove the dipstick tube.

17   Remove the six transmission bellhousing-to-engine bolts **(see illustration)**.

18   Slowly lower both jacks and pull the transmission to the rear, separate it from the engine block dowel pins and make sure the torque converter is detached from the driveplate. Lower the transmission on the jack and wheel it out from under the vehicle.

Installation

19   Make sure the torque converter hub is securely engaged with the pump.

20   If you've removed the converter, spread transmission fluid on the torque converter rear hub, where the transmission front seal rides. With the front of the transmission facing up, rotate the converter back and forth. It should drop down into the transmission front pump in stages. To ensure that the converter is fully engaged, lay a straightedge across the transmission-to-engine mating surface and make sure the converter hub is at least 1/2-inch below the straightedge. Lubricate the front hub of the converter with multipurpose grease.

21   With the transmission secured to the jack, wheel it into position, raise it up into position and connect the oil cooler lines.

22   Turn the torque converter to line up the holes with the holes in the driveplate. The white paint mark made on the torque converter and driveplate in Step 13 must line up.

23   With the aid of a helper, carefully move the transmission forward until the dowel pins and the torque converter are engaged. Make sure the transmission mates with the engine with no gap. If there's a gap, make

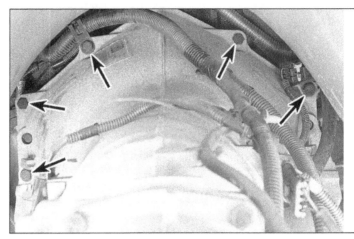

10.17 To detach the transmission from the engine, remove these six bolts (arrows) (right lower bolt not visible in this photo)

sure there are no wires or other objects pinched between the engine and transmission and also make sure the torque converter is completely engaged in the transmission front pump. Try to rotate the converter - if it doesn't rotate easily, it's probably not fully engaged in the pump. If necessary, lower the transmission and install the converter fully.

24   Install the transmission-to-engine bolts and tighten them to the torque listed in this Chapter's Specifications. As you're tightening the bolts, verify that the engine and transmission mate completely at all points. If not, find out why. Never try to force the engine and transmission together with the bolts or you'll break the transmission case.

25   Install the transmission mount and crossmember. Tighten all nuts and bolts securely.

26   Install the torque converter-to-driveplate bolts. Tighten them to the torque listed in this Chapter's Specifications.

27   Install the dipstick tube and the oil cooler lines.

28   Remove the jack stands supporting the engine and transmission.

29   Reconnect all electrical connectors.

30   Reattach the catalytic converters and brackets.

31   Reattach the shift linkage (see Section 4) or the shift cable (see Section 5).

32   Reconnect the driveshaft (see Chapter 8).

33   Check and refill the transmission with the specified fluid (see Chapter 1).

34   Remove the jackstands and lower the vehicle.

Chapter 8 Drivetrain

Contents

Specifications

Torque specifications

	Ft-lbs (unless otherwise indicated)
U-joint strap bolts...	16
Pinion shaft lock bolt..	96 in-lbs

1 Driveshaft and universal joints - general information

Warning: *Because nearly all the procedures covered in this Chapter involve working under the vehicle, make sure that it's securely supported on sturdy jackstands or on a hoist where the vehicle can be easily raised and lowered.*

Note: *This section deals with the driveshaft and universal joints; Section 5 offers general descriptions and checking procedures for rear-axle components.*

1   The driveshaft is a tube that transmits power between the transmission and the differential. Universal joints are located at either end of the driveshaft.

2   The driveshaft employs an internally splined slip yoke at the front, which engages the externally splined output shaft inside the end of the transmission extension housing. This setup allows the driveshaft to slide in and out of the extension housing during operation. An oil seal in the end of the extension housing prevents transmission fluid from leaking out and dirt from entering the transmission. If leakage is evident at the front of the driveshaft, replace the oil seal (see Chapter 7).

3   Generally speaking, the driveshaft assembly requires little service. The universal joints are lubricated for life. If U-joint problems develop, the driveshaft must be removed from the vehicle and the U-joints replaced.

4   Since the driveshaft is a balanced unit, it's important that no undercoating, mud, etc. be allowed to remain on it. When the vehicle is raised for service it's a good idea to clean the driveshaft and inspect it for any obvious damage. Also, make sure the small weights used to originally balance the driveshaft are

in place and securely attached. Whenever the driveshaft is removed it must be reinstalled in the same relative position to preserve the balance.

5   Problems with the driveshaft are usually indicated by a noise or vibration while driving the vehicle. A road test should verify if the problem is the driveshaft or another vehicle component. Refer to *Troubleshooting* at the front of this manual. If you suspect trouble, inspect the driveline (see Section 2).

2 Driveline inspection

1   Raise the rear of the vehicle and support it securely on jackstands. Block the front wheels to keep the vehicle from rolling off the stands.

2   Crawl under the vehicle and visually inspect the driveshaft. Look for any dents or cracks in the tubing. If any are found, the driveshaft must be replaced.

3   Check for oil leakage at the front and rear of the driveshaft. Leakage where the driveshaft enters the transmission indicates a defective transmission extension housing seal (see Chapter 7). Leakage where the driveshaft enters the differential indicates a defective pinion seal (see Section 9).

4   While under the vehicle, have an assistant rotate a rear wheel so the driveshaft will rotate. As it does, make sure the universal joints are operating properly without binding, noise or looseness.

5   The universal joint can also be checked with the driveshaft motionless, by gripping your hands on either side of the joint and attempting to twist the joint. Any movement at all in the joint is a sign of considerable wear. Lifting up on the shaft will also indicate movement in the universal joints.

6   Finally, check the driveshaft mounting bolts at the ends to make sure they're tight.

3 Driveshaft - removal and installation

Removal

Refer to illustrations 3.3 and 3.4

1   Raise the vehicle and support it securely on jackstands. Block the front wheels to prevent the vehicle from rolling.

2   Place the transmission in Park or Neutral with the parking brake off.

3   Make reference marks on the driveshaft and the pinion flange in line with each other **(see illustration)**. This is to make sure the driveshaft is reinstalled in the same position to preserve the balance.

3.3 Before disconnecting the driveshaft from the differential, mark the pinion flange and the U-joint to ensure proper balance after reassembly

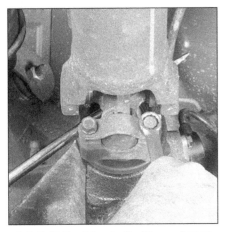

3.4 Hold the driveshaft with a prybar or large screwdriver while breaking loose the clamp bolts

4.2 Remove the inner retainers from the U-joint by tapping them off with a screwdriver and hammer

4.3 Press out the bearing cups as shown

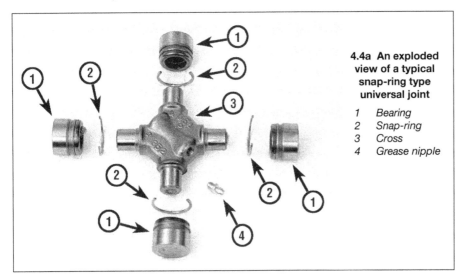

4.4a An exploded view of a typical snap-ring type universal joint

1 *Bearing*
2 *Snap-ring*
3 *Cross*
4 *Grease nipple*

4.4b After it has been pressed out, remove the bearing cup from the yoke ear

4   Remove the rear universal joint bolts and clamps **(see illustration)**. Turn the driveshaft as necessary to bring the bolts into the most accessible position.

5   Tape the bearing caps to the spider to prevent the caps from coming off during removal.

6   Lower the rear of the driveshaft, then slide the front of the driveshaft out of the transmission.

7   Wrap a plastic bag over the transmission extension housing and secure it in place with a rubber band. This will prevent loss of fluid and protect against contamination while the driveshaft is out.

Installation

8   Remove the plastic bag from the transmission and wipe the area clean. Inspect the oil seal carefully. Procedures for replacement of this seal can be found in Chapter 7.

9   Slide the splined front end of the driveshaft into the transmission.

10   Raise the rear of the driveshaft into position, checking to be sure the marks are in alignment. If not, turn the rear wheels to match the pinion flange and the driveshaft.

11   Remove the tape securing the bearing caps and install the clamps and bolts. Tighten all bolts to the torque listed in this Chapter's Specifications. Lower the vehicle.

4   Universal joints - replacement

Refer to illustrations 4.2, 4.3, 4.4a, 4.4b, 4.11 and 4.13

Note: *Always purchase a universal joint service kit(s) for your model vehicle before starting the procedure which follows. Also, read through the entire procedure before beginning work.*

1   Remove the driveshaft (refer to Section 3).

2   Place the shaft on a workbench equipped with a vise. If equipped, remove the snap-rings **(see illustration)**. **Note:** *If no snap-rings are present, the bearing cups are retained by injected-nylon rings which will break as the cups are pressed out. Snap-rings will be used on reassembly.*

3   Place the universal joint in the vise with a 1-1/8 inch socket against one ear of the shaft yoke and a 5/8-inch socket placed on the opposite bearing cup **(see illustration)**.

Caution: *Never clamp the driveshaft tubing itself in a vise, as the tube may be bent.*

4   Press the bearing cup out of the yoke ear, shearing the plastic retaining ring on the bearing, if equipped **(see illustrations)**. **Note:** *If the cup does not come all the way out of the yoke, it may be pulled free with adjustable pliers, then removed.*

5   Turn the driveshaft 180-degrees and press the opposing bearing cup out of the yoke.

6   Disengage the cross from the yoke and remove the cross. If you're replacing the front universal joint, repeat Steps 3 through 5 to remove the remaining cups.

7   If the remaining universal joint is being replaced, press the bearing cups from the slip yoke as detailed above.

8   When reassembling the driveshaft, always install all parts included in the U-joint service kit.

9   If necessary, remove all remnants of the plastic bearing retainers from the grooves in the yokes. Failure to do so may keep the bearing cups from being pressed into place and prevent the bearing retainers from seating properly.

10   Using multi-purpose grease to retain the needle bearings, assemble the bearings, cups and washers. Make sure the bearings do not

4.11 Assemble the new cross and bearing cups

4.13 Install snap-rings in the snap-ring grooves

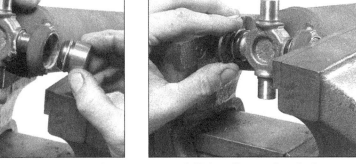

5.6 The identification code and manufacturer's code are stamped on the front of the right axle tube

become dislodged during the assembly and installation procedures.

11   In the vise, assemble the cross and cups in the yoke, installing the cups as far as possible by hand **(see illustration)**.

12   Move the cross back and forth horizontally to ensure alignment, then press the cups into place a little at a time, continuing to center the cross to keep the proper alignment.

13   As soon as one snap-ring groove clears the inside of the yoke, stop pressing and install the snap-ring **(see illustration)**.

14   Continue to press on the bearing cup until the opposite snap-ring can be installed. If difficulty is encountered, strike the yoke sharply with a hammer. This will spring the yoke ears slightly and allow the snap-ring groove to move into position.

15   Install the driveshaft (see Section 3).

5   Rear axle - description, check and identification

Description

1   The rear axle assembly is a hypoid, semi-floating type (the centerline of the pinion gear is below the centerline of the ring gear). When the vehicle goes around a corner, the differential allows the outer rear tire to turn more quickly than the inner tire. The axleshafts are splined to the differential side gears, so when the vehicle goes around a corner, the inner tire, which turns more slowly than the outer tire, turns its side gear more slowly than the outer tire turns its side gear. The differential pinion (or "spider") gears roll around the slower side gear, driving the outer side gear - and tire - more quickly. The differential is housed within a casting with a pressed steel cover, known as the "carrier." The steel axle tubes are pressed into and welded to the carrier.

2   An optional locking limited-slip rear axle is also available. This differential allows for normal operation until one wheel loses traction. A limited-slip unit is similar in design to a conventional differential, except for the addi-

tion of a pair of clutch "cones" which slow the rotation of the differential case when one wheel is on a firm surface and the other on a slippery one. The difference in wheel rotational speed produced by this condition applies additional force to the pinion gears and through the cone, which is splined to the axleshafts, equalizes the rotation speed of the axleshaft driving the wheel with traction.

Check

3   Often, a suspected "axle" problem lies elsewhere. Do a thorough check of other possible causes before assuming the axle is the problem.

4   The following noises are those commonly associated with axle diagnosis procedures:

a)   *Road noise is often mistaken for mechanical faults. Driving the vehicle on different surfaces will show whether the road surface is the cause of the noise. Road noise will remain the same if the vehicle is under power or coasting.*

b)   *Tire noise is sometimes mistaken for mechanical problems. Tires which are worn or low on pressure are particularly susceptible to emitting vibrations and noises. Tire noise will remain about the same during varying driving situations, where axle noise will change during coasting, acceleration, etc.*

c)   *Engine and transmission noise can be deceiving because it will travel along the driveline. To isolate engine and transmission noises, make a note of the engine speed at which the noise is most pronounced. Stop the vehicle and place the transmission in Neutral and run the engine to the same speed. If the noise is the same, the axle is not at fault.*

5   Because of the special tools needed, overhauling the differential isn't cost effective for a do-it-yourselfer. The procedures included in this Chapter describe axleshaft removal and installation, axleshaft oil seal replacement, axleshaft bearing replacement and removal of the entire unit for repair or

replacement. Any further work should be left to a dealer service department or other qualified repair facility.

Identification

Refer to illustration 5.6

6   If the rear axle must be replaced, refer to the identification code and manufacturer's code stamped on the front of the right axle tube **(see illustration)**. This number contains information on the rear axle ratio, differential type, manufacturer and build date information, all of which are necessary to ensure that you get the right axle.

6   Rear axleshaft - removal and installation

Refer to illustrations 6.3 and 6.4

1   Raise the rear of the vehicle, support it securely and remove the wheel and brake drum or caliper and disc (refer to Chapter 9).

2   Remove the cover from the differential carrier and allow the oil to drain into a container.

3   Remove the lock bolt from the differential pinion shaft. Remove the pinion shaft **(see illustration)**.

6.3 Remove the lock bolt and pull out the differential pinion gear shaft (arrow)

6.4 Remove the C-lock from the axleshaft

7.2 You can pry out the old seal with the axleshaft

7.3 Install the new axleshaft seal with a seal driver

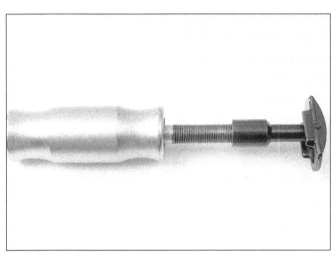

8.3a A typical slide hammer and axleshaft bearing removal accessory

8.3b Use a slide hammer and bearing removal attachment to remove the old axleshaft bearing

4   Push the outer (flanged) end of the axleshaft in and remove the C-lock from the inner end of the shaft **(see illustration)**.
5   Withdraw the axleshaft, taking care not to damage the oil seal in the end of the axle

8.4 Install the new axleshaft bearing with a bearing driver, large socket or section of pipe

housing as the splined end of the axleshaft passes through it.
6   Installation is the reverse of removal. Tighten the lock bolt to the torque listed in this Chapter's Specifications.
7   Always use a new cover gasket and tighten the cover bolts to the torque listed in the Chapter 1 Specifications.
8   Refill the axle with the correct quantity and grade of lubricant (see Chapter 1).

7   Rear axleshaft oil seal - replacement

Refer to illustrations 7.2 and 7.3
1   Remove the axleshaft (see Section 6).
2   Pry the oil seal out of the end of the axle housing with a large screwdriver or the inner end of the axleshaft **(see illustration)**.
3   Apply multi-purpose grease to the oil seal lips and recess, then tap the new seal evenly into place with a hammer and seal installation tool **(see illustration)**, large socket or piece of pipe so the lips are facing in and the metal face is visible from the end

of the axle housing. When correctly installed, the face of the oil seal should be flush with the end of the axle housing.

8   Rear axleshaft bearing - replacement

Refer to illustrations 8.3a, 8.3b and 8.4
1   Remove the axleshaft (see Section 6) and the oil seal (see Section 7).
2   A bearing puller which grips the bearing from behind will be required for this job.
3   Attach a slide hammer to the puller and extract the bearing from the axle housing **(see illustrations)**.
4   Clean out the bearing recess and drive in the new bearing with a bearing installer tool, a big socket or a piece of pipe positioned against the outer bearing race **(see illustration)**. Make sure the bearing is driven in to the full depth of the recess and the numbers on the bearing are visible from the outer end of the axle housing.
5   Install a new oil seal (see Section 7), then install the axleshaft (Section 6).

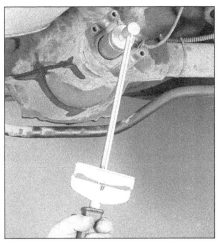

9.3 Use an inch-pound torque wrench to check the torque necessary to rotate the pinion shaft

9.4 Mark the relative positions of the pinion nut and flange before removing the nut

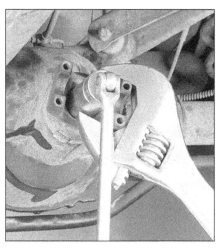

9.6 Hold the pinion flange with a large wrench while removing the pinion nut

9.9 Use a seal removal tool (shown) or a large screwdriver to pry out the old pinion seal - have a drain pan ready to catch the fluid

9.10 Lubricate the lips of the new pinion seal and seat it squarely in the bore, then drive it into the carrier with a seal driver, a large socket (shown) or a block of wood

13   Install the washer (if equipped) and pinion nut. Tighten the nut carefully until the original number of threads are exposed.

14   Measure the torque required to rotate the pinion and tighten the nut in small increments until it matches the figure recorded in Step 4. **Note:** *Rotate the pinion several times after each tightening.* In order to compensate for the drag of the new oil seal, the nut should be tightened more until the rotational torque of the pinion exceeds the earlier recording by 3 to 5 in-lbs. Be very careful and do not over-tighten the pinion nut and do not back-off the pinion nut after it has been tightened. **Caution:** *If the pinion nut is over-tightened, the differential must be disassembled and a new crush sleeve installed.*

15   Connect the driveshaft and lower the vehicle.

10   Rear axle assembly - removal and installation

Refer to illustrations 10.7a, 10.7b, 10.7c and 10.8

1   Loosen the rear wheel lug nuts, raise the rear of the vehicle and support it securely on jackstands placed under the frame rails. Block the front wheels and remove the rear wheels.

2   Mark the driveshaft and companion flange, unbolt the driveshaft and support it out of the way on a wire hanger (see Section 3).

3   Position a floor jack under the differential housing. Raise the jack just enough to take up the weight, but not far enough to take the weight of the vehicle off the jackstands.

4   Remove the rear brake assemblies (see Chapter 9).

5   Remove the cover from the differential and drain the lubricant into a container (see Chapter 1).

6   Remove the axleshafts (see Section 6).

9   Pinion oil seal - replacement

Refer to illustrations 9.3, 9.4, 9.6, 9.9 and 9.10

1   Raise the rear of the vehicle and support it securely on jackstands. Block the front wheels to keep the vehicle from rolling off the stands.

2   Disconnect the driveshaft and fasten it out of the way.

3   Use an inch-pound torque wrench to check the torque required to rotate the pinion. Record the torque value for use later **(see illustration)**.

4   Scribe or punch alignment marks on the pinion shaft, nut and flange **(see illustration)**.

5   Count the number of threads visible between the end of the nut and the end of the pinion shaft and record the number for use later.

6   A special tool or a large wrench can be used to hold the companion flange while loosening the pinion nut **(see illustration)**.

7   Remove the pinion nut.

8   Using a small puller, pull off the companion flange. Do not attempt to pry behind the flange or hammer on the flange or the end of the pinion shaft.

9   Pry out the old seal **(see illustration)**.

10   Lubricate the lips of the new seal with multi-purpose grease and tap it evenly into position with a seal installation tool or a large socket. Make sure it enters the housing squarely and is tapped in to its full depth **(see illustration)**.

11   Align the mating marks made before disassembly and install the companion flange. If necessary, tighten the pinion nut to draw the flange into place. Do not try to hammer the flange into position.

12   Apply non-hardening sealant to the ends of the splines visible in the center of the flange so oil will be sealed in.

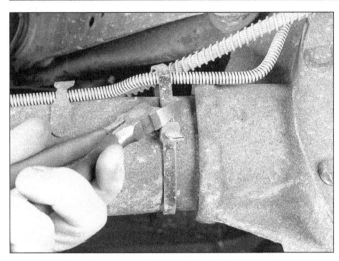

10.7a To detach the metal brake lines from the rear axle,
cut these bands on the axle tube . . .

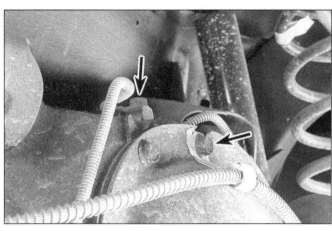

10.7b . . . remove this bolt (upper arrow) on top of the differential
housing to detach the brake line junction block from the axle,
remove this bolt (lower arrow) from the differential cover
to detach the parking brake cable from the axle

7 Detach the brake lines from the axle housing and remove the brake line junction block bolt on top of the differential housing **(see illustrations)**. On models with ABS, unbolt the ABS sensor and the bracket for the sensor lead **(see illustration)**.
8 Remove the brake backing plates **(see illustration)**.

9 Disconnect both shock absorbers (see Chapter 10).
10 Lower the jack under the differential housing enough to remove the coil springs.
11 Disconnect the lower and upper control arms from the axle housing (see Chapter 10).
12 Remove the rear axle assembly from under the vehicle.

13 Installation is the reverse of the removal procedure. Be sure to tighten all fasteners to the torque listed in the Chapter 9 and 10 Specifications.
14 When installation is complete, fill the differential with the recommended oil and bleed the brake system.

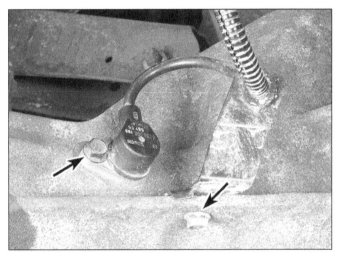

10.7c To detach the ABS wheel speed sensor from the differential
housing, remove this bolt (left arrow) and to detach the sensor
lead bracket from the housing, remove this bolt (right arrow)

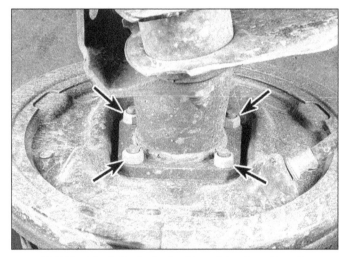

10.8 To detach the brake backing plates from the axle
tube flange, remove these four nuts (arrows)

Chapter 9 Brakes

Contents

Specifications

Disc brakes

Disc thickness after resurfacing (minimum)	
Front	1.250 inches
Rear	0.733 inch
Discard thickness*	
Front	1.209 inches
Rear	0.724 inch
Disc runout (maximum)	
Front	0.005 inch
Rear	
Disc (installed on vehicle)	0.006 inch
Rear axle shaft flange runout	0.002 inch
Disc thickness variation (maximum)	0.0005 inch

* Refer to the marks stamped on the disc (they supersede information printed here)

Rear drum brakes

Drum diameter	
Standard	9.50 inches
Service limit*	9.59 inches
Drum taper (maximum)	0.003 inch
Out-of-round (maximum)	0.002 inch

* Refer to marks stamped on the drum (they supersede information printed here)

Torque specifications

	Ft-lbs (unless otherwise indicated)
Caliper mounting bolts	
Front caliper guide pins	38
Rear caliper guide pins	27
Rear caliper mounting plate bolts	74
Front brake hose-to-caliper banjo bolt	32
Rear brake hose-to-caliper banjo bolt	
1993 and 1994	22
1995 on	33
Rear wheel cylinder bolts	
1993 and 1994	115 in-lbs
1995 on	156 in-lbs

1   General information

All vehicles covered by this manual are equipped with hydraulically operated front and rear brake systems. All front brake systems are disc type; rear brake systems are drum brakes on some models and disc type on others. A quick visual inspection will tell you what type of brakes you have at the rear of your vehicle. All brakes are self-adjusting.

The hydraulic system consists of two separate front and rear circuits. The master cylinder has separate reservoirs for the two circuits and in the event of a leak or failure in one hydraulic circuit, the other circuit will remain operative. A visual warning of circuit failure or air in the system is given by a warning light activated by displacement of the piston in the brake distribution (pressure differential warning) switch from its normal "in balance" position.

The parking brake system is mechanically operated and cable-actuated. It's applied by a foot-operated pedal adjacent to the left kick panel and released by a pull-handle under the left end of the dash. When the parking brake pedal is applied on models with rear drum brakes, each rear cable pulls on a parking brake lever attached to the rear shoe of each brake assembly. When the parking brake pedal is applied on models with rear discs, each rear cable activates a lever that expands a horseshoe-shaped one-piece rear parking brake shoe against the inside of the rear disc.

A combination valve, located in the engine compartment, consists of three sections providing the following functions: The metering section limits pressure to the front brakes until a predetermined front input pressure is reached and until the rear brakes are activated. There is no restriction at inlet pressures below 3 psi, allowing pressure equalization during non-braking periods. The proportioning section proportions outlet pressure to the rear brakes after a predetermined rear input pressure has been reached, preventing early rear wheel lock-up under heavy brake loads. The valve is also designed to assure full pressure to the rear brakes should the front brakes fail, and vice-versa.

The pressure differential warning switch is designed to continuously compare the front and rear brake pressure from the master cylinder and energize the dash warning light in the event of either front or rear brake system failure. The design of the switch and valve are such that the switch will stay in the "warning" position once a failure has occurred. The only way to turn the light off is to repair the cause of the failure and apply a brake pedal force of 450 psi.

The power brake booster, utilizing engine manifold vacuum and atmospheric pressure to provide assistance to the hydraulically operated brakes, is located in the engine compartment, mounted on the driver's side of the firewall.

After completing any operation involving the disassembly of any part of the brake system, always test drive the vehicle to check for proper braking performance before resuming normal driving. When testing the brakes, perform the tests on a clean, dry, flat surface. Conditions other than these can lead to inaccurate test results. Test the brakes at various speeds with both light and heavy pedal pressure. The vehicle should stop evenly without pulling to one side or the other. Avoid locking the brakes because this slides the tires and diminishes braking efficiency and control.

Tires, vehicle load and front-end alignment are factors which also affect braking performance.

Torque values given in the Specifications Section are for dry, unlubricated fasteners.

2   Anti-lock Brake System (ABS) - general information

A four-wheel Anti-lock Brake System (ABS) maintains vehicle maneuverability, directional stability, and optimum deceleration under severe braking conditions on most road surfaces. It does so by monitoring the rotational speed of the wheels and controlling the brake line pressure to the wheels during braking. This prevents the wheels from locking up on slippery roads or during hard braking.

Hydraulic modulator/motor pack assembly

The hydraulic modulator/motor pack assembly, mounted in the left front corner of the engine compartment, controls hydraulic pressure to the front calipers and rear wheel cylinders or calipers by modulating hydraulic pressure to prevent wheel lock-up.

The Electronic Brake Control Module (EBCM) is located above the left (driver's side) kick panel. The EBCM monitors the ABS system and controls the anti-lock valve solenoids. It accepts and processes information received from the brake switch and wheel speed sensors to control the hydraulic line pressure and avoid wheel lock up. It also monitors the system and stores fault codes which indicate specific problems.

Each sensor assembly consists of a variable reluctance sensor mounted adjacent to a "toothed ring" with an air gap between them. A wheel speed sensor and toothed ring are mounted in the hub/bearing unit of each front wheel. A third sensor is mounted on the rear differential housing; its toothed ring is located within the rear differential. The air gap between the sensors and the rings is not adjustable.

A wheel speed sensor measures wheel speed by monitoring the rotation of the toothed ring. As the teeth of the ring move through the magnetic field of the sensor, an AC voltage is generated. This signal frequency increases or decreases in proportion to the speed of the wheel. The EBCM monitors these three signals for changes in wheel

speed; if it detects the sudden deceleration of a wheel, i.e. wheel lockup, the EBCM activates the ABS system.

The ABS system has self-diagnostic capabilities. Each time the vehicle is started, the EBCM runs a self-test. The red BRAKE warning light should come on briefly then go out. The EBCM also monitors the ABS system continuously during vehicle operation. If the ABS INOP light on the dash comes on while you're driving, there is a fault somewhere in the ABS system and the ABS system may be inoperative, but the brakes should still function in their non-ABS mode. Take the vehicle to a dealer service department or other qualified repair shop immediately and have the ABS serviced.

If the ABS INOP light *flashes*, however, there is a more serious fault in the ABS system which may have affected the regular braking system. Pull over immediately and have the vehicle towed to a dealer service department or other qualified repair shop for service.

Although a special electronic tester is necessary to properly diagnose the system, the home mechanic can perform a few preliminary checks before taking the vehicle to a dealer service department which is equipped with this tester.

a) *Make sure the brake calipers are in good condition.*
b) *Check the electrical connector at the controller.*
c) *Check the fuses.*
d) *Follow the wiring harness to the speed sensors and brake light switch and make sure all connections are secure and the wiring isn't damaged.*

If the above preliminary checks don't rectify the problem, the vehicle should be diagnosed by a dealer service department.

3   Disc brake pads - replacement

Refer to illustrations 3.3, 3.4a through 3.4t and 3.4u through 3.4cc

Warning: *Disc brake pads must be replaced on both front wheels at the same time - never replace the pads on only one wheel. Also, the dust created by the brake system may contain asbestos, which is harmful to your health. Never blow it out with compressed air and don't inhale any of it. An approved filtering mask should be worn when working on the brakes. Do not, under any circumstances, use petroleum-based solvents to clean brake parts. Use brake system cleaner only!*

1   Loosen the front wheel lug nuts, raise the front of the vehicle and support it securely on jackstands. Apply the parking brake. Remove the front wheels.

2   Remove about two-thirds of the fluid from the master cylinder reservoir and discard it. **Caution:** *Do not spill brake fluid on painted surfaces.* Position a drain pan under the brake assembly and clean the caliper and surrounding area with brake system cleaner.

3.3 Depress the piston into the caliper with a large C-clamp

3   Position a large C-clamp over the caliper and squeeze the piston back into in its bore **(see illustration)** to provide room for the new brake pads. As the piston is depressed to the bottom of its caliper bore, the fluid in the master cylinder will rise. Make sure it doesn't overflow. If necessary, siphon off some of the fluid.

4   Work on one brake assembly at a time. If you're replacing the front brake pads, follow the accompanying photos, **beginning with illustration 3.4a; if you're replacing the rear pads, start at illustration 3.4u.** Be sure to stay in order and read the caption under each illustration.

5   While the pads are removed, inspect the caliper for seal leaks (evidenced by moisture around the cavity) and for any damage to the piston dust boots. If excessive moisture is evident, rebuild the caliper (see Section 4). When you're done, make sure you tighten the caliper bolts to the torque listed in this Chapter's Specifications.

6   Install the brake pads on the opposite wheel, then install the wheels and lower the vehicle. Tighten the lug nuts to the torque listed in the Chapter 1 Specifications. Add brake fluid to the reservoir until it's full (see Chapter 1).

Front brake pads

3.4a Before getting started, wash the front brake assembly with brake cleaner to protect yourself from brake dust, which might contain asbestos, a known carcinogen

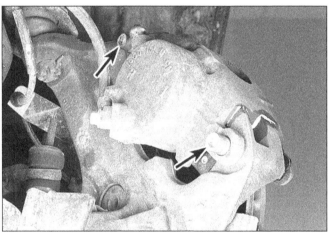

3.4b To detach the front caliper from the steering knuckle, remove these two bolts (arrows); don't remove the banjo bolt for the brake hose unless you intend to overhaul the caliper

3.4c Lift off the caliper . . .

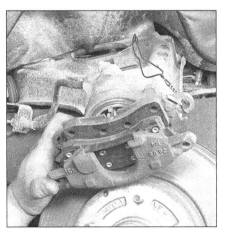

3.4d . . . and suspend it with a piece of wire or a coat hanger to protect the brake hose; NEVER hang the caliper by the brake hose!

3.4e Remove the outer brake pad

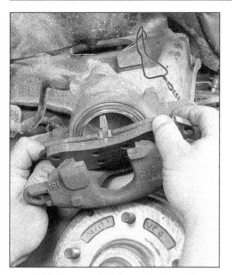

3.4f Remove the inner brake pad

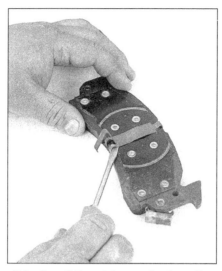

3.4g Pry off the retainer spring from the old inner brake pad . . .

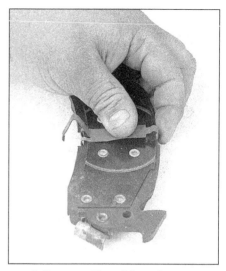

3.4h . . . and install it on the new inner pad

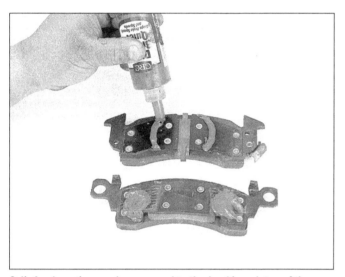

3.4i Apply anti-squeal compound to the backing plates of the new brake pads

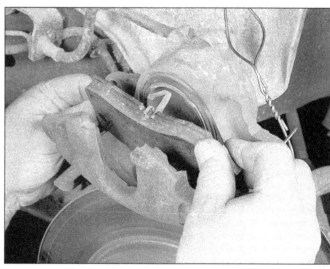

3.4j Install the new inner pad

3.4k Make sure the inner pad retainer spring is fully seated as shown

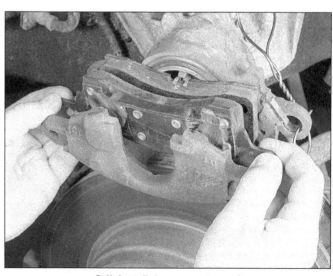

3.4l Install the new outer pad

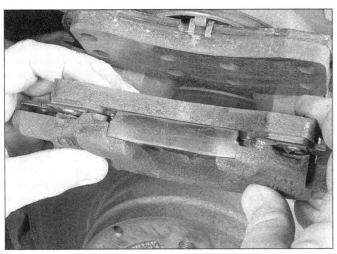

3.4m Make sure this lip on the outer pad is fully seated
as shown . . .

3.4n . . . and these tangs are hooked over the caliper
ears as shown

3.4o Lubricate the lower ways . . .

3.4p . . . and the upper ways of the
steering knuckle with
high-temperature grease

3.4q Install the caliper onto the knuckle

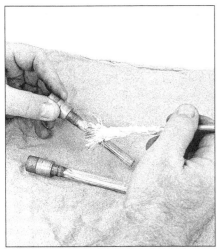

3.4r Clean off the caliper bolts and
lubricate them with
high-temperature grease

3.4s Install the caliper bolts and tighten
them to the torque listed in this
Chapter's Specifications

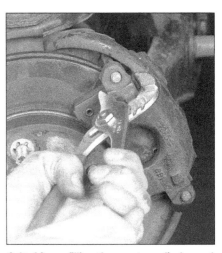

3.4t After refilling the master cylinder and
pumping the brake pedal to seat the pads,
use a pair of large adjustable (water
pump) pliers to bend the upper ears of the
outer pad until the ears are flush with the
caliper housing, with no clearance

Rear brake pads

3.4u Before beginning, wash the disc and caliper assembly with brake cleaner to protect yourself from brake dust, which might contain asbestos

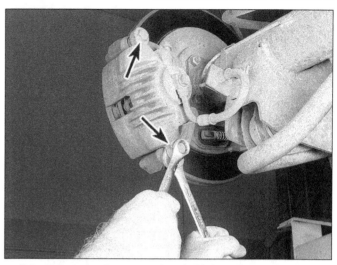

3.4v Using a back-up wrench to hold the caliper guide pins, remove the caliper bolts (arrows)

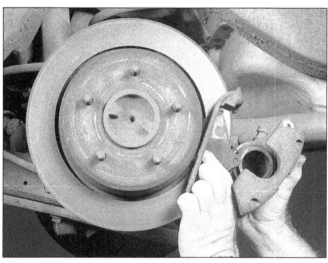

3.4w Remove the caliper assembly . . .

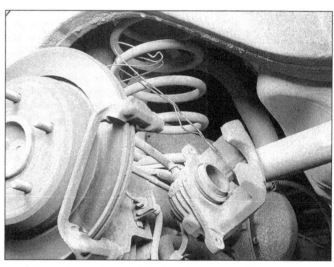

3.4x . . . and hang it out of the way with a coat hanger or a piece of wire

3.4y Remove the outer brake pad

3.4z Remove the inner pad

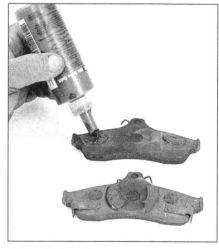

3.4aa Apply anti-squeal compound to the backing plates of the new pads

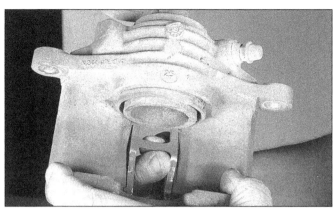

3.4bb Make sure the anti-rattle clip is in good shape; if it's damaged, replace it

3.4cc Install the pads and the caliper, then tighten the caliper bolts to the torque listed in this Chapter's Specifications

7   Pump the brakes several times to seat the pads against the disc, then check the fluid level again.

8   Check the operation of the brakes before driving the vehicle in traffic. Try to avoid heavy brake applications until the brakes have been applied lightly several times to seat the pads.

4   Disc brake caliper - removal, overhaul and installation

Warning: *The dust created by the brake system may contain asbestos, which is harmful to your health. Never blow it out with compressed air and don't inhale any of it. An approved filtering mask should be worn when working on the brakes. Do not, under any circumstances, use petroleum-based solvents to clean brake parts. Use brake system cleaner only!*

Note: *If an overhaul is indicated (usually because of fluid leaks, a stuck piston or broken bleeder screw) explore all options before beginning this procedure. New and factory rebuilt calipers are available on an exchange basis, which makes this job quite easy. If you*

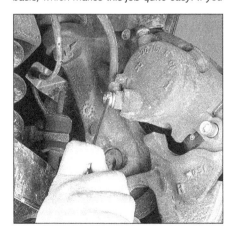

4.5 To detach the brake flex hose from the caliper, remove this banjo bolt; discard the old sealing washers and use new ones when reattaching the brake hose

decide to rebuild the calipers, make sure rebuild kits are available before proceeding. Always rebuild or replace the calipers in pairs - never rebuild just one of them.

Removal

Refer to illustration 4.5

1   Remove the cover from the brake fluid reservoir and siphon off two-thirds of the fluid into a container and discard it.

2   Loosen the wheel lug nuts, raise the front or rear of the vehicle and place it securely on jackstands. Remove the front or rear wheel.

3   If you're removing a rear caliper, reinstall one wheel lug nut to hold the disc in place. Don't remove both calipers at the same time.

4   Push the caliper piston back into its bore with a C-clamp **(see illustration 3.3)**. As the piston is depressed to the bottom of the caliper bore, the fluid in the master cylinder will rise. Make sure that it does not overflow.

5   Remove the banjo fitting bolt holding the brake hose **(see illustration)**, then remove and discard the sealing washers found on either side of the banjo fitting. Always use new sealing washers when reinstalling the brake hose.

6   To prevent brake fluid leakage and contamination, plug the openings in the caliper and brake hose. **Note:** *If you're just removing*

4.10 Using a wood block as a cushion, ease the piston out of the caliper bore with compressed air

the caliper for access to other components, don't disconnect the hose.

7   Remove the caliper (front caliper, **see illustration 3.4a**; rear caliper, **see illustration 3.4i**) and separate the caliper from the disc.

8   Remove the brake pads from the caliper (see Section 3).

Overhaul

Refer to illustrations 4.10, 4.11, 4.12, 4.16, 4.17, 4.18 and 4.19

Note: *Purchase a brake caliper overhaul kit for your particular vehicle before beginning this procedure.*

9   Clean the exterior of the brake caliper with brake system cleaner (never use gasoline, kerosene or any petroleum-based solvents), then place the caliper on a clean workbench.

10   Place a wooden block or shop rag in the caliper as a cushion, then use compressed air to remove the piston from the caliper **(see illustration)**. Use only enough air pressure to ease the piston out of the bore. If the piston is blown out, even with the cushion in place, it may be damaged. **Warning:** *Never place your fingers in front of the piston in an attempt to catch or protect it when applying compressed air - serious injury could occur.*

11   Carefully pry the dust boot out of the caliper bore **(see illustration)**.

4.11 Pry the dust boot out of the caliper bore with a screwdriver - make sure you don't nick or gouge anything

4.12 To remove the seal from the caliper bore, use a plastic or wooden tool, such as a pencil

4.16 Lubricate the piston bore and seal with clean brake fluid before placing the seal in the caliper bore groove

4.17 Install a new dust boot in the piston groove - note that the folds (the accordion-like pleats) are facing out

12   Using a wooden or plastic tool, remove the piston seal from the groove in the caliper bore **(see illustration)**. Metal tools may cause bore damage.

13   Remove the caliper bleeder valve, then remove and discard the sleeves and bushings from the caliper ears. Also discard all rubber parts.

14   Clean the remaining parts with brake fluid or brake system cleaner. Allow them to drain and then shake them vigorously to remove as much fluid as possible.

15   Carefully examine the piston for nicks and burrs and loss of plating. If surface defects are present, parts must be replaced. Check the caliper bore in a similar way, but light polishing with crocus cloth is permissible to remove light corrosion and stains. Discard the mounting bolts if they are corroded or damaged.

16   When assembling, lubricate the piston bores and seal with clean brake fluid; position the seal in the caliper bore groove. Make sure the seal seats properly and isn't twisted **(see illustration)**.

17   Lubricate the piston with clean brake fluid, then install a new boot in the piston groove with the fold toward the open end of the piston **(see illustration)**.

18   Insert the piston squarely into the caliper bore, then apply force to bottom the piston in the bore **(see illustration)**.

19   Position the dust boot in the caliper counterbore, then use a drift to drive it into position **(see illustration)**. Make sure that the boot is evenly installed below the caliper face.

20   Install the bleeder valve.

Installation

21   Installation is the reverse of removal. Always use new sealing washers when connecting the brake hose. Tighten the caliper guide pins or mounting bolts to the torque listed in this Chapter's Specifications. Be sure to fill the master cylinder and bleed the brakes (see Section 10).

4.18 Lubricate the piston with clean brake fluid, insert it squarely in the bore and carefully push it into the caliper

5   Brake disc - inspection, removal and installation

Refer to illustrations 5.5, 5.6a, 5.6b, 5.7a and 5.7b

1   Loosen the wheel lug nuts, raise the vehicle and place it securely on jackstands. Remove the wheel. If you're inspecting a rear disc, release the parking brake and block the front wheels to prevent the vehicle from rolling.

2   Remove the caliper assembly (see Section 4). It is not necessary to disconnect the brake hose. After removing the caliper mounting bolts, hang the caliper out of the way on a piece of wire. Never hang the caliper by the brake hose because damage to the hose could occur.

3   Inspect the disc surfaces. Light scoring or grooving is normal, but deep grooves or severe erosion is not. If pulsating has been noticed during application of the brakes, suspect disc runout.

4   If you're inspecting a rear disc, reinstall the wheel lug nuts - flat side toward the disc

4.19 Use a hammer and driver to seat the boot in the caliper housing counterbore

- to hold the disc in a flat, vertical plane during the following inspection. It may be necessary to install washers under the lug nuts in order for them to apply pressure to the disc.

5   Attach a dial indicator to the caliper mounting bracket, turn the disc and note the amount of runout **(see illustration)**. Check

5.5 Check the runout of the brake disc with a dial indicator

5.6a The minimum thickness is cast into the disc

5.6b Use a micrometer to check the thickness of the disc and compare this measurement to the minimum allowable thickness cast into the disc

5.7a To remove a rear caliper anchor bracket, remove these bolts (arrows) (the front caliper doesn't use an anchor bracket; it's bolted directly to the steering knuckle)

5.7b Once the anchor bracket is removed, simply pull the disc straight off

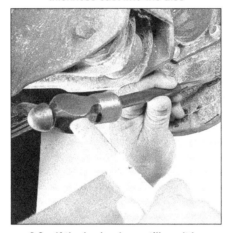

6.3a If the brake drum still can't be removed even after applying penetrating oil to the center hub area, open the adjustment slot by knocking out this cutout with a hammer and punch (buy a rubber grommet at an auto parts store to plug the hole) . . .

6.3b . . . insert a screwdriver and a brake adjusting tool (or another screwdriver) through the adjustment slot . . .

both inner and outer surfaces. If the runout is more than the specified allowable maximum, the disc must be removed from the vehicle and taken to an automotive machine shop for resurfacing.

6   The minimum thickness of the disc is cast into the hub **(see illustration)**. Using a micrometer, measure the thickness of the disc **(see illustration)** and compare it to this specification. If it is less than the specified minimum, replace the disc. Also measure the disc thickness at several points to determine thickness variations in the surface. Any variation over 0.0005-inch may cause pedal pulsations during brake application. If this condition exists and the disc thickness is not below the minimum refinish thickness, the disc can be removed and taken to an automotive machine shop for resurfacing.

7   If a front disc must be removed for repair or replacement, refer to Chapter 1, *Front wheel bearing check, repack and adjustment* for removal of the disc (it's an integral part of the hub). If a rear disc needs to be removed for repair or replacement, remove the caliper anchor bracket, then remove the disc **(see illustrations)**.

6   Drum brake shoes - replacement

Refer to illustrations 6.3a, 6.3b, 6.3c, 6.4a through 6.4z and 6.6

Warning: *Drum brake shoes must be replaced on both wheels at the same time - never replace the pads on only one wheel. Also, the dust created by the brake system may contain asbestos, which is harmful to your health. Never blow it out with compressed air and don't inhale any of it. An approved filtering mask should be worn when working on the brakes. Do not, under any circumstances, use petroleum-based solvents to clean brake parts. Use brake system cleaner only!*

Caution: *Whenever the brake shoes are replaced, the return and hold-down springs should be replaced. Due to the continuous heating/cooling cycle the springs are subjected to, they lose their tension over a period*

of time and may allow the shoes to drag on the drum and wear at a much faster rate than normal.

1   Release the parking brake.

2   Loosen the rear wheel lug nuts, raise the vehicle and place it securely on jackstands. Remove the wheel.

3   If the brake drum cannot be easily pulled off the axle and shoe assembly, make sure that the parking brake is completely released, then squirt some penetrating oil around the center hub area. Allow the oil to soak in and try to pull the drum off again. If the drum still cannot be pulled off, the brake shoes will have to be retracted. Remove the lanced cutout in the backing plate with a hammer and chisel **(see illustration)**. With this lanced area punched in, pull the lever off the adjusting screw wheel with one small screwdriver while turning the adjusting wheel with another small screwdriver, moving the shoes away from the drum **(see illustrations)**. The drum may now be pulled off. **Caution:** *Be sure to retrieve the lanced portion after removing the drum.*

6.3c ... push the actuator off the star
wheel with the screwdriver and hold it off
while turning the star wheel counterclockwise with the brake adjusting tool
(or another small screwdriver) to retract
the brake shoes (turn the star wheel
clockwise to expand the shoes)

6.4a Pry off and discard the drum
retaining washers, if present

4   Wash the brake assembly with brake
cleaner. Do NOT blow it out with compressed
air. To replace the brake shoes, refer to the
accompanying photographs, beginning with

6.4b Mark the relationship of the axle and
brake drum

illustration 6.4a.
5   All four rear shoes must be replaced at
the same time, but to avoid mixing up parts,
work on only one brake assembly at a time.

6.4c Remove the brake drum (if it can't be pulled off, see
illustrations 6.3a, 6.3b and 6.3c)

6.4d Wash the brake assembly with brake system cleaner; DO
NOT blow it out with compressed air

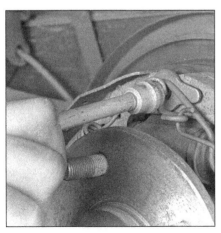

6.4e Remove the shoe return springs -
the spring tool shown here is available at
most auto parts stores and makes this job
much easier and safer

6.4f Pull the bottom of the actuator lever
toward the secondary brake shoe,
compressing the lever return spring - the
actuator link can now be removed from
the top of the lever

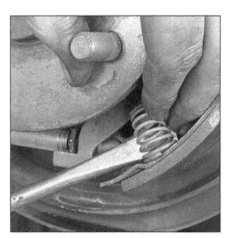

6.4g Pry the actuator lever spring out with
a large screwdriver

6.4h Slide the parking brake strut out from between the axle flange and primary shoe

6.4i Remove the hold-down springs and pins - the hold-down spring tool shown here is available at most auto parts stores

6.4j Remove the actuator lever and pivot - be careful not to let the pivot fall out of the lever

6.4k Spread the top of the shoes apart and slide the assembly around the axle

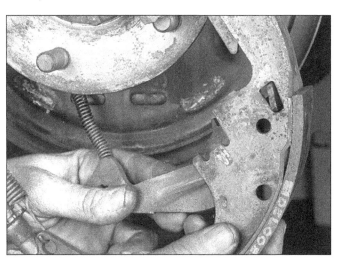

6.4l Unhook the parking brake lever from the secondary shoe

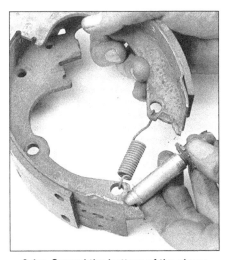

6.4m Spread the bottom of the shoes apart and remove the adjusting screw assembly

6.4n Clean the adjusting screw with solvent. Dry it off and lubricate the threads and end with multi-purpose grease, then reinstall the adjusting screw between the new brake shoes

6.4o Lubricate the shoe contact points on the backing plate with high-temperature brake grease

6.4p Insert the parking brake lever into the slot in the secondary brake shoe

6.4q Spread the shoes apart and slide them into position on the backing plate

6.4r Install the hold-down pin and spring through the backing plate and primary shoe

6.4s Insert the lever pin into the actuator lever, place the lever over the secondary shoe hold-down pin and install the hold-down spring

6.4t Guide the parking brake strut behind the axle flange and engage the rear end of it in the slot on the parking brake lever - spread the shoes enough to allow the other end to seat against the primary shoe

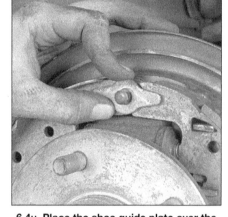

6.4u Place the shoe guide plate over the anchor pin

6   Before reinstalling the drum, it should be checked for cracks, score marks, deep scratches and "hard spots," which will appear as small discolored areas. If the hard spots cannot be removed with fine emery cloth and/or if any of the other conditions listed above exist, the drum must be taken to an automotive machine shop to have it resurfaced. If the drum will not "clean up" before the maximum drum diameter is reached in the machining operation, the drum will have to be replaced with a new one. **Note:** *The maximum diameter is cast into each brake drum* **(see illustration)**.

7   Install the brake drum, lining up the marks made before removal if the old brake

6.4v Hook the lower end of the actuator link to the actuator lever, then loop the top end over the anchor pin

6.4w Install the lever return spring over the tab on the actuator lever, then push the spring up onto the brake shoe

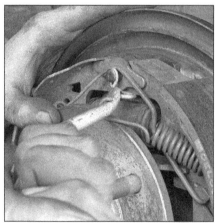

6.4x Install the primary and secondary shoe return springs

6.4y With all parts installed, rock the assembly back and forth with your hands to ensure that all parts are seated

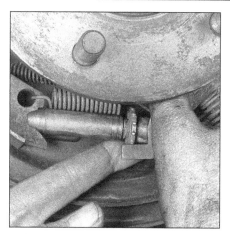

6.4z Pull out on the actuator lever to disengage it from the adjusting screw wheel, turn the wheel to adjust the shoes in or out as necessary - the brake drum should slide over the shoes and turn with a very slight amount of drag (at which time you'll back-off the adjuster until they don't drag) - (to complete the job, refer to Steps 6 through 9)

6.6 The maximum allowable diameter is cast inside the brake drum

drum is used. Turn the adjuster wheel until the brake shoes just drag on the drum when the drum is rotated, then back-off the adjuster wheel until the shoes don't drag.

8   Mount the wheel and tire, install the wheel lug nuts and tighten them to the torque listed in the Chapter 1 Specifications, then lower the vehicle.

9   Make a number of forward and reverse stops to adjust the brakes until a satisfactory pedal action is obtained.

7   Wheel cylinder - removal, overhaul and installation

Refer to illustrations 7.2 and 7.5

1   Remove the brake shoes (see Section 6).

2   Remove the brake line fitting from the rear of the wheel cylinder **(see illustration)**. Cap the brake line to prevent contamination and excessive fluid loss.

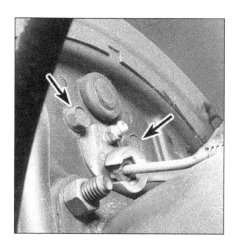

7.2 Unscrew the brake line fitting from the wheel cylinder with a flare nut wrench to prevent rounding off the corners of the nut, then remove the wheel cylinder retaining bolts (arrows)

3   Remove the wheel cylinder retaining bolts from the rear of the backing plate.

4   Remove the cylinder and place it on a clean workbench.

5   Remove the bleeder valve, seals, pistons, boots and spring assembly from the cylinder body **(see illustration)**.

6   Clean the wheel cylinder with brake fluid or brake system cleaner. Do not, under any circumstances, use petroleum-based solvents to clean brake parts.

7   Use filtered, unlubricated compressed air to remove excess fluid from the wheel cylinder and to blow out the passages.

8   Check the cylinder bore for corrosion and scoring. Crocus cloth may be used to remove light corrosion and stains, but the cylinder must be replaced with a new one if the defects cannot be removed easily, or if the bore is scored.

9   Lubricate the new seals with clean brake fluid.

10   Assemble the brake cylinder, making sure the boots are properly seated.

11   Place the wheel cylinder in position on the backing plate. Thread the brake line fitting into the cylinder, being careful not to cross-thread it. Don't tighten the fitting yet.

12   Install the wheel cylinder retaining bolts and tighten them to the torque listed in this Chapter's Specifications. Now tighten the brake line fitting securely.

13   Bleed the brake system (see Section 10).

8   Master cylinder - removal, overhaul and installation

Removal

Refer to illustrations 8.3 and 8.4

Caution: *Have some plugs ready to cap the metal lines that connect the master cylinder to the combination valve. Failure to do so will allow air into the ABS modulator and can allow dirt and moisture to enter the system.*

Note: *A master cylinder overhaul kit should be purchased before beginning this procedure. The kit will include all the replacement parts necessary for the overhaul procedure. The rubber replacement parts, particularly the seals, are the key to fluid control within the master cylinder. As such, it's very important that they*

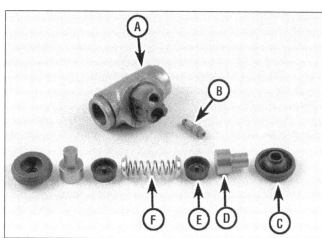

7.5 A typical wheel cylinder assembly

A   *Wheel cylinder body*
B   *Bleeder screw*
C   *Boot*
D   *Piston*
E   *Seal*
F   *Spring*

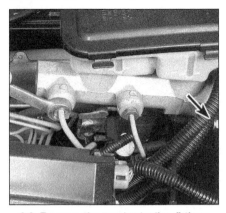

8.3 Remove the two brake line fittings from the master cylinder with a flare-nut wrench; to detach the master cylinder from the power brake booster, remove the left nut (arrow) . . .

8.4 . . . and the right nut

8.5a Pry off the diaphragm from the reservoir cover and inspect it carefully for cracks and tears; if it's damaged, replace it

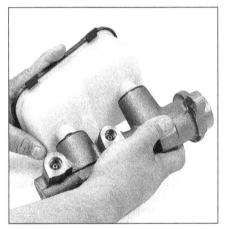

8.5b To separate the reservoir from the master cylinder body, pull it straight out; if the two components are hard to pull apart, use a screwdriver to pry them apart

8.5c Pry out the old grommets

8.6 To remove the lock ring, depress the primary piston assembly and pry out the lock ring

be installed securely and facing in the proper direction. Be careful during the rebuild procedure that no grease or mineral-based solvents come in contact with the rubber parts.

1   Completely cover the front fender and cowling area of the vehicle; brake fluid can ruin painted surfaces if it is spilled.

2   Remove the brake fluid from the master cylinder reservoir and discard it. **Caution:** *Do*

not spill brake fluid on painted surfaces. Place a drain pan under the master cylinder assembly and clean the area around the brake line fittings with brake system cleaner.

3   Disconnect the brake line fittings **(see illustration)**. Rags or newspapers should be placed under the master cylinder to soak up the fluid that will drain out.

4   Remove the two master cylinder mounting nuts **(see illustration)**, and remove the master cylinder from the vehicle. Do not to bend the hydraulic lines running to the combination valve. Plug the ends of both of these lines immediately to prevent air from entering the ABS modulator and to protect the system from moisture and dirt.

Overhaul

Refer to illustrations 8.5a, 8.5b, 8.5c, 8.6, 8.7, 8.8a, 8.8b, 8.9, 8.10, 8.11a, 8.11b, 8.11c, 8.11d, 8.12, 8.13, 8.14a, 8.14b, 8.14c, 8.15, 8.16a and 8.16b

5   Remove the reservoir cover and diaphragm, then discard any remaining fluid in the reservoir. Separate the diaphragm from the reservoir cover **(see illustration)** and inspect it carefully for rips, tears and cracks. Replace the diaphragm if it's damaged.

Remove the reservoir from the master cylinder body and the old grommets between the reservoir and the master cylinder **(see illustrations)**.

6   Remove the primary piston lock ring by depressing the piston and prying out the ring with a screwdriver **(see illustration)**.

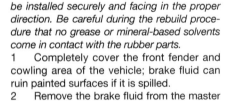

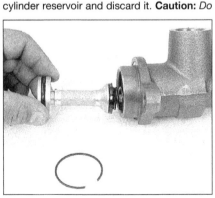

8.7 Pull out the primary piston assembly

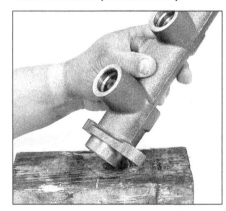

8.8a Tap the master cylinder body on a block of wood to dislodge the secondary piston assembly . . .

8.8b ... then remove the secondary piston assembly from the master cylinder

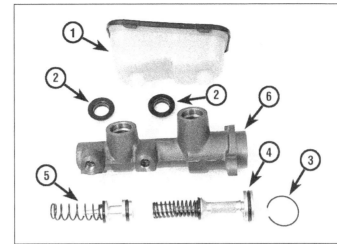

8.9 After washing the master cylinder parts, lay them out for inspection

1   Reservoir assembly (cover, diaphragm and reservoir body)
2   Grommets
3   Lock ring
4   Primary piston and spring
5   Secondary piston and spring
6   Master cylinder body

8.10 Install the secondary seals with the lips facing out

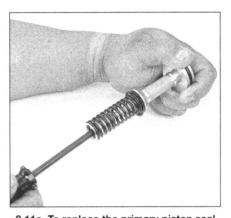

8.11a To replace the primary piston seal (on the spring end of the piston), remove the spring retainer screw and remove the spring

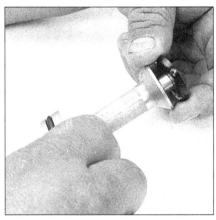

8.11b Remove the seal ...

8.11c ... make sure the primary seal is installed with the lip facing away from the piston ...

8.11d ... install the spring and retainer and install the retainer screw

8.12 This is how the primary piston (upper) and secondary piston (lower) assemblies should look before they're installed

7   Remove the primary piston assembly **(see illustration)**.

8   Remove the secondary piston assembly **(see illustrations)**.

9   Wash all parts with brake system cleaner or denatured alcohol, blow them dry with filtered, unlubricated compressed and lay them out for inspection **(see illustra-**

tion). Inspect the cylinder bore for corrosion and damage. If any corrosion or damage is found, replace the master cylinder body with a new one, as abrasives cannot be used on the bore.

10   Remove the old seals from the secondary piston assembly and install the new seals so that the cups face out **(see illustration)**.

11   Disassemble the primary piston assembly, noting the position of the parts, then remove the old O-ring and seal from the piston, install the new O-ring and seal and reassemble the piston and spring assembly **(see illustrations)**.

12   Make sure the new seals are oriented as shown **(see illustration)**.

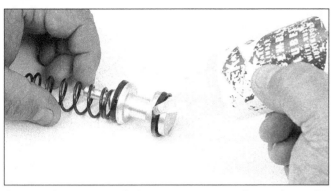

8.13 Lubricate the new O-ring and seal on the primary piston with brake assembly lube or clean brake fluid; lubricate the seals on the secondary piston as well

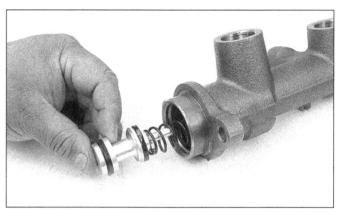

8.14a Install the secondary piston . . .

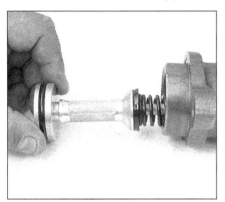

8.14b . . . the primary piston . . .

8.14c . . . push down on the primary piston assembly and install the lock ring

8.15 Lubricate the new grommets with brake assembly lube or clean brake fluid and push them into place

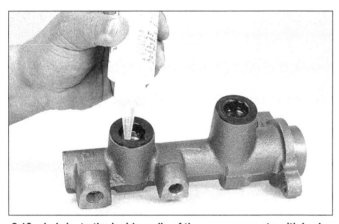

8.16a Lubricate the inside walls of the new grommets with brake assembly lube or clean brake fluid . . .

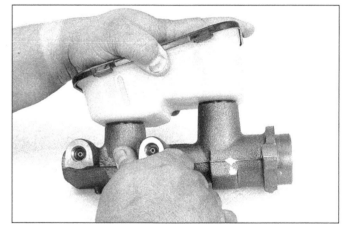

8.16b . . . and install the reservoir

13   Lubricate the new O-ring and seals with brake assembly lube or clean brake fluid **(see illustration)**.

14   Install the secondary and primary piston assemblies in the cylinder bore, depress them and install the lock ring **(see illustrations)**.

15   Lubricate the new grommets with brake assembly lube or clean brake fluid and push them into place **(see illustration)**.

16   Lubricate the insides of the new grommets with brake assembly lube and install the reservoir on the master cylinder body **(see illustrations)**.

17   Make sure the diaphragm is attached to the underside of the reservoir cover.

18   Anytime the master cylinder is removed, the brake hydraulic system must be bled. It's much easier to bleed the rest of the system quickly and effectively if you "bench bleed" the master cylinder before installing it on the vehicle. Bench bleed the master cylinder as follows.

19   Insert threaded plugs of the correct size into the cylinder outlet holes and fill the reservoirs with brake fluid (the master cylinder should be supported in such a manner that brake fluid will not spill out of it during the bench bleeding procedure).

20   Loosen one plug at a time and push the piston assembly into the bore to force air from the master cylinder. To prevent air from being drawn back into the cylinder, the appropriate plug must be tightened before allowing the piston to return to its original position.

21   Stroke the piston three or four times for each outlet to assure that all air has been expelled.

22   Refill the master cylinder reservoirs and install the diaphragm and cap assembly. **Note:** *The reservoir should only be filled to the top of the reservoir divider to prevent*

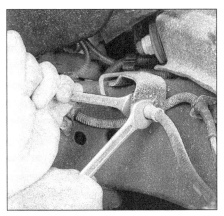

9.2a Using a back-up wrench, unscrew the brake line tube nut from the hose fitting with a flare nut wrench, being careful not to bend the frame bracket or twist the metal brake line

9.2b The rear brake line/brake hose connection, which is located at a bracket on the rear of the axle tube, is basically the same setup as the front and is disconnected the same way

9.3 Use pliers to remove the U-clip from the female fitting at the bracket, then remove the hose from the bracket (front brake hose/line connection shown, rear setup similar)

overflowing when the cover and diaphragm are installed.

Installation

23   The remainder of installation is the reverse of removal. Make sure you bleed the rest of the system when you're done (see Section 10).

---

9   Brake hoses and lines - check and replacement

Inspection

1   About every six months, loosen the wheel lug nuts, raise the vehicle, place it securely on jackstands, remove the wheels, and inspect the flexible hoses which connect the steel brake lines with the front and rear brake assemblies. Look for cracks, chafing, leaks, blisters and any other damage. These are important and vulnerable parts of the brake system, so your inspection should be thorough. You'll need a flashlight and mirror to do the job right. If a hose exhibits any of the above conditions, replace it as follows.

Brake hoses

Front caliper hose (and rear caliper hose on models with rear disc brakes)

Refer to illustrations 9.2a, 9.2b and 9.3

2   Using a back-up wrench, disconnect the brake line from the hose fitting by unscrewing the tube nut with a flare nut wrench (if you have one), being careful not to bend the frame bracket or brake line **(see illustration)**. On front brake hoses, this bracket is located on the frame, behind the upper control arm; on rear hoses, on rear brake hoses, the bracket is located on the back of the axle tube **(see illustration)**.

3   Use pliers to remove the U-clip from the female fitting at the bracket **(see illustration)**, then remove the hose from the bracket.

9.12 Use a flare nut wrench to unscrew the tube nuts, then disconnect the two metal lines from the rear hose junction block on top of the differential; after the metal lines are disconnected, remove the bolt from the top of the junction block and detach the block from the differential housing

4   At the caliper end of the hose, remove the banjo bolt **(see illustration 4.5)**, then disconnect the hose, remove the sealing washers and discard them (always use new sealing washers when reattaching the brake hose to the caliper).

5   When installing the hose, always use new sealing washers on either side of the fitting and lubricate all bolt threads with clean brake fluid before installing them.

6   Attach the hose to the caliper and tighten the banjo bolt to the torque listed in this Chapter's Specifications.

7   Being careful not to twist the new hose, insert the female hose fitting through the bracket on the frame or the axle.

8   Install the U-clip retaining the female fitting to the frame or axle bracket.

9   Using a back-up wrench, attach the brake line to the hose fitting and tighten it securely.

10   When the brake hose installation is com-

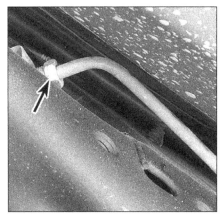

9.13 The forward end of the rear brake hose is attached to the metal line (arrow) from the combination valve at a bracket on top of the rear crossmember; disconnect it the same way you would disconnect a front hose from a metal line

plete, there should be no kinks in the hose. Also make sure that the hose does not contact any part of the suspension. If you're replacing a front brake hose, verify this by turning the wheels to the extreme left and right positions. If the hose contacts anything, disconnect it and correct the installation as necessary.

11   Fill the master cylinder reservoir and bleed the system (see Section 10).

Rear hose between crossmember and junction block on rear axle

Refer to illustrations 9.12 and 9.13

12   Locate the junction block at the rear end of the brake hose, on top of the differential housing. Disconnect the metal brake lines with a flare nut wrench **(see illustration)**, then unbolt the junction block from the differential (it's attached by one bolt running through the top of the block).

13   The forward end of the hose is attached to the metal line running under the pan at a bracket on top of the crossmember **(see illustration)**. Unscrew the tube nut with a

flare nut wrench, and remove the U-clip retaining the hose to the bracket (same setup as a front hose).

14   Being careful not to twist the new hose, insert the female hose fitting through the bracket on top of the crossmember.

15   Install the U-clip retaining the female fitting to the bracket.

16   Using a back-up wrench, attach the brake line to the hose fitting and tighten it securely.

17   Reattach the junction block to the top of the differential and tighten the bolt securely. Reconnect the two metal lines to the junction block and tighten the tube nuts securely.

18   When the brake hose installation is complete, there should be no kinks in the hose. Also make sure that the hose does not contact any part of the suspension. If you're replacing a front brake hose, verify this by turning the wheels to the extreme left and right positions. If the hose contacts anything, disconnect it and correct the installation as necessary.

19   Fill the master cylinder reservoir and bleed the system (see Section 10).

Steel brake lines

20   When it becomes necessary to replace steel lines, use only double-walled steel tubing. Never substitute copper tubing because copper is subject to fatigue cracking and corrosion. The outside diameter of the tubing is used for sizing.

21   Auto parts stores and brake supply houses carry various lengths of prefabricated brake line. These sections can be bent with a tubing bender.

22   When installing the brake line, leave at least 3/4-inch clearance between the line and any moving parts.

10   Brake hydraulic system - bleeding

Refer to illustration 10.8

Warning: *Wear eye protection when bleeding the brake system. If the fluid comes in contact with your eyes, immediately rinse them with water and seek medical attention.*

Note: *Bleeding the hydraulic system is necessary to remove any air that manages to find its way into the system when it's been opened during removal and installation of a hose, line, caliper or master cylinder.*

1   It will probably be necessary to bleed the system at all four brakes if air has entered the system due to low fluid level, or if the brake lines have been disconnected at the master cylinder.

2   If a brake line was disconnected only at a wheel, then only that caliper or wheel cylinder must be bled.

3   If a brake line is disconnected at a fitting located between the master cylinder and any of the brakes, that part of the system served by the disconnected line must be bled.

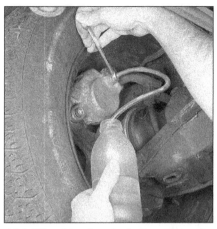

10.8 When bleeding the brakes, a hose is connected to the bleeder screw and then submerged in brake fluid - air will be seen as bubbles in the tube and container (all air must be expelled before moving to the next brake or component)

4   Remove any residual vacuum from the brake power booster by applying the brake several times with the engine off.

5   Remove the master cylinder reservoir cover and fill the reservoir with brake fluid. Reinstall the cover. **Note:** *Check the fluid level often during the bleeding operation and add fluid as necessary to prevent the fluid level from falling low enough to allow air bubbles into the master cylinder.*

6   Have an assistant on hand, as well as a supply of new brake fluid, a clear container partially filled with clean brake fluid, a length of plastic, rubber or vinyl hose to fit over the bleeder valve and a wrench to open and close the bleeder valve.

7   Beginning at the right rear wheel, loosen the bleeder valve slightly, then tighten it to a point where it is snug but can still be loosened quickly and easily.

8   Place one end of the hose over the bleeder valve and submerge the other end in brake fluid in the container **(see illustration)**.

9   Have the assistant push the brake pedal slowly to the floor, then hold the pedal firmly depressed.

10   While the pedal is held depressed, open the bleeder valve just enough to allow a flow of fluid to leave the valve. Watch for air bubbles to exit the submerged end of the tube. When the fluid flow slows after a couple of seconds, close the valve and have your assistant release the pedal.

11   Repeat Steps 9 and 10 until no more air is seen leaving the tube, then tighten the bleeder valve and proceed to the left rear wheel, the right front wheel and the left front wheel, in that order, and perform the same procedure. Be sure to check the fluid in the master cylinder reservoir frequently.

12   Never use old brake fluid. It contains moisture which will deteriorate the brake system components.

13   Refill the master cylinder with fluid at the end of the operation.

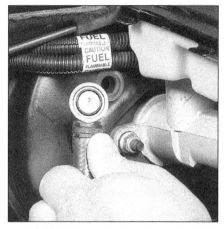

11.6 To detach the intake manifold vacuum hose from the power brake booster, pull it straight out

14   Check the operation of the brakes. The pedal should feel solid when depressed, with no sponginess. If necessary, repeat the entire process. **Warning:** *Do not operate the vehicle if you are in doubt about the effectiveness of the brake system.*

15   As soon as you have completed this procedure, test drive the vehicle and verify that the brake system is in good working order.

11   Power brake booster - check, removal and installation

Operating check

1   Depress the brake pedal several times with the engine off and make sure that there is no change in the pedal reserve distance.

2   Depress the pedal and start the engine. If the pedal goes down slightly, operation is normal.

Airtightness check

3   Start the engine and turn it off after one or two minutes. Depress the brake pedal several times slowly. If the pedal goes down farther the first time but gradually rises after the second or third depression, the booster is airtight.

4   Depress the brake pedal while the engine is running, then stop the engine with the pedal depressed. If there is no change in the pedal reserve travel after holding the pedal for 30 seconds, the booster is airtight.

Removal

Refer to illustrations 11.6, 11.7a, 11.7b, 11.7c and 11.8

Note: *Dismantling of the power brake unit requires special tools. If a problem develops, it is recommended that a new or factory-exchange unit be installed rather than trying to overhaul the original booster.*

5   Remove the mounting nuts which hold the master cylinder to the power brake unit

11.7a The power brake booster pushrod is attached to a pin at the top of the brake pedal; to disconnect the pushrod from the pin, remove the spring clip . . .

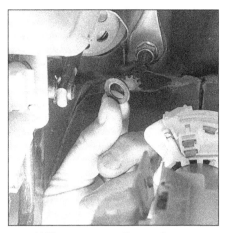

11.7b . . . remove the washer . . .

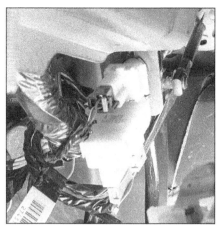

11.7c . . . and pry the end of the pushrod off the pin with a screwdriver

(see Section 8). Carefully move the master cylinder slightly forward, being extremely careful not to bend or kink the lines leading to the master cylinder. If there is any strain on the lines, disconnect them at the master cylinder and plug the ends.

6   Disconnect the intake manifold vacuum hose from the power brake booster **(see illustration)**. Cover the end of the hose.

7   Disconnect the power brake pushrod from the brake pedal **(see illustrations)**. Do not force the pushrod to the side when disconnecting it.

8   Inside the vehicle, loosen the four nuts that secure the booster to the firewall **(see illustration)**. Do not remove these nuts at this time.

9   Now remove the four booster mounting nuts and carefully lift the unit out of the engine compartment.

10   When installing, loosely install the four mounting nuts, then connect the pushrod to the brake pedal. Tighten the nuts to the torque listed in this Chapter's Specifications and reconnect the vacuum hose and master cylinder. If the brake lines were disconnected from the master cylinder, bleed the brake system to eliminate any air which has entered the system (see Section 10).

12   Parking brake - adjustment

Refer to illustration 12.6

1   Generally, the parking brake cables need no regular adjustment. However, they must be adjusted after new rear brake shoes (models with rear drum brakes), new parking brake shoes (models with disc brakes) or new parking brake cables (all models) are installed. It's also a good idea to check cable adjustment whenever the vehicle is raised.

2   Verify that the drums or discs rotate freely without drag when the parking brake lever is released. Then depress the parking brake pedal and verify that the parking brake system holds the vehicle in place. If it doesn't,

adjust the parking brake cable.

3   Raise the vehicle and support it securely on jackstands. Block the front wheels to prevent the vehicle from rolling.

4   If the vehicle is equipped with rear drum brakes, make sure the brake shoes are adjusted (see Section 6, Step 7).

5   Apply the parking brake pedal six ratchet clicks (1991 through 1995 models) or ten clicks (1996 models).

6   Tighten the equalizer adjusting nut **(see illustration)** until the right rear wheel can barely be turned counterclockwise - but not clockwise - with two hands.

7   Remove the jackstands and lower the vehicle.

8   Release the parking brake pedal and test the operation of the parking brake on an incline.

13   Parking brake cables - replacement

All models

1   Fully release the parking brake lever.

2   Loosen the left or right rear wheel lug nuts if you're replacing a rear parking brake

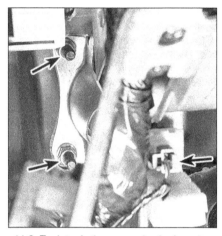

11.8 To detach the power brake booster from the firewall, remove these nuts (arrows - right two nuts not visible)

cable. Raise the rear of the vehicle and place it securely on jackstands. Remove the left or right rear wheel if you're replacing a rear cable.

3   Loosen the equalizer nut **(see illustration 12.6)** to provide sufficient cable slack.

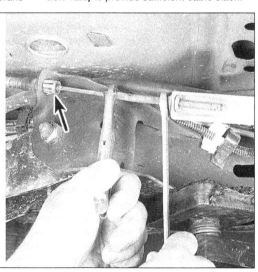

12.6 To adjust the parking brake cable, adjust the brake shoes (drum brake models), apply the parking brake pedal six ratchet clicks (ten clicks on 1996 models), then, holding the rear cable with a pair of small vise grips to prevent it from turning, tighten the equalizer adjusting nut until the right rear wheel can barely be turned counterclockwise - but not clockwise - with two hands. To detach the right rear parking brake cable from the bracket behind the equalizer, squeeze the cable ferrule (arrow) with a pair of pliers and pull the cable through the bracket

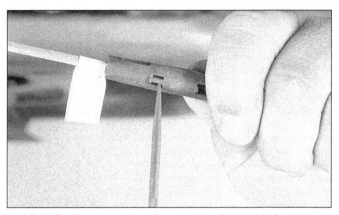

13.4a To disconnect the left rear or the front cable from the retainer, use an awl to pry out the small locking tab for the cable you wish to detach . . .

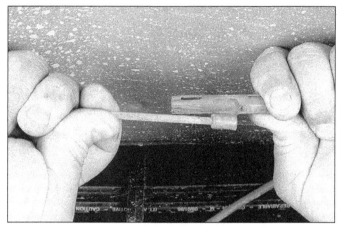

13.4b . . . then disengage the cable from the retainer

13.4c To detach the left rear cable from the equalizer, squeeze the ferrule with a pair of pliers as shown and pull the cable through the equalizer

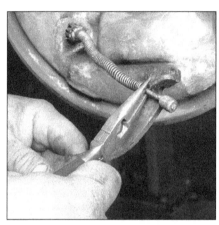

13.5a To disengage a rear cable from its parking brake lever on a model with rear drum brakes, you'll have to disassemble the rear brake assembly far enough to access the parking brake lever, which is attached to the secondary brake shoe

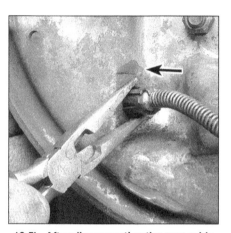

13.5b After disconnecting the rear cable from the parking brake lever, squeeze this cable ferrule with a pair of pliers and pull the cable through the backing plate

Rear cables

Left rear cable

Refer to illustrations 13.4a, 13.4b, 13.4c, 13.5a, 13.5b, 13.6a, 13.6b

4   Disengage the left rear cable from the

front cable retainer **(see illustrations)** and from the equalizer **(see illustration)**.
5   On models with rear drum brakes, remove the brake drum and shoe assembly

(see Section 6), then disengage the cable from the parking brake lever **(see illustration)**, squeeze the cable ferrule at the backing plate with a pair of pliers **(see illustration)** and pull the cable through the backing plate (see Section 6).
6   On models with rear disc brakes, disengage the cable from the parking brake lever **(see illustration)**, then squeeze the cable ferrule at the bracket with a pair of pliers **(see illustration)** and slide the cable through the bracket.

Right rear cable

Refer to illustration 13.8

7   Remove the equalizer nut **(see illustration 12.6)** and disconnect the right rear cable from the equalizer.
8   Squeeze the cable ferrule at the bracket behind the equalizer **(see illustration 12.6)** with a pair of pliers, then cut the two cable ties **(see illustration)** that attach the cable to the axle housing.
9   On models with rear drum brakes, refer to Step 5.
10   On models with rear disc brakes, refer to Step 6.

13.6a To disengage a rear parking brake cable from the parking brake lever on models with rear disc brakes, simply pull it off

13.6b To detach a rear parking brake cable from the bracket just forward of the parking brake lever on models with rear disc brakes, squeeze the cable ferrule together with a pair of pliers

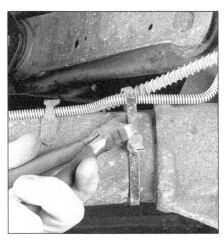

13.8 To detach the right rear parking brake cable from the axle housing, cut the cable ties (only one cable tie shown here, other tie not visible in this photo)

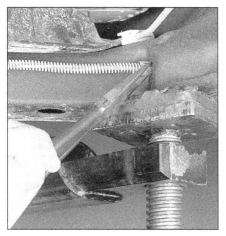

13.12 To disengage the front parking brake cable from the frame, squeeze this cable ferrule with a pair of pliers

13.15 To disengage the front parking brake cable from the parking brake pedal assembly, squeeze the cable ferrule and pull the cable out of the pedal bracket, then disengage the plug (arrow) on the end of the cable from the pedal clevis

Front cable

Refer to illustrations 13.12 and 13.15

11   Disengage the front parking brake cable from the retainer **(see illustrations 13.4a and 13.4b)** that connects it to the left rear cable.
12   Squeeze the cable ferrule at the frame with a pair of pliers **(see illustration)**.
13   Lower the vehicle.
14   Remove the left (driver's side) kick panel (see Chapter 11).
15   Disconnect the front cable from the parking brake pedal bracket and disengage the cable end plug from the pedal clevis **(see illustration)**.
16   Pry out the grommet where the cable goes through the body and remove the cable.

All models

17   Installation is the reverse of removal.
18   Be sure to adjust the parking brake when you're done (see Section 12).

---

14   Parking brake shoes - removal and installation

Refer to illustrations 14.3, 14.4, 14.5a and 14.5b

Note: *The following procedure applies to models with rear disc brakes, which use a small parking brake shoe assembly and drum inside the rear disc. The parking brake system on models with rear drum brakes uses the rear shoes as the parking brakes (see Section 6).*
1   Loosen the rear wheel lug nuts, raise the vehicle and place it securely on jackstands. Remove the rear wheels.
2   Remove the rear brake calipers (see Section 4) and discs (see Section 5).
3   Lift the parking brake shoe assembly straight up **(see illustration)**.
4   Lubricate the shoe guide pads with high temperature grease **(see illustration)**.

14.3 To remove a rear parking brake shoe assembly, lift it straight up

5   Install the new shoe assembly **(see illustrations)**.
6   Repeat steps 3 through 5 for the other side.
7   Install the brake discs and calipers.

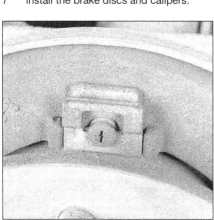

14.5a When installing the new shoe assembly, make sure it's fully seated into this retaining clip . . .

14.4 Lubricate the shoe guide pads (the raised areas on the backing plate) with high-temperature grease

8   Install the rear wheels, remove the jackstands and lower the vehicle. Tighten the wheel lug nuts to the torque listed in the Chapter 1 Specifications.

14.5b . . . and both ends of the shoe are properly engaged with the parking brake lever housing

15.1 The brake light switch (arrow) is located at the upper end of the brake pedal

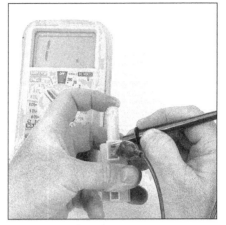

15.6a When the brake pedal is released, it depresses the brake light switch plunger, opening the circuit, so there should be no continuity

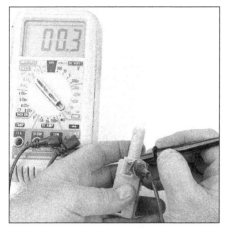

15.6b When the brake pedal is applied, the switch plunger is released, closing the circuit, so there should be continuity

15   Brake light switch - check, replacement and adjustment

Check

Refer to illustrations 15.1, 15.6a and 15.6b

1   The brake light switch **(see illustration)** is located on the brake pedal bracket. You'll need to remove the knee bolster (the trim panel beneath the steering column), and the metal plate behind it, to get to the switch and connector (see Chapter 11).

2   With the brake pedal in the fully released position, the switch plunger is pressed into the switch housing. When the brake pedal is depressed, the plunger protrudes from the switch, which closes the circuit and sends current to the brake lights.

3   If the brake lights are inoperative, check the fuse first (see Chapter 12).

4   If the fuse is okay, verify that voltage is available at the switch.

5   If there's no voltage to the switch, use a test light to find the open circuit condition between the fuse panel and the switch. If there is voltage to the switch, close the switch (depress the brake pedal) and verify that there's voltage on the other terminal of the switch.

6   If there's no voltage on the other terminal of the switch with the brake pedal depressed, replace the switch (see Step 7). If voltage is available, check for voltage at the brake lights. If no power is present, look for an open circuit condition between the switch and the brake lights. Also check the brake light bulbs, even though it's unlikely that both of them would fail at the same time. As a final test, remove the switch (see below) and test the switch on the bench **(see illustrations)**. There should be no continuity when the switch plunger is depressed (brake pedal in the released position) and continuity when the plunger is released (brake pedal applied).

Replacement

Refer to illustrations 15.8 and 15.9

7   Remove the knee bolster (see Chapter 11).

8   Remove the switch from the retainer **(see illustration)**.

9   Unplug the electrical connector from the switch **(see illustra-tions 15.1 and 15.9)**.

10   The switch must be adjusted as it's installed (see below).

Adjustment

11   With the brake pedal released, insert the switch into the retainer and push it in until it stops (don't be afraid to use a firm hand to push the switch barrel into the retainer).

12   Slowly pull the brake pedal to the rear; the switch barrel will automatically ratchet itself back to its adjusted position. Pull the pedal back until you no longer hear any "clicking" sounds. The switch is now adjusted.

13   You can check your work with an ohm-meter or continuity tester by verifying that the switch contacts are open at one inch or less of brake pedal travel, and closed there-after.

14   Installation is otherwise the reverse of removal.

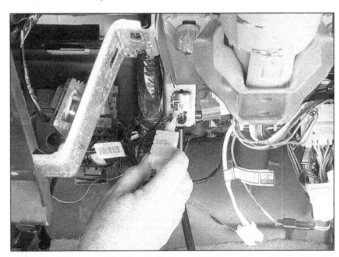

15.8 To disengage the brake light switch from its retainer, simply pull it straight back

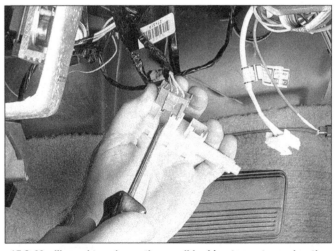

15.9 You'll need to release the small locking tangs to unplug the electrical connectors from the brake light switch

Chapter 10
Suspension and steering systems

Contents

Specifications

General
Power steering fluid type .... See Chapter 1

Torque specifications
Front suspension

Ft-lbs (unless otherwise indicated)

Balljoints
    Lower ballstud nuts 83*
    Upper ballstud nuts 61*
Lower control arms
    Pivot bolts 114
    Pivot bolt nuts 107
Shock absorbers
    Upper nut .... 97 in-lbs
    Lower bolts 20
Stabilizer bar
    Clamp bolts 24
    Link bolt nuts .... 156 in-lbs
Upper control arms
    Pivot shaft retaining nuts 70

Rear suspension
Lower control arms
    Lower control arm-to-rear axle bracket
        Nut 91
        Bolt 114
    Lower control arm-to-frame bracket
        Nut 91
        Bolt 114
Shock absorbers
    Upper nuts/bolts
        Nut .... 156 in-lbs
        Bolt 18
    Lower nuts 63
Stabilizer bar
    Clamp bolts 18
    Link bolts/nuts 16
Upper control arms
    Upper control arm-to-frame crossmember
        Nut 95
        Bolt 114

Torque specifications

Ft-lbs (unless otherwise indicated)

Rear suspension (continued)

Upper control arms (continued)

Upper control arm-to-rear axle bracket

Nut	74
Bolt	83

Steering

Idler arm

Idler arm-to-center link nut	35
Idler arm-to-frame nuts	61

Pitman arm

Pitman arm-to-center link nut	35
Pitman arm-to-steering gear nut	179

Steering damper

Steering damper-to-center link nut	53
Steering damper-to-frame nut	40
Steering gear bolts/nuts	63
Steering shaft U-joint pinch bolt/nut	35
Steering wheel nut	32

Tie-rod assembly

Tie-rod adjuster clamp pinch bolts/nuts	168 in-lbs
Tie-rod end castle nuts (inner and outer)	35*

*Do not back off castle nut for cotter pin insertion.

1.1a Front suspension and steering components

1	Stabilizer bar	6	Center link	10	Lower shock absorber mounting bolts
2	Stabilizer bar clamps	7	Inner tie-rod ends	11	Lower control arms
3	Steering gear	8	Tie-rod adjuster tube and clamps	12	Lower control arm pivot bolts/nuts
4	Pitman arm	9	Outer tie-rod ends	13	Lower control arm balljoints
5	Idler arm				

1 General information

Suspension

Refer to illustrations 1.1a, 1.1b and 1.2

The front suspension **(see illustrations)** is fully independent; it allows each wheel to compensate for road surface irregularities without a significant effect on the other wheel. The suspension at each front wheel consists of an upper and lower control arm, a steering knuckle between the two arms, a shock absorber and a coil spring. The coil spring is positioned between the spring housing and the lower control arm. The shock absorber, which is positioned inside the coil spring, is bolted to the lower control arm and to the roof of the spring housing. The steering knuckle is connected to the upper and lower control

arms by a pair of balljoints, one in each arm. The lower balljoints are pressed into the lower control arms; the upper balljoints are riveted to the upper control arms. Both upper and lower balljoints are replaceable. A stabilizer bar controls vehicle roll during cornering. The stabilizer bar is attached to the frame by a pair of steel clamps and to the lower control arms by link bolts.

The semi-independent rear suspension **(see illustration)** consists of a solid rear axle housing suspended by a pair of shock absorbers and coil springs and is located by two upper and two lower control arms. The springs are positioned between brackets on the rear axle and the frame spring seats; insulators are installed between the springs and their upper and lower seats. The shocks are bolted to brackets on the axle tube and the frame spring seats. A sta-

bilizer bar controls vehicle roll during cornering. The stabilizer bar is bolted to the lower control arms.

Steering

The power-assisted steering system **(see illustration 1.1a)** consists of the steering gearbox, the Pitman arm, the center link, the idler arm, and the tie-rod assemblies (inner tie-rod, adjuster tube and outer tie-rod end).

2 Shock absorber (front) - removal and installation

Refer to illustrations 2.2 and 2.3

1   Loosen the wheel lug nuts, raise the front of the vehicle and support it securely on jackstands, then remove the wheel.

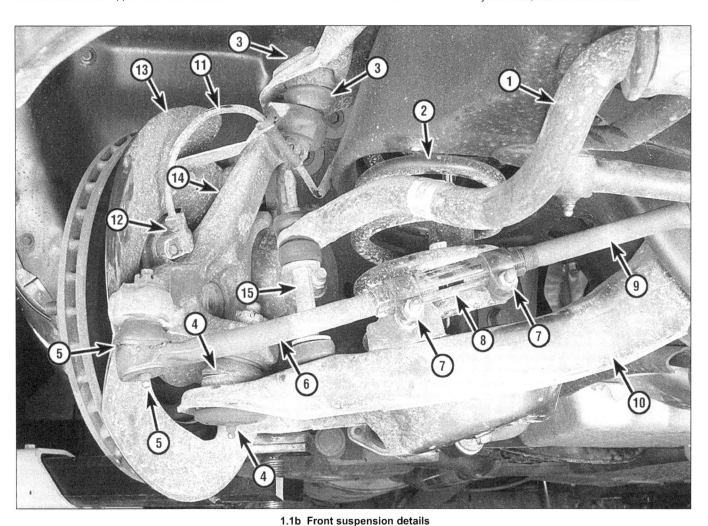

1.1b Front suspension details

1	*Stabilizer bar*	*6*	*Outer tie-rod end*	*11*	*ABS sensor electrical lead*
2	*Coil spring*	*7*	*Adjuster tube clamps*	*12*	*ABS wheel speed sensor*
3	*Upper control arm balljoint*	*8*	*Adjuster tube*	*13*	*Brake backing plate/splash shield*
4	*Lower control arm balljoint*	*9*	*Inner tie-rod end*	*14*	*Steering knuckle*
5	*Outer tie-rod end balljoint*	*10*	*Lower control arm*	*15*	*Stabilizer bar link*

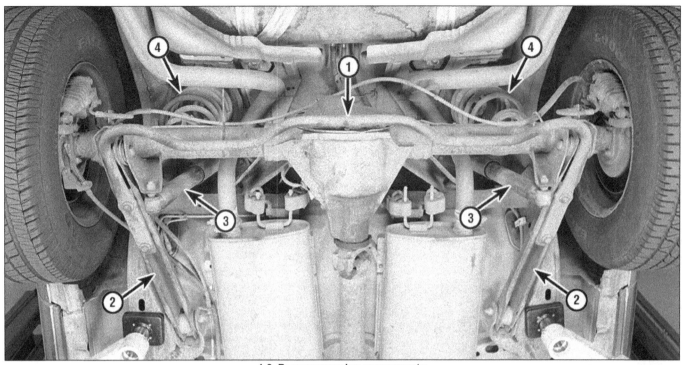

1.2 Rear suspension components

1 Stabilizer bar	2 Lower control arms	3 Shock absorbers	4 Coil springs
(not on all models)			

2.2 Using a small mirror, locate the upper shock absorber mounting nut inside this hole in the upper control arm. Note the two flat sides on the end of the damper rod; put a small wrench on these flats while breaking loose the upper shock nut with another wrench

2.3 To detach the lower end of the shock absorber from the lower control arm, remove these two bolts (arrows)

3.2 To remove the stabilizer bar link, remove this nut (arrow) and remove the bolt, noting the order of the bushings, washers and spacer

2   Holding the flats of the shock absorber damper rod with one wrench, remove the upper shock absorber-to-body nut **(see illustration)**, washer and rubber bushing (don't forget the other washer and bushing underneath, after you have removed the shock).

3   Unbolt the shock absorber from the lower control arm **(see illustration)**.

4   Remove the shock absorber.

5   Installation is the reverse of removal. Be sure to tighten all fasteners to the torque listed in this Chapter's Specifications.

3   Stabilizer bar (front) - removal and installation

Refer to illustrations 3.2 and 3.3

1   Raise the vehicle and support it securely on jackstands.

2   Remove the nuts **(see illustration)** from the link bolts that attach each end of the stabilizer bar to the lower control arms. Note the order in which the washers, rubber bushings and sleeve are removed to ensure proper reassembly.

3   Remove the stabilizer bar clamps and rubber bushings **(see illustration)**.

4   Remove the stabilizer bar.

5   Inspect the stabilizer bar and link bolt rubber parts for cracks and tears. Replace all damaged rubber parts.

6   Apply multi-purpose grease to the bushing installation areas on the stabilizer bar before installing the bushings and clamps.

7   Make sure the slits in the bushings are facing toward the front of the vehicle.

8   Make sure the stabilizer bar is centered in the bushings and clamps before tightening the mounting bolts.

9   Installation is otherwise the reverse of removal procedure. Be sure to tighten all bolts to the torque listed in this Chapter's Specifications.

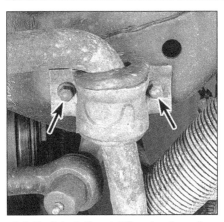

3.3 To detach the stabilizer bar from the frame, remove these bolts (arrows) from each bushing clamp; note how the slit in the bushing faces forward - it must be oriented the same way when you reattach the stabilizer bar

4 Coil spring (front) - removal and installation

Removal

Refer to illustrations 4.6 and 4.8

1 Loosen the front wheel lug nuts, raise the vehicle and place it securely on jackstands. Remove the wheel.
2 Remove the wheel speed sensor **(see illustration 5.2)**.
3 Remove the shock absorber (see Section 2).
4 Remove the stabilizer bar link bolt **(see illustration 3.2)**.
5 Disconnect the outer tie-rod end from the steering knuckle (see Section 16).
6 Install a suitable internal type spring compressor in accordance with the tool manufacturer's instructions **(see illustration)**. Compress the spring enough to relieve all pressure from the spring seats (but don't compress it any more than necessary; if the spring is compressed to 9.6 inches or less, it could be ruined). When you can wiggle the spring, it's compressed enough. (You can buy a suitable spring compressor at most auto parts stores or rent one from a tool rental yard.)
7 Support the lower control arm with a floor jack.
8 Remove the control arm pivot bolts and nuts **(see illustration)**.
9 Pull the lower control arm down and to the rear, then guide the compressed coil spring out.
10 If the coil spring is being replaced, carefully unscrew the spring compressor.

Installation

11 Inspect the upper and lower spring insulators. If either insulator is cracked or excessively worn, replace it. Inspect the coil spring for chips in the corrosion protection coating. If the coating has been chipped or damaged, replace the spring.

4.6 A typical aftermarket internal-type spring compressor tool: The hooked arms grip the upper coils of the spring, the plate is inserted below the lower coils, and when the nut on the threaded rod is turned, the spring is compressed

12 If the coil spring is being replaced, install the spring compressor and compress the spring.
13 With the lower spring insulator in place, position the spring on the lower control arm with the flat end of the spring facing up and the tapered end facing down. Make sure the tapered end seats on the lower control arm with the lower end of the spring seated in the lowest part of the spring seat. The end of the spring must cover all or part of one of the drain holes in the lower control arm, but the other hole must not be covered.
14 Put the floor jack under the lower control arm and raise the arm into position in the frame. Install the control arm pivot bolts and nuts. Tighten the nuts until they're snug but don't torque them yet.
15 Remove the spring compressor.
16 Reattach the outer tie-rod end to the steering knuckle (see Section 16).
17 Reattach the stabilizer bar link to the lower control arm (see Section 3).
18 Install the shock absorber (see Section 2).

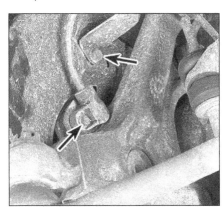

5.2 To detach the ABS wheel speed sensor and wiring harness bracket from the steering knuckle, remove these two bolts (arrows)

4.8 Support the lower control arm with a floor jack, then remove the pivot bolts and nuts (arrow indicates front pivot bolt)

19 Install the ABS wheel speed sensor, if equipped. Tighten the sensor bolt and harness bracket bolts securely.
20 Position the floor jack under the lower control arm balljoint and raise the arm to simulate normal ride height. Tighten the lower control arm pivot bolt nuts to the torque listed in this Chapter's Specifications.
21 Install the wheel, remove the jackstands and lower the vehicle. Tighten the wheel lug nuts to the torque listed in the Chapter 1 Specifications.

5 Steering knuckle - removal and installation

Refer to illustrations 5.2, 5.4, 5.6 and 5.9

1 Loosen the wheel lug nuts, raise the front of the vehicle and support it securely on jackstands. Remove the wheel.
2 If the vehicle is equipped with ABS, remove the wheel speed sensor and wire harness bracket **(see illustration)**.
3 Remove the brake caliper and disc (see Chapter 9).
4 Remove the brake disc shield **(see illustration)**.

5.4 To detach the disc brake shield from the steering knuckle, remove these three bolts (arrows)

5.6 Remove the hub seal and inspect it for cracks and tears; replace it if it's damaged

5.9 A special tool is recommended for pushing the balljoint ballstuds out of the steering knuckle, but you can fabricate your own tool (shown here) from a large bolt, nut, washer and socket. In this photo, our setup is being used to push the upper ballstud out of the knuckle; to push out the lower ballstud, simply reverse the tool, with the socket seated against the upper ballstud nut

7.3 Support the lower control arm with a jackstand positioned between the lower spring seat and the balljoint (the farther out the arm you place the jackstand, the greater the leverage ratio on the arm)

5   Disconnect the tie-rod end from the steering knuckle (see Section 16).
6   Remove the hub seal **(see illustration)**. Inspect the seal for cracks and tears. If it's damaged, replace it.
7   Disconnect the lower end of the shock absorber from the lower control arm (see Section 3).
8   Support the lower control arm with a floor jack under the balljoint area and raise the jack just enough to support the lower arm **(see illustration 7.3)**. **Warning:** *Do NOT remove the floor jack while the steering knuckle is removed! If you do, the coil spring will jump off its lower seat with considerable force, enough force to cause serious injury if it should strike you.*
9   To separate the lower and upper control arms from the steering knuckle, remove the ballstud nut cotter pins, loosen the ballstud nuts and force the ballstuds out of the knuckle with a special tool, available at most auto parts stores, or a suitable alternative such as the tool we fabricated from hardware store nuts and bolts and a deep socket **(see illustration)**. Remove the ballstud nuts and remove the knuckle.

10   If you intend to reuse the old steering knuckle, inspect the tapered ballstud holes in the knuckle for dirt and unusual wear such as elongation or out-of-roundness. If any wear or damage is evident, replace the steering knuckle.
11   Installation is the reverse of removal. Don't forget the hub seal, and be sure to tighten the ballstud nuts to the torque listed in this Chapter's Specifications.

---

6   Lower control arm (front) - removal and installation

1   Loosen the wheel lug nuts, raise the vehicle and place it securely on jackstands. Remove the wheel.
2   Remove the shock absorber (see Section 2).
3   Disconnect the stabilizer bar link bolt

from the lower control arm (see Section 3).
4   Remove the coil spring (see Section 4).
5   To disconnect the lower control arm from the steering knuckle, remove the ballstud nut cotter pin, loosen the ballstud nut and force the ballstud out of the steering knuckle with a tool like the one shown in **illustration 5.9**.
6   Remove the lower control arm.
7   If you intend to reinstall the same control arm, be sure to inspect the control arm bushings for cracks, tears and other damage. If either bushing is damaged, take the arm to an automotive machine shop and have new bushings installed.
8   Installation is the reverse of removal. Be sure to tighten all fasteners to the torque listed in this Chapter's Specifications.

---

7   Upper control arm - removal and installation

Refer to illustrations 7.3, 7.5a and 7.5b
1   If you're removing the upper control arm from the left (driver's side):

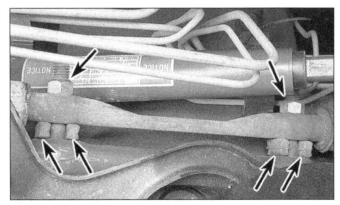

7.5a Remove the upper control arm pivot shaft retaining nuts (arrows), remove the shims (arrows), count and label them to ensure that they're installed in the same place during reassembly, then put them in a plastic bag for safekeeping (left side shown)

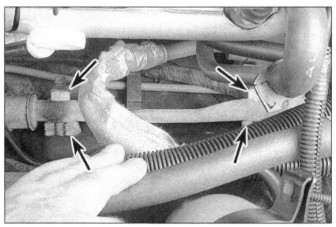

7.5b The upper control arm pivot shaft on the right (passenger's side) is the same, but a little more difficult to access

8.5 It's easy to check the lower control arm balljoint for wear: The shoulder (arrow) protrudes 0.050 inch (approximately 3/64-inch) from the surface of the cover. As the balljoint wears, this shoulder moves up into the balljoint assembly; if the shoulder is flush with the cover, or is up inside the hole, replace the balljoint

a) *Remove the air cleaner assembly and the intake duct* (see Chapter 4).
b) *It may also be helpful to remove the intermediate steering shaft* (see Section 15). *If you do this, put the ignition key lock cylinder in the Lock position to prevent damage to the SIR coil (clockspring) for the airbag system.*

2   Loosen the front wheel lug nuts, raise the front of the vehicle and support it securely on jackstands, then remove the wheel.
3   Support the lower control arm with a floor jack placed under the lower balljoint area **(see illustration)**. The jack must remain here, to keep the coil spring and lower control arm in position, until the upper control arm has been reinstalled.
4   If the vehicle is equipped with ABS, remove the wheel speed sensor **(see illustration 5.2)**.
5   Loosen the retaining nuts **(see illustrations)** from the upper control arm pivot shaft and remove the alignment shims from each bolt. Tape the shims together and clearly label them to ensure that they're returned to the same location during reassembly, then store them in labeled plastic bags so you don't lose them or mix them up.
6   Remove the cotter pin for the upper ballstud nut, loosen the nut, force the ballstud out of the steering knuckle with a tool such as the one shown in **illustration 5.9**, remove the nut and separate the ballstud and knuckle.
7   Remove the pivot shaft nuts **(see illustrations 7.5a or 7.5b)**.
8   Remove the upper control arm.
9   Installation is the reverse of removal. Be sure to put the shims back where they were and tighten all fasteners to the torque listed in this Chapter's Specifications.

8.9a To replace an upper balljoint, drill out the four rivet heads with progressively larger drill bits; start with a 1/8-inch drill bit, then proceed to a 1/4-inch, then switch to a 1/2-inch bit to cut off the rivet heads (stop drilling once the heads are removed)

8   Balljoints - check and replacement

Check
1   Inspect the control arm balljoints for looseness whenever either of them is separated from the steering knuckle. See if you can turn the ballstud in its socket with your fingers.
2   If the balljoint is loose, or if the ballstud can be turned, replace the balljoint. You can also check the balljoints with the suspension assembled as follows.
3   Raise the front of the vehicle and support it securely on jackstands placed under the lower control arms. Position the stands as close to each balljoint as possible. Make sure the vehicle is stable. It should not rock on the stands.

Lower balljoints
Refer to illustration 8.5

4   Wipe each balljoint clean and inspect the seal for cuts and tears. If the seal is damaged, replace the balljoint.
5   The lower balljoints employ a visual wear indicator **(see illustration)** that allows easy diagnosis. The shoulder at the base of the grease fitting protrudes about 0.050-inch (3/64-inch) from the lower surface of the balljoint's lower cover when the balljoint is new. As the balljoint wears, this shoulder slowly moves up into the balljoint. If the shoulder is flush with or inside the surface of the lower cover, replace the lower balljoint.

Upper balljoints
6   Position a dial indicator against the wheel rim, grasp the top and bottom of the tire and "rock" the tire, alternately pushing the top and pulling the bottom, and vice-versa. The dial indicator should indicate no more than 0.125-

8.9b Use a punch to knock out the rivet shanks

8.9c If any rivet material remains, the balljoint may have to be knocked loose with a chisel and hammer

inch deflection. If the indicated reading exceeds this figure, replace the upper balljoint.

Replacement
Refer to illustrations 8.9a through 8.9h

7   If you're replacing a lower balljoint, remove the lower control arm (see Section 6). If you're replacing an upper balljoint, removing the upper control arm (see Section 7) is desirable, but not essential; the upper balljoint *can* be replaced without removing the arm (although it's easier to do with the arm secured in a bench vise).
8   To replace the lower balljoint, take the control arm and a new balljoint to an automotive machine shop. The machine shop will press out the old balljoint and press in the new unit. You cannot do this at home unless you have a hydraulic press or a balljoint replacement tool with the proper adapters.
9   To replace the upper balljoint, drill out the rivets as follows: Using a 1/8-inch drill bit, drill a 1/4-inch deep hole in the center of each rivet. Then switch to a 1/2-inch drill bit and finish the job; drill just deep enough to remove the rivet head. Insert a punch through the rivet holes and knock out the old balljoint **(see illustrations)**. Install the new balljoint;

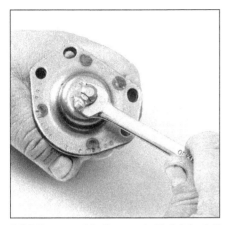

8.9d To assemble the new balljoint, install the grease fitting . . .

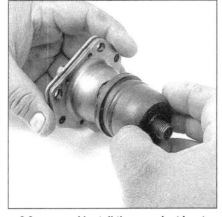

8.9e . . . and install the new dust boot

8.9f Install the balljoint in the upper control arm . . .

some assembly may be required **(see illustrations)**. Tighten the nuts to the torque specified in the instructions that come with the kit.

10   Install the lower or upper control arm (see Section 6 or 7).

11   Have the front end alignment checked by a dealer service department or alignment shop.

9   Stabilizer bar (rear) - removal and installation

1   Raise the vehicle and support it securely on jackstands.

2   Remove the nuts and washers **(see illustration 1.2)** from the bolts.

3   It's really not necessary to remove the retainers, brackets or sleeves unless you're planning to replace a rear lower control arm. If so, remove these parts and install them on the new lower arm.

4   Installation is the reverse of removal. Be sure to tighten the rear stabilizer retaining nuts to the torque listed in this Chapter's Specifications.

8.9g . . . install the metal dust shield (if equipped) . . .

10   Shock absorber (rear) - removal and installation

Refer to illustrations 10.2, 10.4 and 10.5

1   Raise the rear of the vehicle and support it on jackstands.

2   Support the rear axle assembly with a floor jack under the side being worked on

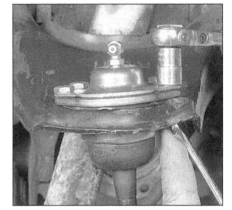

8.9h . . . install the bolts and nuts, tightening the nuts to the torque specified in the kit instructions

(see illustration). Warning: *Failure to support the rear axle could allow a rear coil spring to fly out and cause injury. Damage to the brake hose or the driveshaft U-joint could also result.*

3   On models with air suspension, disconnect the air lines from the shocks by turning the spring clip 90-degrees and carefully pulling out the air line housing.

10.2 Before removing a rear shock absorber, support the end of the rear axle from which you're removing the shock with a jack

10.4 To detach the upper end of the shock absorber from the frame, remove these two bolts (arrows)

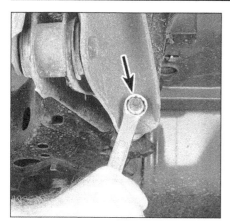

10.5 To detach the lower end of the shock absorber from the rear axle, remove this nut (arrow) and washer

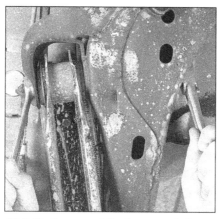

12.6 To disconnect the lower control arm from the frame bracket, remove the nut and bolt as shown

13.4 To disconnect the upper control arm from the rear axle bracket, remove this nut and bolt

4   Remove the upper mounting nuts and bolts **(see illustration)**.
5   Remove the lower shock mounting nut **(see illustration)** and washer from the bracket on the rear axle.
6   Remove the shock absorber.
7   Remove the shock boot heat shield from the old shock and switch it to the new shock.
8   Installation is the reverse of removal. Be sure to tighten all fasteners to the torque listed in this Chapter's Specifications.
9   Now move the jack over to the other end of the rear axle and repeat this procedure for the other rear shock absorber.

11 Coil spring (rear) - removal and installation

Note: *If both coil springs are to be removed, remove and install them one at a time to prevent the axle assembly from slipping sideways or rocking forward, complicating reassembly.*
1   Raise the rear of the vehicle and support it on jackstands. Place the jackstands under the vehicle jacking points, not under the rear axle.
2   Support the side of the rear axle assembly being serviced with a floor jack **(see illustration 10.2)**. **Warning:** *Failure to support the rear axle could allow a rear coil spring to fly out and cause injury.* **Caution:** *Failure to support the rear axle could result in damage to the brake hose or the driveshaft U-joint.*
3   It's a good idea to chain the coil spring to the rear axle to prevent a spring from popping out when the axle is lowered. Bolt the chain together, but leave enough slack in the chain to allow the coil spring to extend fully.
4   On models with air suspension, disconnect the automatic level control sensor link from the upper control arm.
5   Remove the rear brake line junction block bolt from the axle. Don't disconnect the metal lines from the junction block.
6   Remove the shock absorber lower mounting nut and disconnect the shock from the rear axle bracket **(see illustration 10.5)**.

7   Carefully lower the jack until all tension is removed from the coil spring, then remove the spring and the upper and lower spring insulators.
8   Inspect the upper and lower spring insulators. If either insulator is cracked or excessively worn, replace it. Inspect the coil spring for chips in the corrosion protection coating. If the coating has been chipped or damaged, replace the spring.
9   Installation is the reverse of removal. Don't forget to install the upper and lower insulators. The tape on the spring should be at the top, and the upper end of the spring should face toward the left side of the vehicle, perpendicular to the centerline of the vehicle (within five degrees to the rear and 15 degrees to the front). Before tightening the fasteners to the torque listed in this Chapter's Specifications, raise the rear axle to simulate normal ride height.

12 Lower control arm (rear) - removal and installation

Refer to illustration 12.6

Note: *If both lower control arms are to be removed, remove and install them one at a time to prevent the axle assembly from slipping sideways or rocking forward, complicating reassembly.*
1   Loosen the rear wheel lug nuts. Raise the rear of the vehicle and support it securely on jackstands. Place the jackstands under the vehicle jacking points, not under the rear axle. Remove the rear wheel.
2   Support the rear axle assembly with a floor jack **(see illustration 10.2)**. **Warning:** *The jack must remain in this position throughout the entire procedure. Failure to support the rear axle could allow a coil spring to pop out, which could cause serious bodily injury if it were to strike you or someone nearby.* **Caution:** *Failure to support the rear axle could result in damage to the brake hose or the driveshaft U-joint.*
3   Detach the stabilizer bar, if equipped, from the lower control arm (see Section 9).

4   On models equipped with ABS, detach the wheel speed sensor harness from the lower control arm.
5   Remove the lower control arm-to-axle bracket nut and bolt.
6   Remove the lower control arm-to-frame bracket nut and bolt **(see illustration)**.
7   Remove the control arm.
8   Inspect the bushings at either end of the control arm. If either bushing is cracked or torn, take the lower control arm and a new bushing to an automotive machine shop to have the bushing replaced.
9   Installation is the reverse of the removal procedure. Raise the rear axle with the floor jack to simulate normal ride height, then tighten all fasteners to the torque listed in this Chapter's Specifications.

13 Upper control arm (rear) - removal and installation

Refer to illustrations 13.4 and 13.5

Note: *If both upper control arms are to be removed, remove and install them one at a time to prevent the axle assembly from slipping sideways or rocking forward, complicating reassembly.*
1   Loosen the rear wheel lug nuts. Raise the rear of the vehicle and place it securely on jackstands. Remove the rear wheel.
2   Support the rear axle with a floor jack **(see illustration 10.2)**. **Warning:** *The jack must remain in this position throughout the entire procedure. Failure to support the rear axle could allow a coil spring to pop out, which could cause serious injury if it were to strike you or someone nearby.* **Caution:** *Failure to support the rear axle could result in damage to the brake hose or the driveshaft U-joint.*
3   On models with air suspension, disconnect the automatic level control sensor link from the upper control arm.
4   Remove the nut and bolt that attach the upper control arm to the rear axle bracket **(see illustration)**.
5   Remove the nut and bolt that attach the

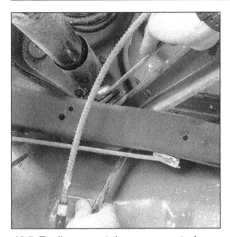

13.5 To disconnect the upper control arm from the frame crossmember, remove this nut and bolt

upper control arm to the frame crossmember **(see illustration)**.

6   Installation is the reverse of removal. Raise the rear axle with the floor jack to simulate normal ride height, then tighten all fasteners to the torque listed in this Chapter's Specifications.

14   Steering wheel -
removal and installation

Refer to illustrations 14.2a, 14.2b, 14.2c, 14.3, 14.4a, 14.4b, 14.4c, 14.4d, 14.6, 14.7 and 14.8

Warning: *These models have airbags. Always turn the steering wheel to the straight ahead position, place the ignition switch in the Lock position and disable the airbag system before working in the vicinity of the impact sensors, steering column or instrument panel to avoid the possibility of accidental deployment of the airbag, which could cause personal injury* (see Chapter 12).

14.3 To detach the airbag module from the steering wheel, remove the retaining screws (left side shown) (on earlier models, there are four retaining screws - two on each side; on later models, there are two - one on each side)

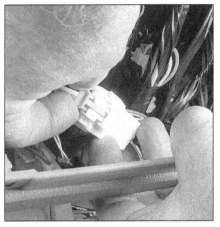

14.2a Before you can unplug either of the yellow electrical connectors for the airbag modules, you must remove the connector position assurance (CPA) retainers (the small locks that prevent the connectors from accidentally unplugging)

Caution: *On models equipped with a Delco Loc II or Theftlock audio system, be sure the lockout feature is turned off before performing any procedure which requires disconnecting the battery.*

1   Disconnect the cable from the negative battery terminal.

2   To disable the airbag system:

a) *Turn the steering wheel so that the front wheels are pointing straight ahead.*

b) *Turn the ignition key to the Lock position.*

c) *Remove the airbag fuse from the instrument panel fuse block (see Chapter 12).*

d) *Remove the left kick panel (see Chapter 11).*

e) *Remove the connector position assurance (CPA) retainers (connector locks) and unplug the yellow connectors at the base of the steering column* **(see illustrations)**. *(There are two yellow connectors: the one with the green CPA is for the steering wheel airbag; the other connector is for the passenger-side airbag. Unplug both connectors just to be safe.)*

14.4a Lift off the airbag module . . .

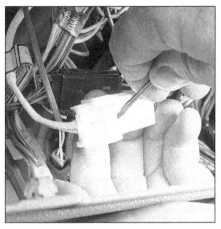

14.2b Use a small screwdriver to unlock the airbag module connectors . . .

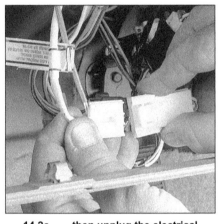

14.2c . . . then unplug the electrical connector for each airbag module

f) *The airbags are now disabled. Don't be alarmed if you note that the AIRBAG warning light comes on when the ignition key is turned to On. This is normal operation; it does not indicate an airbag system malfunction unless it does not turn off.*

3   Remove the airbag module retaining screws **(see illustration)**.

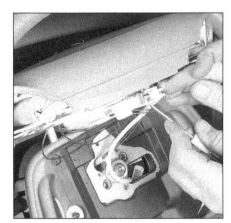

14.4b . . . remove the connector position assurance (CPA) retainer . . .

14.4c ... unlock and unplug the yellow airbag module connector ...

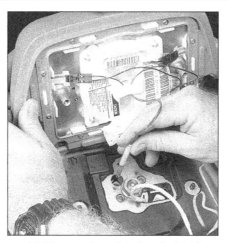

14.4d ... and unplug the electrical connector for the horn lead and ground wire

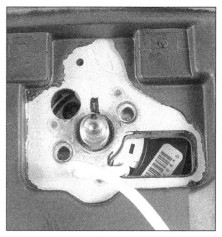

14.6 After removing the steering wheel nut, mark the relationship of the steering wheel to the steering shaft to ensure proper reassembly

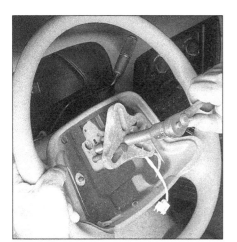

14.7 Use a steering wheel puller to remove the steering wheel

14.8 When properly installed, the airbag coil will be centered with the marks aligned (circle) and the tab fitted between the projections on the top of steering column (arrow)

4   Remove the airbag module and unplug the electrical connectors for the module, and for the horn lead and ground wire **(see illustrations)**. **Warning:** *Carry the airbag module with the trim side facing away from your body. Set the airbag module aside with the trim side facing up.*
5   Remove the steering wheel retaining nut.
6   Mark the relationship of the steering wheel to the steering shaft **(see illustration)**.
7   Remove the steering wheel with a puller **(see illustration)**. **Warning:** *Don't allow the steering shaft to turn with the steering wheel removed. If the shaft turns, the airbag coil assembly (the mechanism which protects the airbag wiring when the steering wheel is turned) will become uncentered, which will cause the airbag harness to break when the vehicle is returned to service.*
8   Installation is the reverse of removal. Before installing the steering wheel, make sure that the airbag coil assembly is centered **(see illustration)**. Be sure to tighten the steering wheel nut and the airbag module retaining screws to the torque values listed in this Chapter's Specifications.

nectors and install the airbag fuse. For more information about the airbag system, refer to Chapter 12.

15 Intermediate steering shaft - removal and installation

Refer to illustrations 15.1a, 15.1b, 15.2 and 15.3
Warning: *Disable the airbag system before working in the vicinity of the steering wheel, instrument panel or any airbag system component. Failure to do so could cause accidental deployment of the airbag resulting in personal injury (see Chapter 12).*
Caution: *Before starting, make sure that the wheels are pointing straight ahead and the ignition key lock cylinder is in the Lock position. Failure to do so might result in damage to the Supplemental Inflatable Restraint (SIR) coil (the airbag module lead).*
1   Detach the shield from the steering gear return pipe nut **(see illustrations)**.

9   To enable the airbag system, plug in the yellow connectors at the base of the steering column, install the CPAs in the con-

15.1a Locate the plastic steering gear shield (arrow) ...

15.1b ... and pry it off the steering gear return pipe nut with a screwdriver (the shield is flexible - prying it off won't damage it)

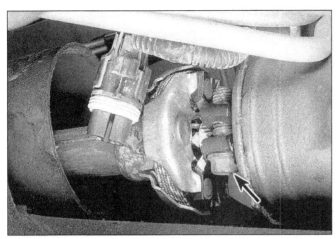

15.2 To detach the lower end of the intermediate shaft from the steering gear, mark the relationship of the shaft to the steering gear, then remove the pinch bolt (arrow) from the lower coupling

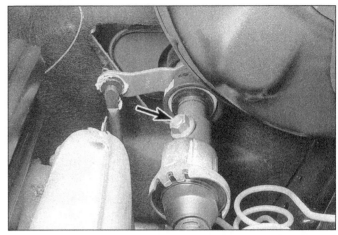

15.3 To detach the upper end of the intermediate shaft from the steering gear, remove the pinch bolt (arrow) and nut from the upper intermediate shaft coupling

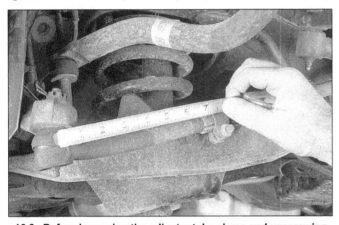

16.8a Before loosening the adjuster tube clamp and unscrewing the tie-rod end, measure the distance from the end of the adjuster tube to the center of the ballstud and record your measurement . . .

16.8b . . . or count the number of threads showing outside the adjuster tube

2   Remove the pinch bolt from the lower coupling **(see illustration)**. Push the intermediate shaft up and to the rear to disengage it from the steering gear.

3   Remove the pinch bolt and nut from the upper coupling **(see illustration)**.

4   Remove the intermediate shaft. **Warning:** *DO NOT allow the steering column shaft to rotate with the intermediate shaft removed or damage to the airbag system could occur.*

5   Installation is the reverse of removal. Be sure to tighten the upper and lower coupling pinch bolts to the torque listed in this Chapter's Specifications.

---

16   Steering linkage - inspection, removal and installation

Inspection

1   The steering linkage **(see illustrations 1.1a and 1.1b)** connects the steering gear to the front wheels and keeps the wheels in proper relation to each other. The linkage consists of the Pitman arm, the idler arm, the center link, and two adjustable tie-rod assem-

blies. The Pitman arm, which is fastened to the steering gear shaft, moves the center link back-and-forth. The center link is supported on the other end by a frame-mounted idler arm. The back-and-forth motion of the center link is transmitted to the steering knuckles through a pair of tie-rod assemblies. Each tie rod consists of an inner and outer tie-rod end, a threaded adjuster tube and two clamps.

2   Set the wheels in the straight-ahead position and lock the steering wheel.

3   Raise one side of the vehicle until the tire is approximately 1-inch off the ground.

4   Mount a dial indicator with the needle resting on the outside edge of the wheel. Grasp the front and rear of the tire and, using light pressure, wiggle the wheel back-and-forth and note the dial indicator reading. The gauge reading should be less than 0.108-inch. If the play in the steering system is more than specified, inspect each steering linkage pivot point and ballstud for looseness and replace parts, if necessary.

5   Raise the vehicle and support it on jackstands. Push up, then pull down on the center link end of the idler arm, exerting a force of approximately 25 pounds each way. Measure

the total distance the end of the arm travels. If the play is greater than 1/4-inch, replace the idler arm.

6   Check for torn ballstud boots, frozen joints and bent or damaged linkage components.

Removal and installation

Warning: *DO NOT allow the steering column shaft to rotate with any of the following steering linkage components removed or damage to the airbag system could occur.*

Tie-rod

Refer to illustrations 16.8a, 16.8b, 16.9a, 16.9b, 16.10a, 16.10b, 16.10c and 16.15

Note: *This procedure covers replacing the tie-rod ends as well as the entire tie-rod. If you'll only be replacing a tie-rod end, ignore the Steps that don't apply.*

7   Loosen the wheel lug nuts, raise the vehicle and support it securely on jackstands. Apply the parking brake. Remove the wheel.

8   If the inner or outer tie-rod end must be replaced, measure the distance from the end of the adjuster tube to the center of the ball-

16.9a Remove the cotter pin . . .

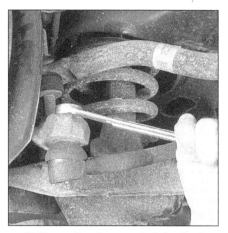

16.9b . . . then loosen - but don't yet remove - the castellated nut on the tie-rod end ballstud

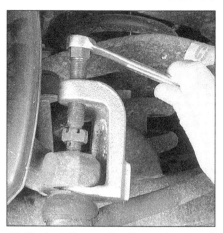

16.10a Install a small puller on the tie-rod end ballstud and separate the ballstud from the steering knuckle; leaving the nut in place will prevent the parts from violently separating

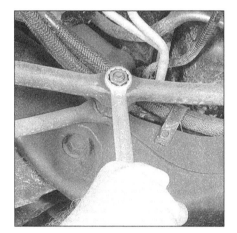

16.10b If you're separating the inner tie-rod end from the center link, loosen - but don't remove - the nut . . .

16.10c . . . then install a small puller and separate the inner tie-rod end and the center link

stud and record it **(see illustration)**. You can also count the number of exposed threads on the tie-rod end (the threads sticking out from the adjuster tube) **(see illustration)**.

9   Remove the cotter pin and loosen, but do not remove, the castellated nut from the ballstud **(see illustrations)**. If only the outer tie-rod end is being replaced, loosen the outer nut only. If only the inner tie-rod end is being replaced, loosen the inner nut only. If the entire tie-rod assembly is being replaced, loosen both nuts.

10   If the outer tie-rod end or the entire tie-rod assembly is being replaced, use a small puller to separate the outer tie-rod end from the steering knuckle **(see illustration)**, then remove the castellated nut and detach the outer tie-rod end ballstud from the knuckle. If the inner tie-rod end or the entire tie-rod is being replaced, separate the inner tie-rod end from the center link **(see illustrations)**.

11   Loosen the adjuster tube clamp bolts and unscrew the outer and/or inner tie-rod end(s).

12   Lubricate the threaded portion of the tie-rod end with chassis grease. Screw the new tie-rod end into the adjuster tube and adjust the distance from the tube to the ballstud to the

previously measured dimension. The number of threads showing on the inner and outer tie-rod ends should be equal within three threads. Don't tighten the clamp yet.

13   Connect the disconnected ballstud nut(s). Tighten the nut(s) to the torque listed in this Chapter's Specifications and install a new cotter pin. If the ballstud spins when attempt-

ing to tighten the nut, force it into the tapered hole with a large pair of pliers. If necessary, tighten the nut slightly to align a slot in the nut with the hole in the ballstud.

14   Insert the inner tie-rod end ballstud into the center link until it's seated. Install the nut and tighten it to the torque listed in this Chapter's Specifications.

15   Tighten the clamp nuts. The center of the bolt should be nearly horizontal and the adjuster tube slot must not line up with the gap in the clamps **(see illustration)**.

16   Install the wheel and lug nuts, lower the vehicle and tighten the lug nuts to the torque listed in the Chapter 1 Specifications. Drive the vehicle to an alignment shop to have the front end alignment checked and, if necessary, adjusted.

Pitman arm

Refer to illustrations 16.19, 16.20 and 16.21

17   Raise the vehicle and support it securely on jackstands.

18   Remove the center link nut from the Pitman arm ballstud. Discard the nut - don't reuse it.

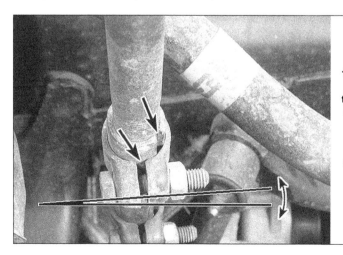

16.15 Make sure the pinch bolts for the tie-rod adjuster tube clamps are roughly horizontal, and the slot in the adjuster tube is not aligned with the gaps in the clamps (arrows)

16.19 Separate the center link from the Pitman arm with a small puller

16.20 Mark the relationship of the Pitman arm to the steering gear shaft

19   Using a puller, separate the Pitman arm ballstud from the center link **(see illustration)**.
20   Remove the Pitman arm-to-steering gear nut and washer. Mark the Pitman arm and the steering gear shaft to ensure proper alignment at reassembly time **(see illustration)**.
21   Remove the Pitman arm with a Pitman arm puller **(see illustration)**.
22   Inspect the ballstud threads for damage. Inspect the ballstud seals for excessive wear. Clean the threads on the ballstud.
23   Installation is the reverse of removal. Make sure the marks you made on the Pitman arm and Pitman shaft are aligned.

Idler arm

Refer to illustration 16.27

24   Raise the vehicle and support it securely on jackstands. Apply the parking brake.
25   Loosen but do not remove the idler arm-to-center link nut.
26   Separate the idler arm from the center link with a small puller. Remove the nut.

27   To detach the idler arm, remove the idler arm-to-frame nuts **(see illustration)**.
28   To install the idler arm, position it on the frame and install the bolts, tightening them to the torque listed in this Chapter's Specifications.
29   Insert the idler arm ballstud into the center link and install the nut. Tighten the nut to the specified torque. If the ballstud spins when attempting to tighten the nut, force it into the tapered hole with a large pair of pliers.

Center link

30   Raise the vehicle and support it securely on jackstands. Apply the parking brake.
31   Separate the two inner tie-rod ends from the center link (see Step 10).
32   Separate the center link from the Pitman arm **(see illustration 16.19)**.
33   Separate the idler arm from the center link.
34   Installation is the reverse of the removal procedure. If the ballstuds spin when attempting to tighten the nuts, force them into the tapered holes with a large pair of pliers. Be sure to tighten all of the nuts to the torque

listed in this Chapter's Specifications.

Steering damper

35   Inspect the steering damper for fluid leakage. A slight film of fluid near the shaft seal is normal, but if there's excessive fluid present and it's obviously coming from the steering damper, replace the damper.
36   Inspect the steering damper bushing for excessive wear. If it's in bad shape, replace the damper.
37   To test the damper itself, disconnect it from the frame or axle end (see next step). Using as much travel as possible, extend and compress the damper. The resistance should be smooth and constant for each stroke. If any binding or unusual noises are present, replace the damper.
38   Remove the damper ballstud-to-center link cotter pin, then remove the nut. Separate the damper from the center link.
39   Remove the steering damper mounting bolt and nut, then remove the damper.
40   Installation is the reverse of removal. Tighten all the fasteners securely.

16.21 Separate the Pitman arm from the steering gear shaft with a Pitman arm puller

16.27 To detach the idler arm, remove these two nuts (arrows)

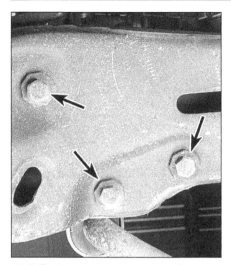

17.7 To detach the steering gear from the frame, remove these bolts (arrows)

17 Steering gear - removal and installation

Removal

Refer to illustration 17.7

Warning: *Disable the airbag system before working in the vicinity of the steering wheel, instrument panel or any airbag system component. Failure to do so could cause accidental deployment of the airbag resulting in personal injury (see Chapter 12).*
Caution: *Before starting, make sure that the wheels are pointing straight ahead and the ignition key lock cylinder is in the Lock position. Failure to do so might result in damage to the Supplemental Inflatable Restraint (SIR) coil (the airbag module lead).*

1   Raise the front of the vehicle and support it securely on jackstands. Apply the parking brake.
2   Place a drain pan under the steering gear. Pry loose the intermediate shaft shield **(see illustrations 15.1a and 15.1b)**, disconnect the power steering hose fittings and cap the ends to prevent excessive fluid loss and contamination.
3   Mark the relationship of the intermediate shaft lower universal joint to the steering gear input shaft. Remove the intermediate shaft lower pinch bolt **(see illustration 15.2)**.
4   Disconnect the Pitman arm from the center link **(see illustration 16.19)**.
5   Remove the Pitman arm nut and washer. Mark the relationship of the Pitman arm to the shaft so it can be installed in the same position **(see illustration 16.20)**.
6   Remove the Pitman arm from the shaft with a special puller **(see illustration 16.21)**.
7   Support the steering gear and remove the mounting bolts **(see illustration)**. Lower the unit, separate the intermediate shaft from the steering gear input shaft and remove the steering gear from the vehicle. **Warning:** *On*

models equipped with an airbag, *DO NOT allow the steering column shaft to rotate with the steering gear removed or damage to the airbag system could occur.*

Installation

8   Slide the Pitman arm onto the shaft. Make sure the marks are aligned. Install the washer and nut and tighten the nut to the torque listed in this Chapter's Specifications.
9   Raise the steering gear and Pitman arm into position, align the marks on the intermediate shaft and steering gear input shaft and connect the intermediate shaft to the steering gear. Reattach the Pitman arm to the center link.
10   Install the mounting bolts and washers and tighten them to the torque listed in this Chapter's Specifications.
11   Tighten the Pitman arm-to-steering gear nut and the Pitman arm-to-center link nut to the torque listed in this Chapter's Specifications.
12   Install the lower intermediate shaft pinch bolt and tighten it to the torque listed in this Chapter's Specifications. Install the plastic shield.
13   Connect the power steering hose fittings to the steering gear and fill the power steering pump reservoir with the recommended fluid (see Chapter 1).
14   Lower the vehicle and bleed the steering system (see Section 19).

18 Power steering pump - removal and installation

Removal

1   Disconnect the cable from the negative battery terminal.
2   Drain the coolant (see Chapter 1).
3   Using a syringe or siphon, remove as much power steering fluid as possible from the power steering pump reservoir to prevent spillage (see Chapter 1).
4   Remove the serpentine drivebelt (see Chapter 1).
5   If the vehicle is equipped with a mechanical fan, remove the pulley bracket for the engine cooling fan.
6   Detach the heater inlet and outlet hoses from the water pump (see Chapter 3). Detach the heater hose clip from the alternator and power steering pump bracket. Set the hoses aside.
7   Disconnect the pressure and return hoses and the reservoir hose from the power steering pump.
8   Remove the power steering pump-to-bracket bolts, then remove the power steering pump, the bracket between the pump and pulley, and the pulley, as a single assembly.
9   Remove the pulley from the power steering pump with a pulley removal tool.
10   Separate the pump from the bracket.

11   Installation is the reverse of removal. Tighten all pump bolts securely. To install the pulley, a pulley installation tool will be required (these are available at most auto parts stores).
12   Prime the pump by turning the pulley in the reverse direction to that of normal rotation (counterclockwise as viewed from the front) until air bubbles cease to emerge from the fluid when observed through the reservoir filler cap.
13   Install the serpentine drivebelt (see Chapter 1).
14   Bleed the power steering system (see Section 19).

19 Power steering system - bleeding

1   This is not a routine operation and normally will only be required when the system has been dismantled and reassembled.
2   Fill the reservoir to the correct level with fluid of the recommended type and allow it to remain undisturbed for at least two (2) minutes.
3   Start the engine and run it for two or three seconds only. Check the reservoir and add more fluid as necessary.
4   Repeat the operations described in the preceding paragraph until the fluid level remains constant.
5   Raise the front of the vehicle until the wheels are clear of the ground.
6   Start the engine and increase the speed to about 1500 rpm. Now turn the steering wheel gently from stop-to-stop. Check the reservoir fluid level.
7   Lower the vehicle to the ground and, with the engine still running, move the vehicle forward sufficiently to obtain full right lock followed by full left lock. Recheck the fluid level. If the fluid in the reservoir is extremely foamy, allow the vehicle to stand for a few minutes with the engine switched off and then repeat the previous operations. At the same time, check the belt tightness and check for a bent or loose pulley. Check also to make sure the power steering hoses are not touching any other part of the vehicle, especially sheet metal or the exhaust manifold.
8   The procedures above will normally remedy an extreme foam condition and/or an objectionably noisy pump (low fluid level and/or air in the power steering fluid are the leading causes of this condition). If, however, either or both conditions persist after a few trials, the power steering system will have to be thoroughly checked. Do not drive the vehicle until the condition(s) have been remedied.

20 Wheels and tires - general information

Refer to illustration 20.1

All vehicles covered by this manual are equipped with metric-sized fiberglass or steel-

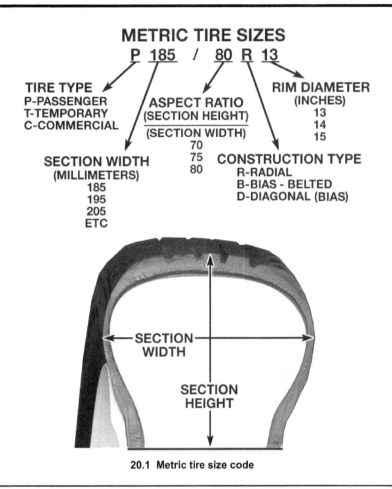

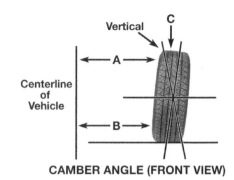

CAMBER ANGLE (FRONT VIEW)

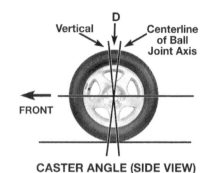

CASTER ANGLE (SIDE VIEW)

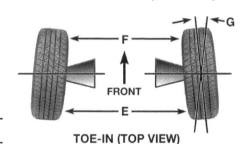

TOE-IN (TOP VIEW)

20.1 Metric tire size code

21.1 Front end alignment details

A minus B = C (degrees camber)
D = Degrees caster
E minus F = toe-in (measured in inches)
G - toe-in (expressed in degrees)

belted radial tires **(see illustration)**. Use of other size or type of tires may affect the ride and handling of the vehicle. Don't mix different types of tires, such as radials and bias belted, on the same vehicle as handling may be seriously affected. It's recommended that tires be replaced in pairs on the same axle, but if only one tire is being replaced, be sure it's the same size, structure and tread design as the other. Because tire pressure has a substantial effect on handling and wear, the pressure on all tires should be checked at least once a month or before any extended trips (see Chapter 1).

Wheels must be replaced if they are bent, dented, leak air, have elongated bolt holes, are heavily rusted, out of vertical symmetry or if the lug nuts won't stay tight. Wheel repairs that use welding or peening are not recommended.

Tire and wheel balance is important to the overall handling, braking and performance of the vehicle. Unbalanced wheels can adversely affect handling and ride characteristics as well as tire life. Whenever a tire is installed on a wheel, the tire and wheel should be balanced by a shop with the proper equipment.

21 Front end alignment - general information

Refer to illustration 21.1

A front end alignment refers to the adjustments made to the front wheels so they are in proper angular relationship to the suspension and the ground **(see illustration)**. Front wheels that are out of proper alignment not only affect steering control, but also increase tire wear. Camber, caster and toe-in can be adjusted on the vehicles covered by this manual.

Getting the proper front wheel alignment is a very exacting process, one in which complicated and expensive machines are necessary to perform the job properly. Because of this, you should have a technician with the proper equipment perform these tasks. We will, however, use this space to give you a basic idea of what is involved with front end alignment so you can better understand the process and deal intelligently with the shop that does the work.

Camber is the tilting of the front wheels from vertical when viewed from the front of the vehicle. Caster is the tilting of the top of the front steering axis from the vertical. A tilt toward the rear is positive caster and a tilt toward the front is negative caster. These two angles are adjusted by altering the relationship of the upper control arm to the crossmember.

Toe-in is the turning in of the front wheels. The purpose of a toe specification is to ensure parallel rolling of the front wheels. In a vehicle with zero toe-in, the distance between the front edges of the wheels will be the same as the distance between the rear edges of the wheels. The actual amount of toe-in is normally only a fraction of an inch. Toe-in adjustment is controlled by the position of the tie-rod end on the tie-rod. Incorrect toe-in will cause the tires to wear improperly by making them scrub against the road surface.

Chapter 11 Body

Contents

1 General information

Warning: *The models covered by this manual are equipped with airbags. Always disable the airbag system (see Chapter 12) before working in the vicinity of the impact sensors, steering column or instrument panel. Failure to follow these procedures may cause accidental deployment of the airbag, which could cause personal injury. The airbag circuits are easily identified by yellow insulation covering the entire wiring harness. Do not use electrical test equipment on any of these wires or tamper with them in any way.*

Caution: *On models equipped with a Delco Loc II or Theftlock audio system, be sure the lockout feature is turned off before performing any procedure which requires disconnecting the battery.*

The vehicles covered by this manual are built with a body-on-frame construction. The frame is a ladder-type, consisting of two box steel side rails joined by crossmembers. These crossmembers are welded to the side rails, with the exception of the transmission crossmember which is bolted into place for easy removal. The vehicle bodies are secured to the chassis by rubber insulated mounts and

can be completely removed from the chassis.

Certain body components are particularly vulnerable to accident damage and can be unbolted and repaired or replaced. Among these parts are the body moldings, front fenders, doors, bumpers, the hood, trunk lid and all glass.

Only general body maintenance practices and body panel repair procedures within the scope of the do-it-yourselfer are included in this Chapter.

2 Body - maintenance

1   The condition of your vehicle's body is very important, because the resale value depends a great deal on it. It's much more difficult to repair a neglected or damaged body than it is to repair mechanical components. The hidden areas of the body, such as the wheel wells, the frame and the engine compartment, are equally important, although they don't require as frequent attention as the rest of the body.

2   Once a year, or every 12,000 miles, it's a good idea to have the underside of the body steam cleaned. All traces of dirt and oil will be removed and the area can then

be inspected carefully for rust, damaged brake lines, frayed electrical wires, damaged cables and other problems. The front suspension components should be greased after completion of this job.

3   At the same time, clean the engine and the engine compartment with a steam cleaner or water-soluble degreaser.

4   The wheel wells should be given close attention, since undercoating can peel away and stones and dirt thrown up by the tires can cause the paint to chip and flake, allowing rust to set in. If rust is found, clean down to the bare metal and apply an anti-rust paint.

5   The body should be washed about once a week. Wet the vehicle thoroughly to soften the dirt, then wash it down with a soft sponge and plenty of clean soapy water. If the surplus dirt is not washed off very carefully, it can wear down the paint.

6   Spots of tar or asphalt thrown up from the road should be removed with a cloth soaked in solvent.

7   Once every six months, wax the body and chrome trim. If a chrome cleaner is used to remove rust from any of the vehicle's plated parts, remember that the cleaner also removes part of the chrome, so use it sparingly.

3   Vinyl trim - maintenance

Don't clean vinyl trim with detergents, caustic soap or petroleum-based cleaners. Plain soap and water works just fine, with a soft brush to clean dirt that may be ingrained. Wash the vinyl as frequently as the rest of the vehicle. After cleaning, application of a high-quality rubber and vinyl protectant will help prevent oxidation and cracks. The protectant can also be applied to weatherstripping, vacuum lines and rubber hoses, which often fail as a result of chemical degradation, and to the tires.

4   Upholstery and carpets - maintenance

1   Every three months remove the floormats and clean the interior of the vehicle (more frequently if necessary). Use a stiff whisk broom to brush the carpeting and loosen dirt and dust, then vacuum the upholstery and carpets thoroughly, especially along seams and crevices.

2   Dirt and stains can be removed from carpeting with basic household or automotive carpet shampoos available in spray cans. Follow the directions and vacuum again, then use a stiff brush to bring back the "nap" of the carpet.

3   Most interiors have cloth or vinyl upholstery, either of which can be cleaned and maintained with a number of material-specific cleaners or shampoos available in auto supply stores. Follow the directions on the product for usage, and always spot-test any upholstery cleaner on an inconspicuous area (bottom edge of a back seat cushion) to ensure that it doesn't cause a color shift in the material.

4   After cleaning, vinyl upholstery should be treated with a protectant. **Note:** *Make sure the protectant container indicates the product can be used on seats - some products may make a seat too slippery.* **Caution:** *Do not use protectant on vinyl-covered steering wheels.*

5   Leather upholstery requires special care. It should be cleaned regularly with saddle-soap or leather cleaner. Never use alcohol, gasoline, nail polish remover or thinner to clean leather upholstery.

6   After cleaning, regularly treat leather upholstery with a leather conditioner, rubbed in with a soft cotton cloth. Never use car wax on leather upholstery.

7   In areas where the interior of the vehicle is subject to bright sunlight, cover leather seating areas of the seats with a sheet if the vehicle is to be left out for any length of time.

5   Body repair - minor damage

Plastic body panels

The following repair procedures are for minor scratches and gouges. Repair of more serious damage should be left to a dealer service department or qualified auto body shop. Below is a list of the equipment and materials necessary to perform the following repair procedures on plastic body panels. Although a specific brand of material may be mentioned, it should be noted that equivalent products from other manufacturers may be used instead.

> *Wax, grease and silicone removing solvent*
> *Cloth-backed body tape*
> *Sanding discs*
> *Drill motor with three-inch disc holder*
> *Hand sanding block*
> *Rubber squeegees*
> *Sandpaper*
> *Non-porous mixing palette*
> *Wood paddle or putty knife*
> *Curved tooth body file*
> *Compoxy repair material, or equivalent (for flexible panels)*
> *Structural bonding epoxy, or equivalent (for rigid panels)*

Flexible panels (front and rear bumper fascia)

1   Remove the damaged panel, if necessary or desirable. In most cases, repairs can be carried out with the panel installed.

2   Clean the area(s) to be repaired with a wax, grease and silicone removing solvent applied with a water-dampened cloth.

3   If the damage is structural, that is, if it extends through the panel, clean the backside of the panel area to be repaired as well. Wipe dry.

4   Sand the rear surface about 1-1/2 inches beyond the break.

5   Cut two pieces of fiberglass cloth large enough to overlap the break by about 1-1/2 inches. Cut only to the required length.

6   Mix the adhesive from the repair kit according to the instructions included with the kit, and apply a layer of the mixture approximately 1/8-inch thick on the backside of the panel. Overlap the break by at least 1-1/2 inches

7   Apply one piece of fiberglass cloth to the adhesive and cover the cloth with additional adhesive. Apply a second piece of fiberglass cloth to the adhesive and immediately cover the cloth with additional adhesive insufficient quantity to fill the weave.

8   Allow the repair to cure for 20 to 30 minutes at 60-degrees to 80-degrees F.

9   If necessary, trim the excess repair material at the edge.

10   Remove all of the paint film over and around the area(s) to be repaired. The repair material should not overlap the painted surface.

11   With a drill motor and a sanding disc (or a rotary file), cut a "V" along the break line approximately 1/2-inch wide. Remove all dust and loose particles from the repair area.

12   Mix and apply the repair material. Apply a light coat first over the damaged area; then continue applying material until it reaches a level slightly higher than the surrounding finish.

13   Cure the mixture for 20 to 30 minutes at 60-degrees to 80-degrees F.

14   Roughly establish the contour of the area being repaired with a body file. If low areas or pits remain, mix and apply additional adhesive.

15   Block sand the damaged area with sandpaper to establish the actual contour of the surrounding surface.

16   If desired, the repaired area can be temporarily protected with several light coats of primer. Because of the special paints and techniques required for flexible body panels, it is recommended that the vehicle be taken to a paint shop for completion of the body repair.

Steel body panels

See photo sequence

Repair of minor scratches

17   If the scratch is superficial and does not penetrate to the metal of the body, repair is very simple. Lightly rub the scratched area with a fine rubbing compound to remove loose paint and built-up wax. Rinse the area with clean water.

18   Apply touch-up paint to the scratch, using a small brush. Continue to apply thin layers of paint until the surface of the paint in the scratch is level with the surrounding paint. Allow the new paint at least two weeks to harden, then blend it into the surrounding paint by rubbing with a very fine rubbing compound. Finally, apply a coat of wax to the scratch area.

19   If the scratch has penetrated the paint and exposed the metal of the body, causing the metal to rust, a different repair technique is required. Remove all loose rust from the bottom of the scratch with a pocket knife, then apply rust inhibiting paint to prevent the formation of rust in the future. Using a rubber or nylon applicator, coat the scratched area with glaze-type filler. If required, the filler can be mixed with thinner to provide a very thin paste, which is ideal for filling narrow scratches. Before the glaze filler in the scratch hardens, wrap a piece of smooth cotton cloth around the tip of a finger. Dip the cloth in thinner and then quickly wipe it along the surface of the scratch. This will ensure that the surface of the filler is slightly hollow. The scratch can now be painted over as described earlier in this section.

Repair of dents

20   When repairing dents, the first job is to pull the dent out until the affected area is as close as possible to its original shape. There is no point in trying to restore the original shape completely as the metal in the damaged area will have stretched on impact and cannot be restored to its original contours. It is better to bring the level of the dent up to a point which is about 1/8-inch below the level of the surrounding metal. In cases where the dent is very shallow, it is not worth trying to pull it out at all.

21   If the back side of the dent is accessible, it can be hammered out gently from behind using a soft-face hammer. While doing this, hold a block of wood firmly against the opposite side of the metal to absorb the hammer blows and prevent the metal from being stretched.

22   If the dent is in a section of the body which has double layers, or some other factor makes it inaccessible from behind, a different technique is required. Drill several small holes through the metal inside the damaged area, particularly in the deeper sections. Screw long, self-tapping screws into the holes just enough for them to get a good grip in the metal. Now the dent can be pulled out by pulling on the protruding heads of the screws with locking pliers.

23   The next stage of repair is the removal of paint from the damaged area and from an inch or so of the surrounding metal. This is done with a wire brush or sanding disk in a drill motor, although it can be done just as effectively by hand with sandpaper. To complete the preparation for filling, score the surface of the bare metal with a screwdriver or the tang of a file, or drill small holes in the affected area. This will provide a good grip for the filler material. To complete the repair, see the subsection on filling and painting later in this Section.

Repair of rust holes or gashes

24   Remove all paint from the affected area and from an inch or so of the surrounding metal using a sanding disk or wire brush mounted in a drill motor. If these are not available, a few sheets of sandpaper will do the job just as effectively.

25   With the paint removed, you will be able to determine the severity of the corrosion and decide whether to replace the whole panel, if possible, or repair the affected area. New body panels are not as expensive as most people think and it is often quicker to install a new panel than to repair large areas of rust.

26   Remove all trim pieces from the affected area except those which will act as a guide to the original shape of the damaged body, such as headlight shells, etc. Using metal snips or a hacksaw blade, remove all loose metal and any other metal that is badly affected by rust. Hammer the edges of the hole in to create a slight depression for the filler material.

27   Wire brush the affected area to remove the powdery rust from the surface of the metal. If the back of the rusted area is accessible, treat it with rust inhibiting paint.

28   Before filling is done, block the hole in some way. This can be done with sheet metal riveted or screwed into place, or by stuffing the hole with wire mesh.

29   Once the hole is blocked off, the affected area can be filled and painted. See the following subsection on filling and painting.

Filling and painting

30   Many types of body fillers are available, but generally speaking, body repair kits which contain filler paste and a tube of resin hardener are best for this type of repair work. A wide, flexible plastic or nylon applicator will be necessary for imparting a smooth and contoured finish to the surface of the filler material. Mix up a small amount of filler on a clean piece of wood or cardboard (use the hardener sparingly). Follow the manufacturer's instructions on the package, otherwise the filler will set incorrectly.

31   Using the applicator, apply the filler paste to the prepared area. Draw the applicator across the surface of the filler to achieve the desired contour and to level the filler surface. As soon as a contour that approximates the original one is achieved, stop working the paste. If you continue, the paste will begin to stick to the applicator. Continue to add thin layers of paste at 20-minute intervals until the level of the filler is just above the surrounding metal.

32   Once the filler has hardened, the excess can be removed with a body file. From then on, progressively finer grades of sandpaper should be used, starting with a 180-grit paper and finishing with 600-grit wet-or-dry paper. Always wrap the sandpaper around a flat rubber or wooden block, otherwise the surface of the filler will not be completely flat. During the sanding of the filler surface, the wet-or-dry paper should be periodically rinsed in water. This will ensure that a very smooth finish is produced in the final stage.

33   At this point, the repair area should be surrounded by a ring of bare metal, which in turn should be encircled by the finely feathered edge of good paint. Rinse the repair area with clean water until all of the dust produced by the sanding operation is gone.

34   Spray the entire area with a light coat of primer. This will reveal any imperfections in the surface of the filler. Repair the imperfections with fresh filler paste or glaze filler and once more smooth the surface with sandpaper. Repeat this spray-and-repair procedure until you are satisfied that the surface of the filler and the feathered edge of the paint are perfect. Rinse the area with clean water and allow it to dry completely.

35   The repair area is now ready for painting. Spray painting must be carried out in a warm, dry, windless and dust free atmosphere. These conditions can be created if you have access to a large indoor work area, but if you are forced to work in the open, you will have to pick the day very carefully. If you are working indoors, dousing the floor in the work area with water will help settle the dust which would otherwise be in the air. If the repair area is confined to one body panel, mask off the surrounding panels. This will help minimize the effects of a slight mismatch in paint color. Trim pieces such as chrome strips, door handles, etc., will also need to be masked off or removed. Use masking tape and several thickness of newspaper for the masking operations.

36   Before spraying, shake the paint can thoroughly, then spray a test area until the spray painting technique is mastered. Cover the repair area with a thick coat of primer. The thickness should be built up using several thin layers of primer rather than one thick one. Using 600-grit wet-or-dry sandpaper, rub down the surface of the primer until it is very smooth. While doing this, the work area should be thoroughly rinsed with water and the wet-or-dry sandpaper periodically rinsed as well. Allow the primer to dry before spraying additional coats.

37   Spray on the top coat, again building up the thickness by using several thin layers of paint. Begin spraying in the center of the repair area and then, using a circular motion, work out until the whole repair area and about two inches of the surrounding original paint is covered. Remove all masking material 10 to 15 minutes after spraying on the final coat of paint. Allow the new paint at least two weeks to harden, then use a very fine rubbing compound to blend the edges of the new paint into the existing paint. Finally, apply a coat of wax.

6   Body repair - major damage

1   Major damage must be repaired by an auto body/frame repair shop with the necessary welding and hydraulic straightening equipment.

2   If the damage has been serious, it is vital that the structure be checked for proper alignment or the vehicle's handling characteristics may be adversely affected. Other problems, such as excessive tire wear and wear in the driveline and steering may occur.

3   Due to the fact that all of the major body components (hood, fenders, etc.) are separate and replaceable units, any seriously damaged components should be replaced rather than repaired. Sometimes these components can be found in a wrecking yard that specializes in used vehicle components, often at considerable savings over the cost of new parts.

7   Hinges and locks - maintenance

Once every 3000 miles, or every three months, the hinges and latch assemblies on the doors, hood and trunk should be given a few drops of light oil or lock lubricant. The door latch strikers should also be lubricated with a thin coat of grease to reduce wear and ensure free movement. Lubricate the door and trunk locks with spray-on graphite lubricant.

8   Windshield and fixed glass - replacement

Replacement of the windshield and fixed glass requires the use of special fast setting adhesive/caulk materials. These operations should be left to a dealer or a shop specializing in glass work.

These photos illustrate a method of repairing simple dents. They are intended to supplement *Body repair - minor damage* in this Chapter and should not be used as the sole instructions for body repair on these vehicles.

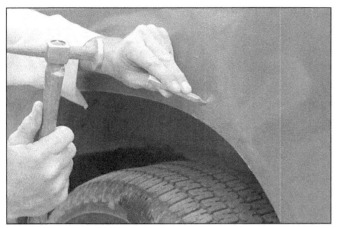

1 If you can't access the backside of the body panel to hammer out the dent, pull it out with a slide-hammer-type dent puller. In the deepest portion of the dent or along the crease line, drill or punch hole(s) at least one inch apart . . .

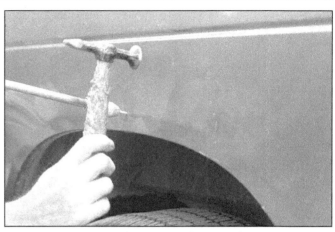

2 . . . then screw the slide-hammer into the hole and operate it. Tap with a hammer near the edge of the dent to help 'pop' the metal back to its original shape. When you're finished, the dent area should be close to its original contour and about 1/8-inch below the surface of the surrounding metal

3 Using coarse-grit sandpaper, remove the paint down to the bare metal. Hand sanding works fine, but the disc sander shown here makes the job faster. Use finer (about 320-grit) sandpaper to feather-edge the paint at least one inch around the dent area

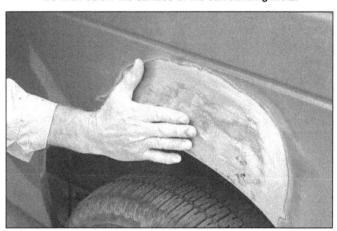

4 When the paint is removed, touch will probably be more helpful than sight for telling if the metal is straight. Hammer down the high spots or raise the low spots as necessary. Clean the repair area with wax/silicone remover

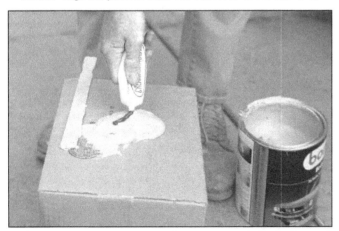

5 Following label instructions, mix up a batch of plastic filler and hardener. The ratio of filler to hardener is critical, and, if you mix it incorrectly, it will either not cure properly or cure too quickly (you won't have time to file and sand it into shape)

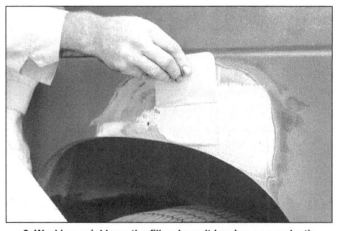

6 Working quickly so the filler doesn't harden, use a plastic applicator to press the body filler firmly into the metal, assuring it bonds completely. Work the filler until it matches the original contour and is slightly above the surrounding metal

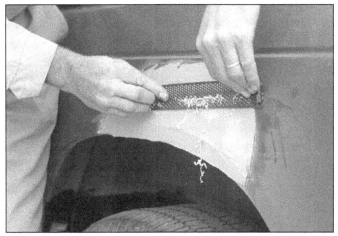

7 Let the filler harden until you can just dent it with your fingernail. Use a body file or Surform tool (shown here) to rough-shape the filler

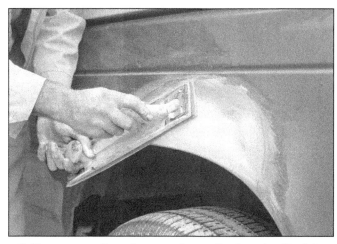

8 Use coarse-grit sandpaper and a sanding board or block to work the filler down until it's smooth and even. Work down to finer grits of sandpaper - always using a board or block - ending up with 360 or 400 grit

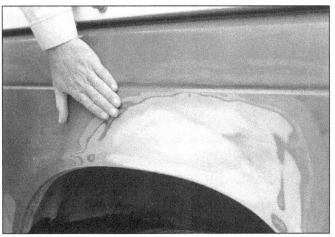

9 You shouldn't be able to feel any ridge at the transition from the filler to the bare metal or from the bare metal to the old paint. As soon as the repair is flat and uniform, remove the dust and mask off the adjacent panels or trim pieces

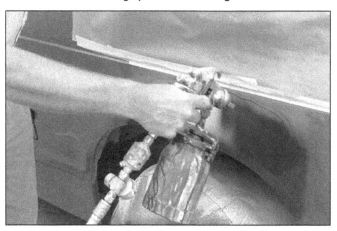

10 Apply several layers of primer to the area. Don't spray the primer on too heavy, so it sags or runs, and make sure each coat is dry before you spray on the next one. A professional-type spray gun is being used here, but aerosol spray primer is available inexpensively from auto parts stores

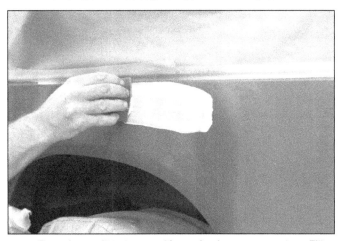

11 The primer will help reveal imperfections or scratches. Fill these with glazing compound. Follow the label instructions and sand it with 360 or 400-grit sandpaper until it's smooth. Repeat the glazing, sanding and respraying until the primer reveals a perfectly smooth surface

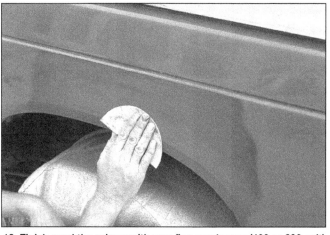

12 Finish sand the primer with very fine sandpaper (400 or 600-grit) to remove the primer overspray. Clean the area with water and allow it to dry. Use a tack rag to remove any dust, then apply the finish coat. Don't attempt to rub out or wax the repair area until the paint has dried completely (at least two weeks)

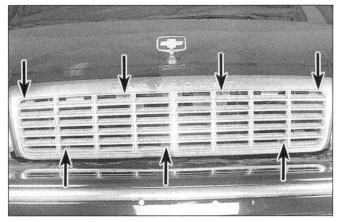

9.1 Radiator grille mounting screw locations (arrows) (Chevrolet shown, Buick similar)

9.2 Open the hood, then remove the screw caps from the back of the lower grille retaining screws

10.3 Use a small screwdriver to pry the retaining clip out of its locking groove, then detach the end of the strut from the mounting stud

11.5 With the help of an assistant to hold the hood, remove the retaining bolts (arrows) from each hinge plate, then lift off the hood

11.10 Loosen the bolts (arrows) and move the hood latch to adjust the hood closed position

9   Radiator grille - removal and installation

Refer to illustrations 9.1 and 9.2

1   Remove the screws securing the top half of the grille **(see illustration)**.
2   Open the hood, and remove the screw caps from the back of the screws securing the lower half of the grille **(see illustration)**.
3   Remove the screws securing the lower half of the grille and detach the grille from the vehicle **(see illustration 9.1)**.
4   Installation is the reverse of removal.

10   Hood and liftgate (station wagon) support struts - removal and installation

Refer to illustration 10.3

Note: *The hood and rear liftgate are heavy and somewhat awkward to hold - at least two people should perform this procedure.*

1   Open the hood or rear liftgate and support it securely.

2   Disconnect any electrical connectors which would interfere with removal.
3   Using a small screwdriver, detach the retaining clips at both ends of the support strut. Then pry or pull sharply to detach it from the vehicle **(see illustration)**.
4   Installation is the reverse of removal.

11   Hood - removal, installation and adjustment

Note: *The hood is somewhat awkward to remove and install, at least two people should perform this procedure.*

Removal and installation

Refer to illustration 11.5

1   Open the hood, then place blankets or pads over the fenders and cowl area of the body. This will protect the body and paint as the hood is lifted off.
2   Make marks or scribe a line around the hood hinge to ensure proper alignment during installation.
3   Disconnect any cables or wires that will interfere with removal.

4   Have an assistant support the weight of the hood and detach the support struts (see Section 10).
5   Remove the hinge-to-hood bolts and lift off the hood **(see illustration)**.
6   Installation is the reverse of removal. Align the hinge bolts with the marks made in step 2.

Adjustment

Refer to illustrations 11.10 and 11.11

7   Fore-and-aft and side-to-side adjustment of the hood is done by moving the hinge plate slot after loosening the bolts or nuts.
8   Scribe a line around the entire hinge plate so you can determine the amount of movement.
9   Loosen the bolts or nuts and move the hood into correct alignment. Move it only a little at a time. Tighten the hinge bolts and carefully lower the hood to check the position.
10   If necessary after installation, the entire hood latch assembly can be adjusted up-and-down as well as from side-to-side on the radiator support so the hood closes securely and flush with the fenders. To make the adjustment, scribe a line or mark around the

11.11 Screw the hood bumpers in or out to adjust the hood flush with the fenders

12.1 Pry out the cable retainer from the backside of the hood latch assembly, then disengage the cable

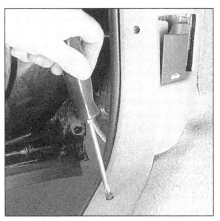

12.6 Remove the driver's side kick panel to access the hood release lever

hood latch mounting bolts to provide a reference point, then loosen them and reposition the latch assembly, as necessary **(see illustration)**. Following adjustment, retighten the mounting bolts.

11   Finally, adjust the hood bumpers on the radiator support so the hood, when closed, is flush with the fenders **(see illustration)**.

12   The hood latch assembly, as well as the hinges, should be periodically lubricated with white, lithium-base grease to prevent binding and wear.

12   Hood latch and release cable - removal and installation

Warning: *The models covered by this manual are equipped with airbags. Always disable the airbag system (see Chapter 12) before working in the vicinity of the impact sensors, steering column or instrument panel. Failure to follow these procedures may cause accidental deployment of the airbag, which could cause personal injury. The airbag circuits are easily identified by yellow insulation covering the entire wiring harness. Do not use electrical test equipment on any of these wires or tamper with them in any way.*

Latch

Refer to illustration 12.1

1   Disconnect the hood release cable by disengaging the cable from the latch assembly **(see illustration)**.

2   Scribe a line around the latch to aid alignment when installing, then remove the retaining bolts to the radiator support **(see illustration 11.10)**. Remove the latch.

3   Installation is the reverse of removal. **Note:** *Adjust the latch so the hood engages securely when closed and the hood bumpers are slightly compressed.*

Cable

Refer to illustrations 12.6 and 12.7

4   Disconnect the hood release cable from the latch assembly as described in step 1.

5   Attach a piece of thin wire or string to

the end of the cable and unclip all remaining cable retaining clips.

6   Working in the passenger compartment, remove the drivers side kick panel **(see illustration)**.

7   Pry open the cable retaining clip, then detach the cable from the hood release lever **(see illustration)**.

8   Pull the cable and grommet rearward into the passenger compartment until you can see the wire or string. Ensure that the new cable has a grommet attached then remove the wire or string from the old cable and fasten it to the new cable.

9   With the new cable attached to the wire or string, pull the wire or string back through the firewall until the new cable reaches the latch assembly.

10   Working in the passenger compartment, reinstall the new cable into the hood release lever. Use pliers to crimp the retaining clip which secures the cable to the release lever.

11   The remainder of the installation is the reverse of removal. **Note:** *Push on the grommet with your fingers from the passenger compartment to seat the grommet in the firewall correctly.*

13   Bumpers - removal and installation

Warning: *The models covered by this manual are equipped with airbags. Always disable the airbag system (see Chapter 12) before working in the vicinity of the impact sensors, steering column or instrument panel. Failure to follow these procedures may cause accidental deployment of the airbag, which could cause personal injury. The airbag circuits are easily identified by yellow insulation covering the entire wiring harness. Do not use electrical test equipment on any of these wires or tamper with them in any way.*

Caution: *On models equipped with a Delco Loc II or Theftlock audio system, be sure the lockout feature is turned off before performing any procedure which requires disconnecting the battery.*

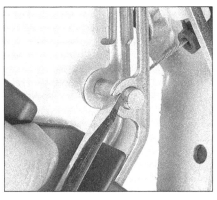

12.7 Using a screwdriver, pry open the retaining clip securing the cable to the hood release lever, then disengage the cable from the lever

Front bumper

Refer to illustrations 13.3, 13.4a, 13.4b, 13.5, 13.6 and 13.8

1   Open the hood and disconnect the negative battery cable.

2   Raise the front of the vehicle and support it securely on jackstands.

3   Working in the front wheel opening, detach the wheel opening moldings from each fender **(see illustration)**.

13.3 Remove the screws and detach the wheel opening moldings

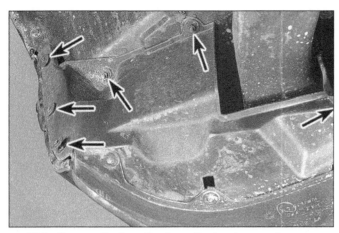

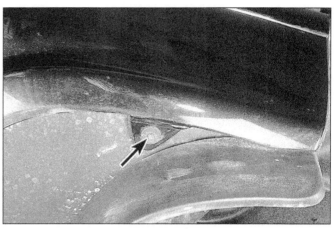

13.4a Remove the inner fenderwell extension panel screws (arrows) from inside the wheel opening . . .

13.4b . . . and from the lower front corner(s) of the bumper cover

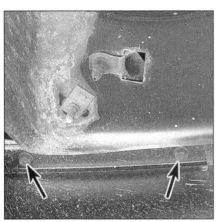

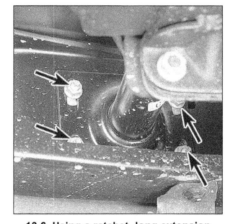

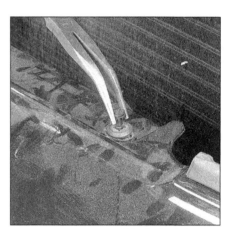

13.5 Detach the bolts (arrows) securing the bumper cover to both fenders

13.6 Using a ratchet, long extension and a socket, remove the front bumper retaining nuts (arrows)

13.8 Pull out on the center pin to release the clips securing the bumper cover to the bumper

4   Detach the retaining screws securing the inner fenderwell extension panels to the bumper cover **(see illustrations)**.

5   Remove the retaining screws securing the bumper cover to each fender **(see illustration)**.

6   Working under the front of the vehicle, remove the retaining nuts securing the bumper to the bumper energy absorbers **(see illustration)**.

7   Separate the bumper cover from the fenders, then pull the bumper assembly straight out and away from the vehicle to remove it.

8   To remove the bumper cover from the bumper, simply detach the plastic clips securing the upper and lower edges of the bumper cover **(see illustration)**.

9   Installation is the reverse of removal.

Rear bumper

Refer to illustrations 13.11, 13.12, 13.14, 13.15 and 13.17

10   Apply the parking brake, raise the vehicle and support it securely on jackstands.

11   Working in the rear wheel opening, detach the wheel opening moldings from each quarter panel. Working under the vehicle, remove the retaining nuts securing the

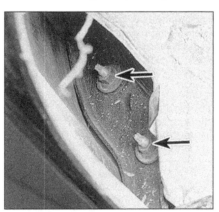

13.11 Working under the vehicle, remove the nuts (arrows) securing the bumper cover to the rear quarter panels

bumper cover to the rear quarter panels **(see illustration)**.

12   Drill out the retaining rivets (if equipped) securing the lower edge bumper cover to the rear quarter panels **(see illustration)**.

13   On Station wagon models, Use a heat gun or blow dryer to loosen the adhesive securing the plastic trim covers at the upper edges of the bumper cover, then carefully pry

13.12 Drill out the retaining rivet (arrow) securing the lower edges of the bumper cover (if equipped)

off the trim covers with a trim removal tool or a putty knife. Drill out the retaining rivets securing the upper edges of the bumper cover. **Note:** *Later models use screws to fasten the upper edges of the bumper cover.*

14   On Sedan models, remove the plastic retaining nuts and clips securing the trunk finishing panels. Peel back the trunk finishing panels and remove the remaining nuts secur-

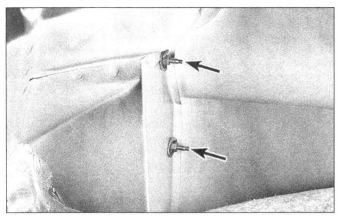

13.14 On Sedan models, peel back the trunk finishing panels to access the remaining bumper cover-to-rear quarter panel retaining nuts (arrows)

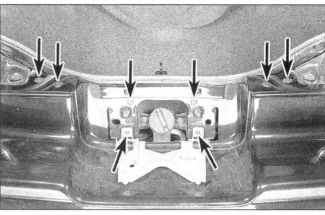

13.15 On Chevrolet models, detach the license plate door assembly and the plastic screws securing the upper edge of the bumper cover (arrows)

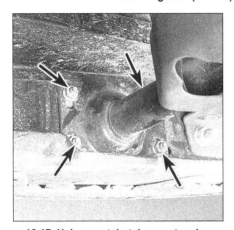

13.17 Using a ratchet, long extension and a socket, remove the rear bumper retaining nuts (arrows)

14.7a Remove the three bolts (arrows) at the front of the fender . . .

14.7b . . . and four more bolts (arrows) located behind the headlight assembly

ing the bumper cover to rear quarter panels **(see illustration)**.

15 On Chevrolet models, remove the rear license plate door assembly and the plastic screws securing the upper edge of the bumper cover **(see illustration)**.

16 On Buick models, disconnect the electrical connectors from the back-up lights.

17 On all models, work under the vehicle and remove the retaining nuts securing the bumper to the bumper energy absorbers **(see illustration)**. Pull the bumper assembly straight out and away from the vehicle to remove it.

18 To remove the bumper cover from the bumper, simply detach the plastic clips securing the upper and lower edges of the bumper cover **(see illustration 13.8)**.

19 Installation is the reverse of removal.

14 Front fender - removal and installation

Refer to illustrations 14.7a through 14.7g

1 Remove the hood (see Section 11).

2 Raise the vehicle, support it securely on jackstands and remove the front wheel.

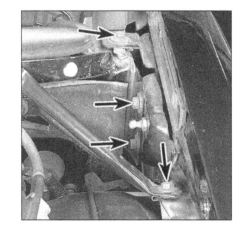

14.7c Remove the four bolts (arrows) securing the fender to the firewall

3 Remove the front side marker light and the headlight housing assembly from the side of the vehicle that the fender is to be removed (see Chapter 12).

4 Remove the front bumper assembly (see Section 13).

5 If your removing the passenger side fender, remove the battery, battery tray (see Chapter 5) and the radio antenna (see Chapter 12).

14.7d Open the front door to access the upper fender-to-door pillar retaining bolt (arrow) . . .

6 If your removing the driver side fender on a later model vehicle, remove the air filter housing assembly (see Chapter 4). Then move the ECM, which is located beneath the air filter housing off to the side to access the fender mounting bolts.

7 Remove the remaining fender mounting bolts **(see illustrations)**.

8   Detach the fender. It's a good idea to have an assistant support the fender while it's being moved away from the vehicle to prevent damage to the surrounding body panels.
9   Installation is the reverse of removal.

15 Trunk lid - removal, installation and adjustment

Note: *The trunk lid is heavy and somewhat awkward to remove and install - at least two people should perform this procedure.*

Removal and installation

Refer to illustration 15.4

1   Open the trunk lid and cover the edges of the trunk compartment with pads or cloths to protect the painted surfaces when the lid is removed.
2   Disconnect any cables or wire harness connectors attached to the trunk lid that would interfere with removal.
3   Use a felt-tip marker or scribe to make alignment marks around the trunk lid hinge.
4   While an assistant supports the lid, remove the hinge bolts from both sides and lift the trunk lid off the vehicle **(see illustration)**.
5   Installation is the reverse of removal.
Note: *When reinstalling the trunk lid, align the hinge with the marks made during removal.*

Adjustment

Refer to illustrations 15.9 and 15.10

6   Fore-and-aft and side-to-side adjustment of the trunk lid is done by moving the hood in relation to the hinge plate after loosening the bolts or nuts.
7   Scribe a line around the entire hinge plate as described earlier in this section so you can judge the amount of movement.
8   Loosen the bolts or nuts and move the trunk lid into correct alignment. Move it only a little at a time. Tighten the hinge bolts or nuts and carefully lower the trunk lid to check the alignment.

14.7e . . . and the lower fender-to-door pillar retaining bolt (arrow)

14.7f Remove the retaining bolt (arrow) securing the fender to the rocker panel - some models have two retaining bolts at this location

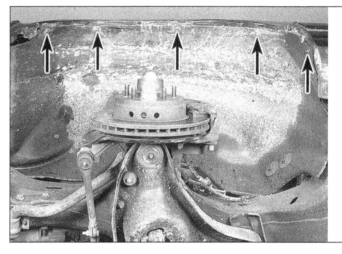

14.7g Detach the remaining bolts (arrows) from the inner fenderwell, then remove the fender from the vehicle

9   If necessary after installation, the entire trunk lid latch assembly can be adjusted up and down as well as from side to side on the trunk lid so the lid closes securely and is flush with the rear quarter panels. To do this, scribe a line around the trunk lid latch mounting bolts to provide a reference point. Then loosen the bolts and reposition the latch assembly as necessary **(see illustra-** **tion)**. Following adjustment, retighten the mounting bolts.
10   Adjust the bumpers on the trunk lid, so that the trunk lid is flush with the rear quarter panels when closed **(see illustration)**.
11   The trunk lid latch assembly, as well as the hinges, should be periodically lubricated with white lithium-base grease to prevent sticking and wear.

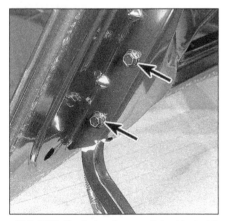

15.4 With the help of an assistant to hold the trunk lid, remove the four retaining bolts and lift off the trunk lid

15.9 Loosen the bolts (arrows) and move the latch assembly as necessary to adjust the trunk lid flush with the quarter panels in the closed position

15.10 Adjust the bumpers so they're slightly compressed when the trunk lid is in the closed position

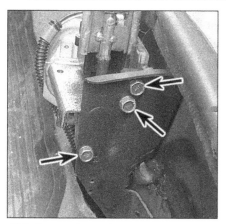

16.3 Automatic closing unit retaining bolt locations (arrows)

16.6 Drill out the two rivets securing the emblem

16.7 Detach the license plate lamp retaining screws (arrows), then disconnect the electrical connector from the lamp assembly

16 Trunk lid latch, latch striker and lock cylinder - removal and installation

Caution: *On models equipped with a Delco Loc II or Theftlock audio system, be sure the lockout feature is turned off before performing any procedure which requires disconnecting the battery.*

Latch and latch striker

Refer to illustration 16.3

1   On models with electrically operated automatic closing units, disconnect the negative cable from the battery. Unplug any electrical connectors and detach any cables from the latch assembly.

2   The trunk lid latch is retained by bolts which can readily be removed with a wrench **(see illustration 15.9)**. For adjustment procedures, see Section 15.

3   On models with manually operated trunk lids the striker is welded to the trunk and can not be adjusted. On models with automatic closing units the striker is retained with bolts **(see illustration)**.

4   Disconnect all electrical connections, then detach the closing unit retaining bolts and remove the striker assembly from the vehicle.

5   Installation is the reverse of removal.

Lock cylinder

Refer to illustrations 16.6, 16.7 and 16.8

6   Remove the trunk lock emblem retaining rivets, then remove the emblem so the lock cylinder can be withdrawn **(see illustration)**.

7   Remove the retaining screws securing the license plate lamp, then pull the lamp assembly down and away to access the lock cylinder retaining clip **(see illustration)**.

8   Using a small drill bit, drill out the rivets securing the lock cylinder retaining clip **(see illustration)**, then pull the retaining clip down or away from the lock cylinder and remove the cylinder from the trunk lid.

9   Installation is the reverse of removal.

17 Liftgate (station wagon) - removal, installation and adjustment

Note: *The liftgate is heavy and somewhat awkward to hold - at least two people should perform this procedure.*

Removal and installation

1   Open the liftgate and support it securely.

2   Remove the liftgate trim panels and disconnect all wiring harness connectors leading to the liftgate.

3   While an assistant supports the liftgate, detach both ends of the support struts. Then pry or pull sharply to remove them from the vehicle.

4   Detach the hinge to liftgate bolts and remove the liftgate from the vehicle.

5   Installation is the reverse of removal.

Adjustment

6   Adjustments are made by loosening the hinge-to liftgate bolts and moving the liftgate. Proper alignment is achieved when the edges of the liftgate are parallel with the rear quarter panel and the top of the tailgate.

7   Finally, adjust the latch striker assembly as necessary (up and down) to provide positive engagement with the latch mechanism.

18 Tailgate (station wagon) - removal, installation and adjustment

Note: *The tailgate is heavy and somewhat awkward to remove and install - at least two people should perform this procedure.*

Removal and installation

1   Open the tailgate so that it rests in the open door position. Wait a few minutes for the tension on the torque rod to relieve itself, then proceed to Step 2.

2   Working just above the left side tailgate hinge, remove the torque rod link retainer and torque rod link.

16.8 Drill out the rivets (arrows) and remove the lock cylinder retaining clip - the lock cylinder can then be removed from the outside of the trunk lid

3   Detach the clips and screws from left rear quarter inner trim panel and remove the trim panel from the vehicle.

4   Disconnect all wiring harness connectors leading to the tailgate that would interfere with removal of the tailgate.

5   Cover the bumper area around the tailgate opening with pads or cloths to protect the painted surfaces when the tailgate is removed.

6   Close the tailgate. Re-open the tailgate so that it rests in the open gate position (lying down).

7   While an assistant supports the tailgate, detach the tailgate support spring.

8   Scribe a line around the hinge to help aid alignment when installing. Working through the inner trim panel opening, remove the hinge retaining nuts.

9   Depress the tailgate door release handle and remove the tailgate from the vehicle.

10   Installation is the reverse of removal.

Adjustment

11   Tailgate adjustments are made by loosening the hinge-to-body or hinge-to-tailgate bolts and moving the tailgate. Proper alignment is achieved when the top of the tail-

19.1a Pull the inside handle lever to access the bezel retaining screw

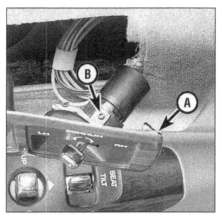

19.1b Use a small screwdriver to release the bezel retaining clip (A), then remove the mirror switch retaining screw (B) (Chevrolet model shown)

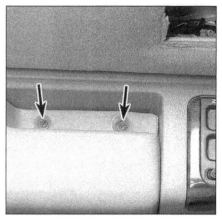

19.2a Remove the screws (arrows) securing the armrest pull handle

19.2b Remove the screw (arrow) located above the map pocket

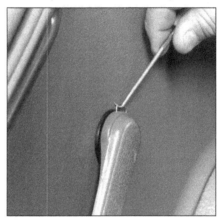

19.3 If your vehicle is equipped with manual windows, use a hooked tool like this to remove the window crank retaining clip

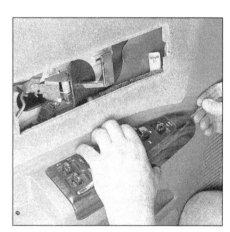

19.4a Use a small screwdriver to disengage the retaining clip at the front of the switch control plate . . .

gate is aligned with the top of the rear quarter panel. If these goals can't be reached by adjusting the hinge to body or hinge to tailgate bolts, body alignment shims may have to be purchased and inserted behind the hinges to achieve correct alignment.

12   To adjust the tailgate closed position, first check that the latch is contacting the center of the latch striker assembly. If not, remove the striker assembly and add or subtract shims to achieve correct alignment.

13   Finally, adjust the latch striker assembly as necessary (up and down or sideways) to provide positive engagement with the latch mechanism and the outside of the tailgate is flush with rear quarter panel.

---

19   Door trim panel - removal and installation

Caution: *On models equipped with a Delco Loc II or Theftlock audio system, be sure the lockout feature is turned off before performing any procedure which requires disconnecting the battery.*

Removal

Refer to illustrations 19.1a, 19.1b, 19.2a, 19.2b, 19.3, 19.4a, 19.4b, 19.6 and 19.8

1   Disconnect the cable from the negative terminal of the battery. Remove the inside door handle trim bezel **(see illustrations)**.

2   Detach the armrest pull handle retaining screws and the lower door trim panel retaining screw **(see illustrations)**

3   On manual window equipped models, remove the window crank, using a hooked tool to remove the retainer clip **(see illustration)**. A special tool is available for this purpose, but it's not essential. With the clip removed, pull off the handle.

4   On power window equipped models, pry out the armrest switch control plate **(see illustrations)** and disconnect the electrical connections.

5   On Buick models, remove the door pull handle retaining screws (if equipped).

6   Insert a wide putty knife, a thin screwdriver or a special trim panel removal tool between the trim panel and the head of the retaining clip to disengage the door panel

19.4b . . . then disconnect the electrical connectors from the backside

retaining clips **(see illustration)**.

7   Once all of the clips and screws are disengaged, detach the trim panel, disconnect any electrical connectors and remove the trim panel from the vehicle by gently pulling it up and out.

19.6 Insert a trim removal tool or a putty knife between the door and the trim panel to disengage the clips along the outer edge of the door trim panel

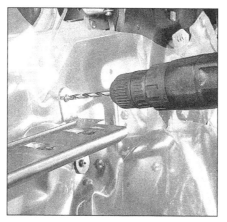

19.8 To access the inner door, drill out the armrest support bracket retaining rivets

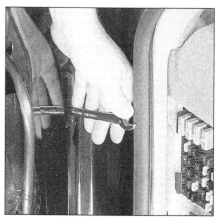

20.7a The front door retaining bolts can be removed using a wrench

8   For access to the inner door remove the door panel support bracket **(see illustration)**. Then peel back the watershield, taking care not to tear it.

Installation

9   If necessary, add more sealant then press the watershield back into place.

10   Install the door panel support bracket. **Note:** *If a rivet gun and rivets are not available, sheet metal screws of the appropriate size may be used to secure the door panel support bracket.*

11   Prior to installation of the door panel, be sure to reinstall any door panel retaining clips which may have come out of the door trim panel during the removal procedure back into the door trim panel.

12   Working through the window opening, engage the hooks on the back of the trim panel onto the door and push down until they are seated, then press the door panel retaining clips into place. **Note:** *When installing door trim panel retaining clips, make sure the clips are lined up with their mating holes first, then gently tap inward with the palm of your hand.*

13   The remainder of the installation is the reverse of removal.

20   Door - removal, installation and adjustment

Caution: *On models equipped with a Delco Loc II or Theftlock audio system, be sure the lockout feature is turned off before performing any procedure which requires disconnecting the battery.*

Note: *The door is heavy and somewhat awkward to remove and install - at least two people should perform this procedure. This procedure applies to both front and rear doors.*

Removal and installation

Refer to illustrations 20.7a and 20.7b

1   Raise the window completely and disconnect the negative cable from the battery if

20.7b Open the front door to access the rear door retaining bolts (arrows)

equipped with power windows.

2   Open the door all the way and support it on jacks or blocks covered with rags to prevent damaging the paint.

3   Remove the door trim panel and water deflector as described in (Section 19).

4   Unplug all electrical connections, ground wires and harness retaining clips from the door. **Note:** *It is a good idea to label all connections to aid the reassembly process.*

5   Working on the door side, detach the rubber conduit between the body and the door. Pull the wiring harness through the conduit hole and remove the wiring from the door.

6   Mark around the door hinges with a pen or a scribe to facilitate realignment during reassembly.

7   Have an assistant hold the door, remove the hinge to door bolts **(see illustrations)** and lift the door off.

8   Installation is the reverse of removal.

Adjustment

Refer to illustration 20.12

9   Having proper door to body alignment is a critical part of a well functioning door assembly. First check the door hinge pins and bushings for excessive play. **Note:** *If the door can*

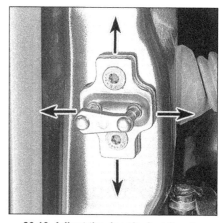

20.12 Adjust the door lock striker by loosening the mounting screws and gently tapping the striker in the desired direction

be lifted (1/16-inch or more) without the car body lifting with it the hinge pins and bushings should be replaced.

10   Door-to-body alignment adjustments are made by loosening the hinge-to-body bolts or hinge-to-door bolts and moving the door. Proper body alignment is achieved when the top of the doors are parallel with the roof section, the front door is flush with the fender, the rear door is flush with the rear quarter panel and the bottom of the doors are aligned with the lower rocker panel. If these goals can't be reached by adjusting the hinge-to-body or hinge-to-door bolts, body alignment shims may have to be purchased and inserted behind the hinges to achieve correct alignment.

11   To adjust the door closed position, first check that the door latch is contacting the center of the latch striker bolt. If not, remove the striker bolt and add or subtract washers to achieve correct alignment.

12   Finally, adjust the latch striker bolt as necessary (up and down or sideways) to provide positive engagement with the latch mechanism **(see illustration)** and the door panel is flush with the center pillar or rear quarter panel.

21.2 Remove the latch retaining screws (arrows) from the end of the door, then detach the actuating rods and pull the latch assembly through the access hole

21.9 To remove the lock cylinder, detach the plastic clip securing the lock rod, then pry off the lock cylinder retaining clip (arrow)

21.11 The outside handle retaining nuts (arrows) can be reached through the access holes in the door frame with a socket, extension and ratchet

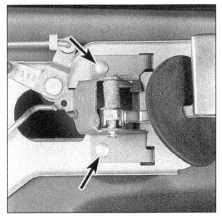

21.14 Drill out the handle retaining rivets (arrows), then rotate the handle out and detach the actuating rods from the backside

22.3 Remove the upper trim molding surrounding the door window

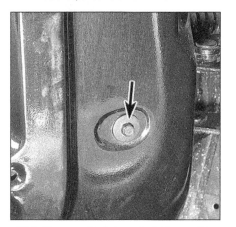

22.4 Remove the forward window guide channel retaining bolts (arrow)

21 Door latch, lock cylinder and handles - removal and installation

Door latch

Refer to illustration 21.2

1   Raise the window then remove the door trim panel and watershield as described in (Section 19).
2   Remove the screws securing the latch to the door **(see illustration)**.
3   Working through the large access hole, position the latch as necessary to disengage the outside door handle and outside lock cylinder to latch rods and the inside handle to latch rod.
4   All door locking rods are attached by plastic clips. The plastic clips can be removed by unsnapping the portion engaging the connecting rod and then by pulling the rod out of its locating hole.
5   Position the latch as necessary to disengage the door lock actuator rod. Then remove the latch assembly from the door.
6   Installation is the reverse of removal.

Door lock cylinder and outside handle

Refer to illustrations 21.9 and 21.11

7   To remove the lock cylinder, raise the window and remove the door trim panel and watershield as described in Section 19.
8   Working through the large access hole, disengage the plastic clip that secures the lock cylinder to the latch rod.
9   Using a pair of pliers, slide the lock cylinder retaining clip out of engagement and remove the lock cylinder from the door **(see illustration)**.
10   To remove the outside handle, work through the access hole and disengage the plastic clip that secures the outside handle-to-latch rod.
11   Remove the outside handle retaining nuts **(see illustration)** and pull the handle from the door.
12   Installation is the reverse of removal.

Inside handle

Refer to illustration 21.14

13   Remove the door trim panel as described in Section 19 and peel away the watershield.

14   Drill out the rivets securing the inside handle **(see illustration)**. Pull forward on the handle to disengage it from the inner door panel.
15   Detach the actuating rods from the backside of the handle and remove it from the vehicle.
16   Installation is the reverse of removal.

22 Door window glass - removal and installation

Refer to illustrations 22.3, 22.4, 22.5a and 22.5b

1   Remove the door trim panel and the plastic watershield (see Section 19).
2   Lower the window glass all the way down into the door.
3   Carefully pry the trim panel out of the door window opening **(see illustration)**.
4   Remove the front guide channel **(see illustration)**.
5   Raise the window just enough to access the window retaining rivets through the hole in the door frame **(see illustrations)**.
6   Place a rag over the glass to help pre-

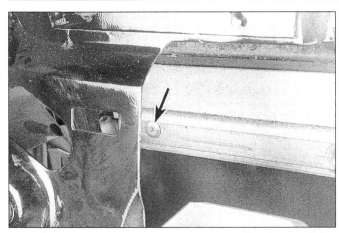

22.5a Raise the window just enough to access the glass retaining rivets (arrow) through the hole in the door frame . . .

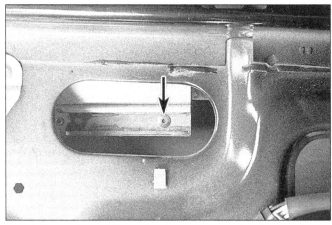

22.5b . . . then, drill out the rivets securing the glass to the equalizer arm

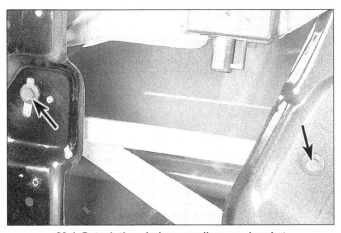

23.4 Detach the window equalizer arm bracket retaining bolts (arrows)

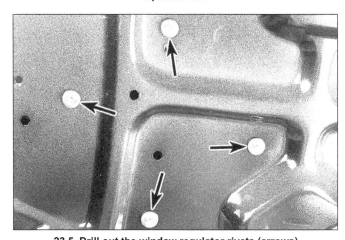

23.5 Drill out the window regulator rivets (arrows)

vent scratching the glass, then drill out the two glass mounting rivets.

7   Remove the glass by pulling it up and out.

8   Installation is the reverse of removal.

23   Door window glass regulator - removal and installation

Refer to illustrations 23.4 and 23.5

Warning: *The regulator arms are under extreme pressure and can cause serious injury if the motor or counter-balance spring is removed without locking the sector gear. This can be done by inserting a bolt and nut through the holes in the backing plate and sector gear to lock them together.*

Caution: *On models equipped with a Delco Loc II or Theftlock audio system, be sure the lockout feature is turned off before performing any procedure which requires disconnecting the battery.*

1   Remove the door trim panel and the plastic watershield (see Section 19).

2   Remove the window glass assembly (see Section 22).

3   On models with power windows, discon-

nect the electrical connector from the window regulator motor.

4   Remove the equalizer arm bracket **(see illustration)**.

5   Drill out the rivets that secure the window regulator to the door frame with a 5/16-inch drill bit **(see illustration)**.

6   Pull the equalizer arm and regulator assemblies through the service hole in the door frame to remove it.

7   Installation is the reverse of removal.

24   Outside mirrors - removal and installation

Refer to illustration 24.5

Caution: *On models equipped with a Delco Loc II or Theftlock audio system, be sure the lockout feature is turned off before performing any procedure which requires disconnecting the battery.*

1   Disconnect the negative cable from the battery.

2   Remove the door trim panel and the plastic watershield (see Section 19).

3   Remove the upper trim molding surrounding the door window.

4   Disconnect the electrical connector from the mirror.

5   Remove the three mirror retaining nuts and detach the mirror from the vehicle **(see illustration)**.

6   Installation is the reverse of removal.

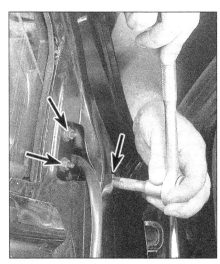

24.5 Outside mirror retaining nut locations (arrows)

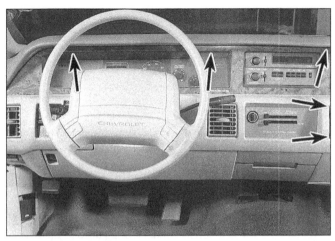

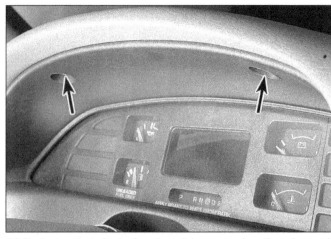

26.2a Instrument cluster bezel retaining screw locations (arrows) (early model Chevrolet shown, Buick similar)

26.2b Instrument cluster bezel retaining screw locations (arrows) (late model Chevrolet shown)

25 Center console (Impala SS) - removal and installation

Warning: *The models covered by this manual are equipped with airbags. Always disable the airbag system before working in the vicinity of the impact sensors, steering column or instrument panel to avoid the possibility of accidental deployment of the airbag(s), which could cause personal injury (see Chapter 12). The yellow wires and connectors routed through the console are for this system. Do not use electrical test equipment on these yellow wires or tamper with them in any way while working around the console.*

Caution: *On models equipped with a Delco Loc II or Theftlock audio system, be sure the lockout feature is turned off before performing any procedure which requires disconnecting the battery.*

1   Disconnect the negative cable from the battery.

2   Apply the parking brake lever and place the gear selector into the neutral position.

3   Pry out the shift lever knob retaining clip and remove the knob.

4   Pry out the gear selector trim bezel by disengaging the clips on the front edge.

5   Peel back the liner and detach the retaining screws securing the front half of the console.

6   Working in the console glove box, detach the retaining screws securing the rear half of the console.

7   Lift the console up and over the shift lever. Disconnect any electrical connections and remove the console from the vehicle.

8   Installation is the reverse of removal.

26 Instrument cluster bezel - removal and installation

Refer to illustrations 26.2a and 26.2b

Warning: *The models covered by this manual are equipped with airbags. Always disable*

the airbag system before working in the vicinity of the impact sensors, steering column or instrument panel to avoid the possibility of accidental deployment of the airbag(s), which could cause personal injury (see Chapter 12). The yellow wires and connectors under the instrument panel are for this system. Do not use electrical test equipment on these yellow wires or tamper with them in any way while working around the instrument panel.

Caution: *On models equipped with a Delco Loc II or Theftlock audio system, be sure the lockout feature is turned off before performing any procedure which requires disconnecting the battery.*

1   Remove the steering column lower trim panel (see Section 27).

2   Remove the bezel-to-instrument panel retaining screws **(see illustrations)**. **Note:** *On Buick and early model Chevrolet vehicles, open the glove box to access the retaining screws located at the far right.*

3   Grasp the bezel securely and pull back sharply to detach the clips from the instrument panel

4   Unplug any electrical connectors that interfere with removal.

5   Installation is the reverse of removal.

27 Dashboard trim panels - removal and installation

Warning: *The models covered by this manual are equipped with airbags. Always disable the air bag system before working in the vicinity of the impact sensors, steering column or instrument panel to avoid the possibility of accidental deployment of the airbag(s), which could cause personal injury (see Chapter 12). The yellow wires and connectors routed through the instrument panel are for this system. Do not use electrical test equipment on these yellow wires or tamper with them in any way while working around the instrument panel.*

Caution: *On models equipped with a Delco Loc II or Theftlock audio system, be sure the lockout feature is turned off before perform-*

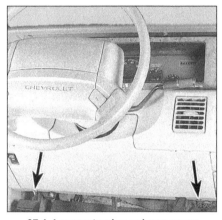

27.1 Lower steering column cover mounting screw locations (arrows) (early Caprice model shown)

ing any procedure which requires disconnecting the battery.

Lower steering column trim cover

Refer to illustration 27.1

1   Remove the bolts at the bottom edge of the knee bolster, then detach the clips or fasteners at the upper edge **(see illustration)**.

2   Unplug any electrical connectors, then lower the trim panel from the instrument panel.

3   Installation is the reverse of removal.

Center trim panel (1994 and later Chevrolet models)

Refer to illustrations 27.6a and 27.6b

4   Remove the instrument cluster bezel (see Section 26).

5   Remove the ashtray.

6   Remove the retaining screws, then grasp the panel securely and detach it from the instrument panel by pulling it straight back **(see illustrations)**.

7   Installation is the reverse of removal.

**27.6a Center trim panel mounting screw locations (arrows)
(1994 and later Caprice)**

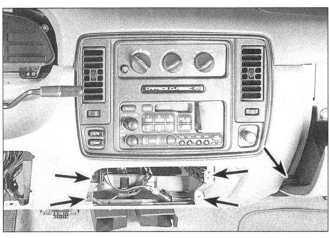

**27.6b Center trim panel mounting screw locations (arrows)
(1994 and later Caprice)**

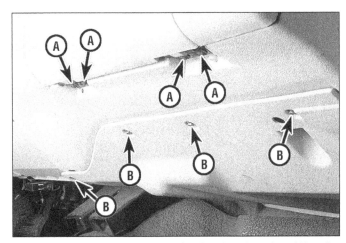

**27.8 Mounting screw locations for the glove box door (A) and
lower right-hand trim panel (B)**

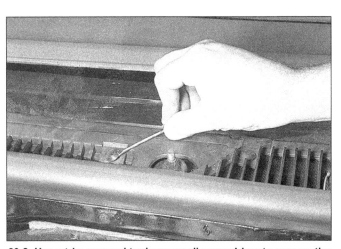

**29.2 Use a trim removal tool or a small screwdriver to remove the
plastic clips securing the center of the cowl cover**

Glove box door and lower right trim panel

Refer to illustration 27.8

8 To remove the glove box door, simply detach the screws from the hinge **(see illustration)**. Open the glove box door, squeeze the plastic sides in and lower the glove box door from the instrument panel.

9 To remove the lower right trim panel, simply detach the screws and lower the trim panel from the instrument panel **(see illustration 27.8)**.

10 Installation is the reverse of the removal.

---

28 Seats - removal and installation

Front seat

1 Position the seat all the way forward or all the way to the rear to access the front seat retaining bolts.

2 Detach any bolt trim covers and remove the retaining nuts.

3 Tilt the seat upward to access the underside, then disconnect any electrical connectors and lift the seat from the vehicle.

4 Installation is the reverse of removal.

Rear seat

5 Push the front edge of the cushion back then lift upward to release the seat cushion retaining tabs. Remove the cushion from the vehicle.

6 Detach the retaining bolts at the lower edge of the seat back.

7 Lift up on the lower edge of the seat back and remove it from the vehicle.

8 Installation is the reverse of removal.

---

29 Cowl cover - removal and installation

Refer to illustrations 29.2 and 29.3

1 Remove the windshield wiper arms (see Chapter 12).

2 Detach the plastic clips securing the cen-

ter of the cowl cover **(see illustration)**.

3 Remove the retaining screws securing the outer edges of the cowl cover **(see illustration)**. Remove the cover from the vehicle.

4 Installation is the reverse of removal.

**29.3 Remove the screws securing the
outer edges of the cowl cover**

Notes

Chapter 12
Chassis electrical system

Contents

1   General information

The electrical system is a 12-volt, negative ground type. Power for the lights and all electrical accessories is supplied by a lead/acid-type battery which is charged by the alternator.

This Chapter covers repair and service procedures for the various electrical components not associated with the engine. Information on the battery, alternator, ignition system and starter motor can be found in Chapter 5. It should be noted that when portions of the electrical system are serviced, the negative battery cable should be disconnected from the battery to prevent electrical shorts and/or fires. **Caution:** *On models equipped with a Delco Loc II or Theftlock audio system, be sure the lockout feature is turned off before performing any procedure which requires disconnecting the battery.*

2   Electrical troubleshooting - general information

A typical electrical circuit consists of an electrical component, any switches, relays, motors, fuses, fusible links or circuit breakers related to that component and the wiring and electrical connectors that link the component to both the battery and the chassis. To help you pinpoint an electrical circuit problem, wiring diagrams are included at the end of this Chapter.

Before tackling any troublesome electrical circuit, first study the appropriate wiring diagrams to get a complete understanding of what makes up that individual circuit. Trouble spots, for instance, can often be narrowed down by noting if other components related to the circuit are operating properly. If several components or circuits fail at one time, chances are the problem is in a fuse or ground connection, because several circuits are often routed through the same fuse and ground connections.

Electrical problems usually stem from simple causes, such as loose or corroded connections, a blown fuse, a melted fusible link or a bad relay. Visually inspect the condition of all fuses, wires and connections in a problem circuit before troubleshooting it.

If testing instruments are going to be utilized, use the diagrams to plan ahead of time where you will make the necessary connections in order to accurately pinpoint the trouble spot.

The basic tools needed for electrical troubleshooting include a circuit tester or voltmeter (a 12-volt bulb with a set of test leads can also be used), a continuity tester, which includes a bulb, battery and set of test leads, and a jumper wire, preferably with a circuit breaker incorporated, which can be used to bypass electrical components. Before attempting to locate a problem with test instruments, use the wiring diagram(s) to decide where to make the connections.

Voltage checks

Voltage checks should be performed if a circuit is not functioning properly. Connect one lead of a circuit tester to either the negative battery terminal or a known good ground. Connect the other lead to an electrical connector in the circuit being tested, preferably nearest to the battery or fuse. If the bulb of the tester lights, voltage is present, which means that the part of the circuit between the electrical connector and the battery is problem free. Continue checking the rest of the circuit in the same fashion. When you reach a point at which no voltage is present, the problem lies between that point and the last test point with voltage. Most of the time the problem can be traced to a loose connection. **Note:** *Keep in mind that some circuits receive voltage only when the ignition key is in the Accessory or Run position.*

Finding a short

One method of finding shorts in a circuit is to remove the fuse and connect a test light or voltmeter in its place to the fuse terminals. There should be no voltage present in the circuit. Move the wiring harness from side-to-side while watching the test light. If the bulb goes on, there is a short to ground somewhere in that area, probably where the insulation has rubbed through. The same test can be performed on each component in the circuit, even a switch.

Ground check

Perform a ground test to check whether a component is properly grounded. Disconnect the battery and connect one lead of a self-powered test light, known as a continuity tester, to a known good ground. Connect the other lead to the wire or ground connection being tested. If the bulb goes on, the ground is good. If the bulb does not go on, the ground is not good.

Continuity check

A continuity check is done to determine if there are any breaks in a circuit - if it is passing electricity properly. With the circuit off (no power in the circuit), a self-powered continuity tester can be used to check the circuit. Connect the test leads to both ends of the circuit (or to the "power" end and a good ground), and if the test light comes on the circuit is passing current properly. If the light doesn't come on, there is a break somewhere in the circuit. The same procedure can be used to test a switch, by connecting the continuity tester to the switch terminals. With the switch turned On, the test light should come on.

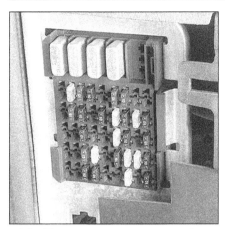

3.1a The passenger compartment fuse block is accessible after removing the cover

3.1b The engine compartment fuse block also contains relays

Finding an open circuit

When diagnosing for possible open circuits, it is often difficult to locate them by sight because oxidation or terminal misalignment are hidden by the electrical connectors. Merely wiggling an electrical connector on a sensor or in the wiring harness may correct the open circuit condition. Remember this when an open circuit is indicated when troubleshooting a circuit. Intermittent problems may also be caused by oxidized or loose connections.

Electrical troubleshooting is simple if you keep in mind that all electrical circuits are basically electricity running from the battery, through the wires, switches, relays, fuses and fusible links to each electrical component (light bulb, motor, etc.) and to ground, from which it is passed back to the battery. Any electrical problem is an interruption in the flow of electricity to and from the battery.

3 Fuses - general information

Refer to illustrations 3.1a, 3.1b, 3.1c and 3.3

1 The electrical circuits of the vehicle are protected by a combination of fuses, circuit breakers and fusible links. The two fuse blocks are located under a cover on the left end of the instrument panel and in the engine compartment **(see illustrations)**.

2 Each of the fuses is designed to protect a specific circuit, and the various circuits are identified on the fuse panel itself.

3 Miniaturized fuses are employed in the fuse block. These compact fuses, with blade terminal design, allow fingertip removal and replacement. If an electrical component fails, always check the fuse first. The easiest way to check fuses is with a test light. Check for power at the exposed terminal tips of each fuse. If power is present on one side of the fuse but not the other, the fuse is blown. A blown fuse can also be confirmed by visually inspecting it **(see illustration)**.

4 Be sure to replace blown fuses with the correct type. Fuses of different ratings are physically interchangeable, but only fuses of the proper rating should be used. Replacing a fuse with one of a higher or lower value than specified is not recommended. Each electrical circuit needs a specific amount of protection. The amperage value of each fuse is molded into the fuse body.

If the replacement fuse immediately fails, don't replace it again until the cause of the problem is isolated and corrected. In most cases, the cause will be a short circuit in the wiring caused by a broken or deteriorated wire.

4 Fusible links - general information

Some circuits are protected by fusible links. Fusible links are circuit protection devices that are part of the wiring harness itself, that are designed to melt and open the circuit when a short causes excessive current flow. Fusible links on these models are on the right (passenger's side) of the engine compartment, on the inner wheelwell. Fusible links are used in circuits which are not ordinarily fused, such as the ignition circuit.

Although the fusible links appear to be a heavier gauge than the wire they are protecting, the appearance is due to the thick insulation. All fusible links are several wire gauges smaller than the wire they are designed to protect.

Fusible links cannot be repaired, but a new link of the same size wire can be put in its place. The procedure is as follows:

a) *Disconnect the negative cable from the battery.*

b) *Disconnect the fusible link from the wiring harness.*

c) *Cut the damaged fusible link out of the wiring just behind the electrical connector.*

d) *Strip the insulation back approximately 1/2-inch.*

e) *Position the electrical connector on the new fusible link and crimp it into place.*

f) *Use rosin core solder at each end of the new link to obtain a good solder joint.*

g) *Use plenty of electrical tape around the soldered joint. No wires should be exposed.*

h) *Connect the battery ground cable. Test the circuit for proper operation.*

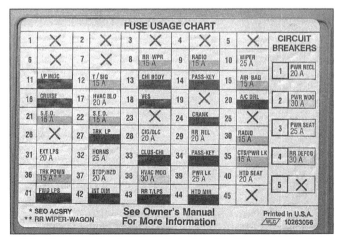

3.1c The fuse circuit can easily be identified by observing the decal located on the inside of each fuse panel cover

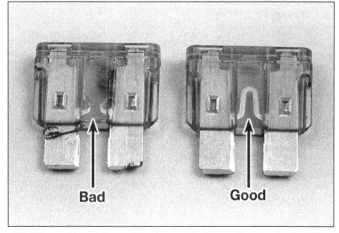

3.3 When a fuse blows, the element between the terminal melts - the fuse on the left is blown, the fuse on the right is good

5 Circuit breakers - general information

Circuit breakers protect components such as power windows, power door locks and headlights. On some models the circuit breaker resets itself automatically, so an electrical overload in a circuit breaker protected system will cause the circuit to fail momentarily, then come back on. If the circuit doesn't come back on, check it immediately. Once the condition is corrected, the circuit breaker will resume its normal function. Some circuit breakers must be reset manually.

6 Relays - general information

General information

1   Several electrical accessories in the vehicle, such as the fuel injection system, horns, starter, and fog lamps use relays to transmit the electrical signal to the component. Relays use a low-current circuit (the control circuit) to open and close a high-current circuit (the power circuit). If the relay is defective, that component will not operate properly. The various relays are mounted in engine compartment **(see illustration 3.1b)** and several locations throughout the vehicle. If a faulty relay is suspected, it can be removed and tested using the procedure below or by a dealer service department or a repair shop. Defective relays must be replaced as a unit.

Testing

2   It's best to refer to the wiring diagram for the circuit to determine the proper hook-ups for the relay you're testing. However, if you're not able to determine the correct hook-up from the wiring diagrams, you may be able to determine the test hook-ups from the information that follows.
3   On most relays, two of the terminals are the relay's control circuit (they connect to the relay coil which, when energized, closes the large contacts to complete the circuit). The other terminals are the power circuit (they are connected together within the relay when the control-circuit coil is energized).
4   Most relays are marked as an aid to help you determine which terminals are the control circuit and which are the power circuit.
5   Connect a fused jumper wire between one of the two control circuit terminals and the positive battery terminal. Connect another jumper wire between the other control circuit terminal and ground. When the connections are made, the relay should click. On some relays, polarity may be critical, so, if the relay doesn't click, try swapping the jumper wires on the control circuit terminals.
6   With the jumper wires connected, check for continuity between the power circuit terminals as indicated by the markings on the relay.
7   If the relay fails any of the above tests, replace it.

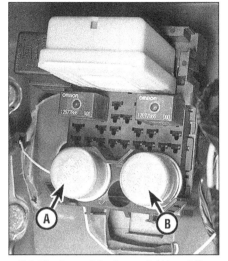

7.1 The hazard flasher unit (A) and the turn signal flasher unit (B) can be pulled straight out from the junction block

7 Turn signal/hazard flasher - check and replacement

Refer to illustration 7.1
Warning: *The models covered by this manual are equipped with airbags. Always disable the airbag system before working in the vicinity of the impact sensors, steering column or instrument panel to avoid the possibility of accidental deployment of the airbag(s), which could cause personal injury (see Section 26). The yellow wires and connectors routed through the instrument panel are for this system. Do not use electrical test equipment on these yellow wires or tamper with them in any way while working under the instrument panel.*
Caution: *On models equipped with a Delco Loc II or Theftlock audio system, be sure the lockout feature is turned off before performing any procedure which requires disconnecting the battery.*
1   The turn signal and hazard flashers are located in a junction block which is mounted to the left of the steering column **(see illustration).**
2   When the flasher unit is functioning properly, an audible click can be heard during its operation. If the turn signals fail on one side or the other and the flasher unit does not make its characteristic clicking sound, a faulty turn signal bulb is indicated.
3   If both turn signals fail to blink, the problem may be due to a blown fuse, a faulty flasher unit, a broken switch or a loose or open connection. If a quick check of the fuse box indicates that the turn signal fuse has blown, check the wiring for a short before installing a new fuse.
4   To replace the flasher, simply unplug it from the junction block.
5   Make sure that the replacement unit is

identical to the original. Compare the old one to the new one before installing it.
6   Installation is the reverse of removal.

8 Turn signal switch assembly - removal and installation

Warning: *The models covered by this manual are equipped with airbags. Always disable the airbag system before working in the vicinity of the impact sensors, steering column or instrument panel to avoid the possibility of accidental deployment of the airbag(s), which could cause personal injury (see Section 26). The yellow wires and connectors routed through the instrument panel are for this system. Do not use electrical test equipment on these yellow wires or tamper with them in any way while working under the instrument panel.*
Caution: *On models equipped with a Delco Loc II or Theftlock audio system, be sure the lockout feature is turned off before performing any procedure which requires disconnecting the battery.*

Removal

Refer to illustrations 8.3a, 8.3b, 8.4, 8.5, 8.6a, 8.6b and 8.8
1   Detach the cable from the negative battery terminal and disable the airbag system (see Section 26).
2   Remove the steering wheel (see Chapter 10).
3   Remove the airbag coil retaining snapring and remove the coil assembly; let the coil hang by the wiring harness. Remove the wave washer. Using a lock plate removal tool, depress the lock plate for access to the retaining ring **(see illustrations)**. Use a small screwdriver to pry the retaining ring out of the groove in the steering column and remove the lock plate.

8.3a Remove the snap-ring and lift the airbag coil off

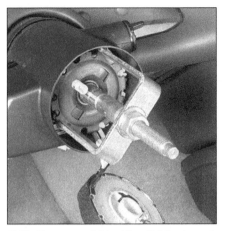

8.3b Use a special tool (available at most auto parts stores) to compress the lock plate for access to the retaining ring

8.4 Remove the hazard warning knob screw (arrow) and the knob

8.5 Remove the cancel cam assembly

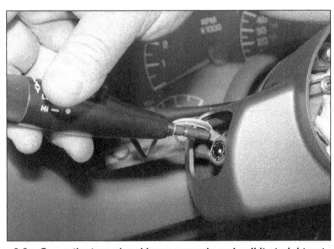

8.6a Grasp the turn signal lever securely and pull it straight out to detach it

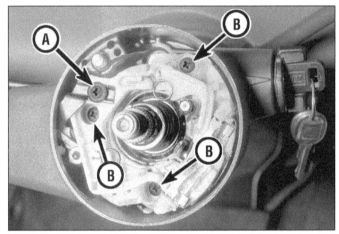

8.6b Remove the turn signal arm screw (A), followed by the switch mounting screws (B) - place the switch in the right turn position to access all the screws

4   Use a small Phillips head screwdriver to remove the hazard warning knob **(see illustration)**.

5   Remove the turn signal cancel cam **(see illustration)**.

6   Pull the multi-function lever straight out to detach the lever, then remove the turn signal switch mounting screws **(see illustrations)**.

7   Remove the knee bolster panel below the steering column.

8   Locate the turn signal switch electrical connector and unplug it **(see illustration)**. Remove the wiring protector.

9   Pull the wiring harness and electrical connector up through the steering column and remove the switch assembly.

Installation

Refer to illustrations 8.14 and 8.15

10   Feed the turn signal switch connector and wiring harness down through the column. Use a section of mechanics wire to pull it through, if necessary.

11   Plug in the connector and replace the wiring protector.

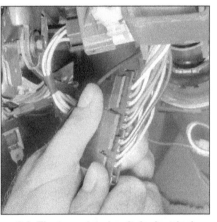

8.8 The turn signal switch electrical connector is located under the dash near the steering column - unplug it as shown

12   Seat the turn signal switch on the column and install the turn signal switch mounting screws and lever arm. Install the hazard knob and multi-function lever. Press the multi-function lever straight in until it snaps in place.

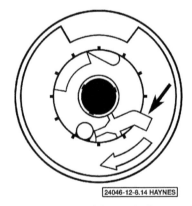

24046-12-8.14 HAYNES

8.14 To center the airbag coil, depress the spring lock (arrow); rotate the hub in the direction of the arrow; back the hub off 2-1/2 turns and release the spring lock

13   Install the cancel cam and the lock plate. Depress the lock plate and install the retaining ring.

14   If necessary, center the airbag coil as follows (it will only become uncentered if the

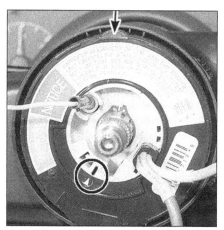

8.15 When properly installed, the airbag coil will be centered with the marks aligned (circle) and the tab fitted between the projections on the top of steering column (arrow)

9.3 Lift off the buzzer switch and clip with needle-nose pliers

9.4 The lock cylinder is held in place by a Torx-head screw (arrow)

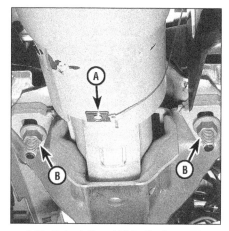

9.11 Detach the shift indicator cable clip (A) and remove the steering column retaining bolts (B)

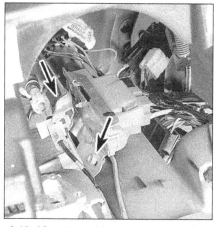

9.13 After the steering column has been lowered, detach the electrical connector and remove the ignition switch retaining screws (arrows)

spring lock is depressed and the hub rotated with the coil off the column) **(see illustration)**:

a) *Turn the coil over and depress the spring lock.*

b) *Rotate the hub in the direction of the arrow until it stops.*

c) *Rotate the hub in the opposite direction 2-1/2 turns and release the spring lock.*

15   Install the wave washer and the airbag coil **(see illustration)**. Pull the slack out of the airbag coil lower wiring harness to keep it tight through the steering column, or it may be cut when the steering wheel is turned. Install the airbag coil retaining snap-ring.

16   The remainder of installation is the reverse of removal.

9   Ignition switch and key lock cylinder - removal and installation

Warning: *The models covered by this manual are equipped with airbags. Always disable the airbag system before working in the vicinity of the impact sensors, steering column or instrument panel to avoid the possibility of accidental deployment of the airbag(s), which could cause personal injury (see Section 26). The yellow wires and connectors routed through the instrument panel are for this system. Do not use electrical test equipment on these yellow wires or tamper with them in any way while working under the instrument panel.*

Caution: *On models equipped with a Delco Loc II or Theftlock audio system, be sure the lockout feature is turned off before performing any procedure which requires disconnecting the battery.*

Lock cylinder

Refer to illustrations 9.3 and 9.4

1   The lock cylinder is located on the upper

right-hand side of the steering column. It should be removed only in the Run position, otherwise damage to the warning buzzer switch may occur.

2   Remove the steering wheel (Chapter 10) and turn signal switch (see Section 8). **Note:** *The turn signal switch need not be fully removed provided that it is pushed to the rear far enough to be slipped over the end of the shaft. Do not pull the harness out of the column.*

3   Insert the key and place the lock cylinder in the Run position, then use needle-nose pliers to remove the buzzer switch **(see illustration)**.

4   Remove the lock cylinder retaining screw **(see illustration)**.

5   Remove the lock cylinder by turning the key to the Start position, then pulling the assembly straight out.

6   To install, rotate the lock cylinder and align the cylinder key with the keyway in the steering column housing.

7   Push the lock all the way in and install the retaining screw.

8   Install the remaining components, referring to the appropriate Sections.

Ignition switch

Refer to illustrations 9.11 and 9.13

9   Disconnect the negative cable at the battery.

10   Place the ignition switch in the Lock position. If the key lock cylinder has been removed, pull the actuating rod up until a definite stop can be felt, then move it down one detent.

11   Remove the lower finish panel from beneath the steering column and remove the shift indicator clip **(see illustration)**. On 1994 and later Chevrolet models, it will be necessary to remove the center trim panel (see Chapter 11).

12   Remove the two nuts that secure the steering column to the lower edge of the dash assembly, then carefully lower the steering column down and rest the steering wheel on the seat **(see illustration 9.11)**.

13   Disconnect the electrical connector at the switch, then remove the ignition switch retaining screws **(see illustration)** and lift the switch out of the steering column jacket.

14   Prior to installation, make sure the ignition switch is in the Lock position.

10.4 Remove the screw (arrow), then detach the headlight switch bezel (1994 and later Chevrolet shown)

11.4a Remove the bolts (arrows) securing the radio

15 Connect the actuating rod to the switch.
16 Press the switch into position and install the screws.
17 Raise the steering column into position and install and tighten the nuts to 20 ft-lbs. Make sure the switch is actuated when the ignition key is turned to the Start position. If it doesn't, loosen the switch screws and adjust the position of the switch on the steering column.

10 Headlight switch - removal and installation

Refer to illustration 10.4

Warning: *The models covered by this manual are equipped with airbags. Always disable the airbag system before working in the vicinity of the impact sensors, steering column or instrument panel to avoid the possibility of accidental deployment of the airbag(s), which could cause personal injury (see Section 26). The yellow wires and connectors routed through the instrument panel are for this system. Do not use electrical test equipment on these yellow wires or tamper with them in any way while working under the instrument panel.*

Caution: *On models equipped with a Delco Loc II or Theftlock audio system, be sure the lockout feature is turned off before performing any procedure which requires disconnecting the battery.*

Note: *To remove the dimmer switch follow the ignition switch removal procedures (see Section 9).*

1 Detach the cable from the negative battery terminal and disable the airbag system (see Section 26).
2 Remove the instrument cluster bezel (see Chapter 11).
3 On 1994 and later Chevrolet models, remove the center trim panel from the dashboard (see Chapter 11).
4 Remove the retaining screw(s) and any retaining clips securing the switch **(see illustration).**

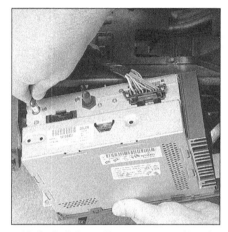

11.4b Pull the radio out, support it and unplug the connectors

5 Disconnect the electrical connector and remove the switch from the vehicle.
6 Installation is the reverse of removal.

11 Radio and speakers - removal and installation

Warning: *The models covered by this manual are equipped with airbags. Always disable the airbag system before working in the vicinity of the impact sensors, steering column or instrument panel to avoid the possibility of accidental deployment of the airbag(s), which could cause personal injury (see Section 26). The yellow wires and connectors routed through the instrument panel are for this system. Do not use electrical test equipment on these yellow wires or tamper with them in any way while working under the instrument panel.*

Caution: *On models equipped with a Delco Loc II or Theftlock audio system, be sure the lockout feature is turned off before performing any procedure which requires disconnecting the battery.*

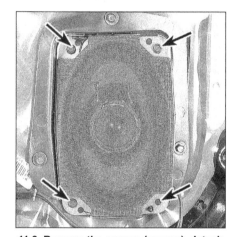

11.8 Remove the screws (arrows), detach the speaker and unplug the electrical connector

Radio

Refer to illustrations 11.4a and 11.4b

1 Detach the cable from the negative battery terminal and disable the airbag system (see Section 26).
2 Remove the instrument cluster bezel (see Chapter 11).
3 On 1994 and later Chevrolet models, remove the center trim panel from the dashboard (see Chapter 11).
4 Remove the screws, pull the radio out and disconnect the electrical connection and antenna lead **(see illustrations).**
5 Remove the radio from the instrument panel.
6 Installation is the reverse of removal.

Front speakers

Refer to illustration 11.8

7 Remove the front door trim panel (see Chapter 11).
8 Remove the speaker retaining bolts. Disconnect the electrical connector and remove the speaker from the vehicle **(see illustration).**
9 Installation is the reverse of removal.

12.3 The antenna base retaining nut can be removed using a pair of snap-ring pliers or similar tool

13.4 When measuring the voltage at the rear window defogger grid, wrap a piece of aluminum foil around the negative probe of the voltmeter and press the foil against the wire with your finger

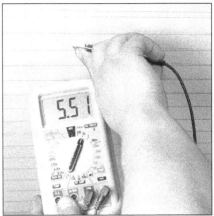

13.5 To determine if a heating element has broken, check the voltage at the center of each element - if the voltage is 6-volts, the element is unbroken - if the voltage is 12-volts, the element is broken between the center and the positive end - if there is no voltage, the element is broken between the center and ground

Rear speakers (Sedan models)

10 Remove the rear seat from the vehicle (see Chapter 11).
11 Detach the rear quarter panel inside trim panels.
12 Pry up the plastic clips securing the rear parcel shelf. Then lift up and out to remove it from the vehicle.
13 Remove the speaker retaining screws. Disconnect the electrical connector and remove the speaker from the vehicle.
14 Installation is the reverse of removal.

12 Antenna - removal and installation

Refer to illustration 12.3

Caution: *On models equipped with a Delco Loc II or Theftlock audio system, be sure the lockout feature is turned off before performing any procedure which requires disconnecting the battery.*

Fender mounted antenna

1 Detach the radio and disconnect the antenna lead from the backside of the radio (see Section 11). Attach a piece of stiff wire to the end of the antenna lead, then remove any retaining clips under the instrument panel securing the antenna lead.
2 On vehicles equipped with fixed type antennas, use a small wrench and remove the antenna mast.
3 Using a pair of snap-ring pliers or similar tool, remove the antenna base retaining nut **(see illustration)**.
4 Remove the right front tire and the right front inner fenderwell as described in the fender removal Section in Chapter 11.
5 On vehicles equipped with a power antenna, remove the antenna motor mounting screws from the inner fenderwell.
6 Pull the antenna motor/base assembly through the inner fenderwell opening until you can see the wire, then remove the old antenna lead from the wire and replace it

with the new antenna lead.
7 Working in the passenger compartment, pull the wire and the antenna lead rearward into the passenger compartment.
8 The remainder of installation is the reverse of removal.

Trunk mounted antenna

9 On vehicles equipped with fixed type antennas, use a small wrench and remove the antenna mast.
10 Remove the antenna mast retaining nut **(see illustration 12.3)**.
11 Working in the trunk, pry out the plastic clips securing the passenger side trunk finishing panels to allow access to the antenna motor.
11 Detach the motor/base assembly retaining bolt(s). Disconnect the antenna lead and the electrical connector (if equipped) and remove the antenna motor/base assembly from the vehicle.
12 Installation is the reverse of removal.

13 Rear window defogger - check and repair

1 The rear window defogger consists of a number of horizontal elements baked onto the glass surface.
2 Small breaks in the element can be repaired without removing the rear window.

Check

Refer to illustrations 13.4, 13.5 and 13.7

3 Turn the ignition switch and defogger system switches to the ON position.
4 When measuring voltage during the next two tests, wrap a piece of aluminum foil around the tip of the voltmeter negative probe and press the foil against the heating element with your finger **(see illustration)**.
5 Check the voltage at the center of each heating element **(see illustration)**. If the voltage is 6-volts, the element is okay (there is no

break). If the voltage is 12-volts, the element is broken between the center of the element and the positive end. If the voltage is 0-volts the element is broken between the center of the element and ground.
6 Connect the negative lead to a good body ground. The reading should stay the same.
7 To find the break, place the voltmeter positive lead against the defogger positive terminal. Place the voltmeter negative lead with the foil strip against the heating element at the positive terminal end and slide it toward the negative terminal end. The point at which the voltmeter deflects from zero to several volts is the point at which the heating element is broken **(see illustration)**.

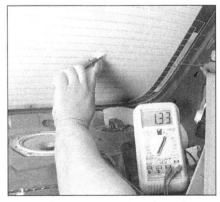

13.7 To find the break, place the voltmeter positive lead against the defogger positive terminal, place the voltmeter negative lead with the foil strip against the heating element at the positive terminal end and slide it toward the negative terminal end - the point at which the voltmeter reading changes abruptly is the point at which the element is broken

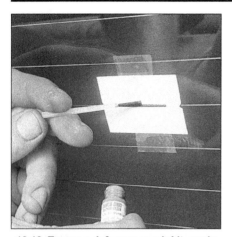

13.13 To use a defogger repair kit, apply masking tape to the inside of the window at the damaged area, then brush on the special conductive coating

Repair

Refer to illustration 13.13

8   Repair the break in the element using a repair kit specifically recommended for this purpose, such as Dupont paste No. 4817 (or equivalent). Included in this kit is plastic conductive epoxy.
9   Prior to repairing a break, turn off the system and allow it to cool off for a few minutes.
10   Lightly buff the element area with fine steel wool, then clean it thoroughly with rubbing alcohol.
11   Use masking tape to mask off the area being repaired.
12   Thoroughly mix the epoxy, following the instructions provided with the repair kit.
13   Apply the epoxy material to the slit in the masking tape, overlapping the undamaged area about 3/4-inch on either end **(see illustration)**.
14   Allow the repair to cure for 24 hours before removing the tape and using the system.

14 Headlights - replacement

Refer to illustrations 14.2 and 14.3

Warning: *Halogen gas filled bulbs are under pressure and may shatter if the surface is scratched or the bulb is dropped. Wear eye protection and handle the bulbs carefully, grasping only the base whenever possible. Do not touch the surface of the bulb with your fingers because the oil from your skin could cause it to overheat and fail prematurely. If you do touch the bulb surface, clean it with rubbing alcohol.*
1   Open the hood.
2   Disconnect the electrical connector from the bulb holder **(see illustration)**.
3   Rotate the headlight bulb retaining ring counterclockwise as viewed from the rear **(see illustration)**.

14.2 Detach the electrical connector from the rear of headlight bulb assembly

4   Withdraw the bulb assembly and retaining ring from the headlight housing.
5   Remove the bulb from the socket assembly by pulling it straight out.
6   Without touching the glass with your bare fingers, insert the new bulb into the socket assembly and then into the headlight housing, install and tighten the retaining ring.
7   Plug in the electrical connector. Test headlight operation, then close the hood.

15 Headlights - adjustment

Refer to illustrations 15.1 and 15.3

Caution: *The headlights must be aimed correctly. If adjusted incorrectly they could blind the driver of an oncoming vehicle and cause a serious accident or seriously reduce your ability to see the road. The headlights should be checked for proper aim every 12 months and any time a new headlight is installed or front end body work is performed. It should be emphasized that the following procedure is only an interim step which will provide temporary adjustment until the headlights can be adjusted by a properly equipped shop.*
1   Headlights have two spring loaded adjusting screws, one on the top controlling up-and-down movement and one on the side controlling left-and-right movement **(see illustration)**.
2   There are several methods of adjusting the headlights. The simplest method requires a blank wall 25 feet in front of the vehicle and a level floor.
3   Position masking tape vertically on the wall in reference to the vehicle centerline and the centerlines of both headlights **(see illustration)**.
4   Position a horizontal tape line in reference to the centerline of all the headlights. **Note:** *It may be easier to position the tape on the wall with the vehicle parked only a few inches away.*
5   Adjustment should be made with the vehicle sitting level, the gas tank half-full and no unusually heavy load in the vehicle.

14.3 Rotate the headlight bulb retaining ring counterclockwise and pull the bulb socket assembly out of the housing - when installing the new bulb, don't touch the surface, but clean it with rubbing alcohol if you do

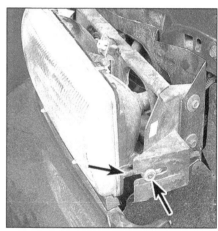

15.1 The headlight vertical adjustment screw is located at the top of the headlight (arrow) and the horizontal screw(s) is on the side of the headlight (arrows) - a Torx-head tool will be required for making headlight adjustments

6   Starting with the low beam adjustment, position the high intensity zone so it is two inches below the horizontal line and two inches to the right of the headlight vertical line. Adjustment is made by turning the top adjusting screw clockwise to raise the beam and counterclockwise to lower the beam. The adjusting screw on the side should be used in the same manner to move the beam left or right.
7   With the high beams on, the high intensity zone should be vertically centered with the exact center just below the horizontal line. **Note:** *It may not be possible to position the headlight aim exactly for both high and low beams. If a compromise must be made, keep in mind that the low beams are the most used and have the greatest effect on safety.*
8   Have the headlights adjusted by a dealer service department or service station at the earliest opportunity.

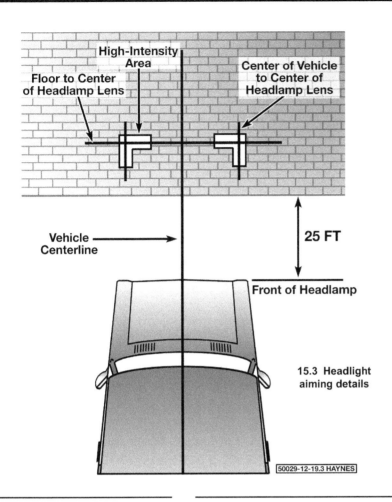

15.3 Headlight aiming details

16.3 Remove the bolts (arrows) located behind the side marker/ turn signal light assembly

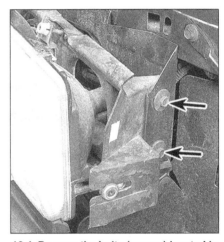

16.4 Remove the bolts (arrows) located in the radiator grille opening

16 Headlight housing - removal and installation

Refer to illustrations 16.3 and 16.4

Warning: *The models covered by this manual are equipped with airbags. Always disable the airbag system before working in the vicinity of the impact sensors, steering column or instrument panel to avoid the possibility of accidental deployment of the airbag(s), which could cause personal injury (see Section 26). The yellow wires and connectors routed through the instrument panel are for this system. Do not use electrical test equipment on these yellow wires or tamper with them in any way while working under the instrument panel.*

1   Remove the headlight bulb(s) (see Section 14).
2   Remove the front side marker/turn signal light assembly (see Section 17).
3   Remove the headlight housing bolts located in the side marker/turn signal light opening **(see illustration)**.
4   Remove the headlight housing bolts located in the radiator grille opening **(see illustration)**.
5   Pull the headlight housing out to remove it.
6   The remainder of the installation is the reverse of removal.

17 Bulb replacement

Front turn signal and side marker lights

Refer to illustrations 17.1 and 17.2

1   Detach the side marker light retaining screw, then pull the side marker light outward **(see illustration)**.
2   Twist the bulb socket a quarter turn coun-

terclockwise, then remove the bulb assembly from the housing **(see illustration)**.
3   The defective bulb can then be twisted out of the socket and replaced.
4   Installation is the reverse of removal.

17.1 Remove the side marker/turn signal housing retaining screw (arrow)

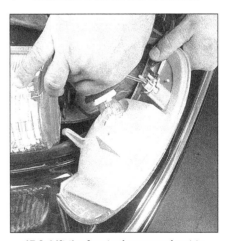

17.2 Lift the front edge up and out to access the bulb(s)

Rear turn signal, brake, tail and back-up lights

Refer to illustrations 17.6, 17.7 and 17.9

5   On sedan models, detach the plastic clips securing the rear finishing panel, then remove the panel from the vehicle. Working from the inside of the trunk compartment, detach the retaining nuts securing the rear tail light housing.

6   On later model sedans, detach the retaining bolt securing the inner edge of the tail light housing **(see illustration)**.

7   Pull the tail light assembly outward to access the tail light bulbs **(see illustration)**.

8   On station wagon models, remove the rear quarter inside trim panel from the affected side, then detach the retaining nut and screws securing the rear tail light housing, then pull the tail light assembly outward to access the tail light bulbs.

9   Twist the bulb(s) socket a quarter turn counterclockwise, then remove the bulb assembly from the housing **(see illustration)**.

10   The defective bulb can then be twisted out of the socket and replaced.

11   Installation of the tail light housing is the reverse of removal.

License plate light

Refer to illustration 17.12

12   Detach the retaining screws which secure the lens to the trunk lid **(see illustration)**.

13   The defective bulb can then be pulled straight out of the socket and replaced.

14   Installation of the lens is the reverse of removal.

High-mounted brake light

15   The high mounted brake light bulbs can be accessed from the cargo compartment on station wagon models and from the rear passenger compartment on sedan models.

16   On station wagon models, remove the trim cover retaining screws and the trim cover from the liftgate, then twist the bulb a quarter turn counterclockwise, then remove the bulb from the housing.

17.6 On later model sedans, remove the bolt (arrow) that secures the inner edge of the tail light housing

17.7 Lift up and out to remove the tail light housing assembly (later model shown)

17.9 Squeeze the clip and lift the bulb holder out of the housing - push in and rotate the bulb to remove it

17   On sedan models, pry off the trim cover and disconnect the electrical connector, then remove the lamp housing retaining screws. Twist the bulb a quarter turn counterclockwise, then remove the bulb from the housing.

Interior light

Refer to illustration 17.18

18   Using a small screwdriver, remove the lens and replace the bulb **(see illustration)**.

Instrument cluster illumination

Refer to illustration 17.19

19   To gain access to the instrument cluster illumination lights, the instrument cluster will have to be removed (see Section 20). The bulbs can then be removed and replaced from the rear of the cluster **(see illustration)**.

17.12 Remove the screws (arrows) to access the license plate light bulb

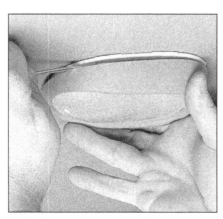

17.18 Pry off the interior light lens to access the bulb

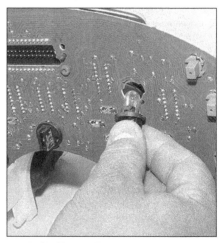

17.19 Rotate the bulb and lift it out of the cluster

19.2 Use a voltmeter or test light to check for battery power at the wiper motor

19.7 Use a small screwdriver to pry off the wiper arm nut cover, then remove the nut (arrow) and pull the arm straight off its splined shaft

18 Daytime Running Lights (DRL) - general information

The Daytime Running Lights (DRL) system used on all Canadian and later US models illuminates the headlights whenever the engine is running. The only exception is with the engine running and the parking brake engaged. Once the parking brake is released, the lights will remain on as long as the ignition switch is on, even if the parking brake is later applied.

The DRL system supplies reduced power to the headlights so they will be bright enough for daytime visibility while prolonging headlight life.

19 Wiper motor - check and replacement

Check

Refer to illustration 19.2

Note: *Refer to the wiring diagrams for wire colors and locations in the following checks. Keep in mind that power wires are generally* larger in diameter and brighter colors, where ground wires are usually smaller in diameter and darker colors. When checking for voltage, probe a grounded 12-volt test light to each terminal at a connector until it lights; this verifies voltage (power) at the terminal.

1   If the wipers work slowly, make sure the battery is in good condition and has a strong charge (see Chapter 1). If the battery is in good condition, remove the wiper motor (see below) and operate the wiper arms by hand. Check for binding linkage and pivots. Lubricate or repair the linkage or pivots as necessary. Reinstall the wiper motor. If the wipers still operate slowly, check for loose or corroded connections, especially the ground connection. If all connections look OK, replace the motor.

2   If the wipers fail to operate when activated, check the fuse. If the fuse is OK, connect a jumper wire between the wiper motor and ground, then retest. If the motor works now, repair the ground connection. If the motor still doesn't work, turn the wiper switch to the HI position and check for voltage at the motor **(see illustration)**. If there's voltage at the motor, remove the motor and check it off the vehicle with fused jumper wires from the battery. If the motor now works, check for binding linkage (see Step 1 above). If the motor still doesn't work, replace it. If there's no voltage at the motor, check for voltage at the wiper control module. If there's voltage at the wiper control module and no voltage at the at the wiper motor, check the switch for continuity (see Section 8). If the switch is OK, the wiper control module is probably bad.

3   If the interval (delay) function is inoperative, check the continuity of all the wiring between the switch and wiper control module. If the wiring is OK, check the resistance of the delay control knob of the multi-function switch (see Section 8). If the delay control knob is within the specified resistance, replace the wiper control module.

4   If the wipers stop at the position they're in when the switch is turned off (fail to park), check for voltage at the park feed wire of the wiper motor connector when the wiper switch is OFF but the ignition is ON. If no voltage is present, check for an open circuit between the wiper motor and the fuse panel.

5   If the wipers won't shut off unless the ignition is OFF, disconnect the wiring from the wiper control switch. If the wipers stop, replace the switch. If the wipers keep running, there's a defective limit switch in the motor; replace the motor.

6   If the wipers won't retract below the hoodline, check for mechanical obstructions in the wiper linkage or on the vehicle's body which would prevent the wipers from parking. If there are no obstructions, check the wiring between the switch and motor for continuity. If the wiring is OK, replace the wiper motor.

Replacement

Refer to illustrations 19.7, 19.9a, 19.9b and 19.10

7   Pry off the cover for the wiper arm nuts, unscrew the nuts and remove both wiper arms **(see illustration)**.

8   Remove the screws and detach the cowl cover (see Chapter 11).

9   Remove the wiper motor shield and motor spindle nut **(see illustrations)**.

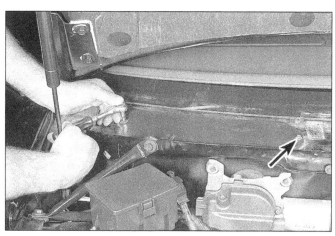

19.9a Detach the wiper motor shield retaining screws . . .

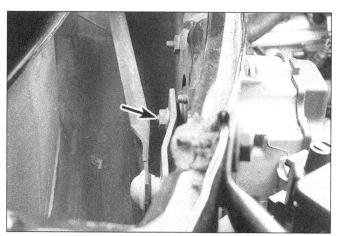

19.9b . . . and the wiper motor spindle nut (arrow)

19.10   Remove the wiper motor retaining bolts (arrows)

10   Remove the wiper motor retaining bolts **(see illustration)**, detach the wiper arm linkage then pull the motor out from the firewall.
11   Disconnect the electrical connector and remove the motor from the vehicle.
12   Installation is the reverse of removal.

20   Instrument cluster - removal and installation

Refer to illustration 20.3

Warning: *The models covered by this manual are equipped with airbags. Always disable the airbag system before working in the vicinity of the impact sensors, steering column or instrument panel to avoid the possibility of accidental deployment of the airbag(s), which could cause personal injury (see Section 26). The yellow wires and connectors routed through the instrument panel are for this system. Do not use electrical test equipment on these yellow wires or tamper with them in any way while working under the instrument panel.*
Caution: *On models equipped with a Delco Loc II or Theftlock audio system, be sure the lockout feature is turned off before performing any procedure which requires disconnecting the battery.*
1   Detach the cable from the negative battery terminal and disable the airbag system (see Section 26).
2   Remove instrument cluster bezel (see Chapter 11).
3   Remove the four bolts, detach the instrument cluster from the electrical connector and lift it from the dash **(see illustration)**.
4   Installation is the reverse of removal.

21   Horn - check and replacement

Check

Refer to illustration 21.3

Note: *Check the fuses before beginning electrical diagnosis.*
1   Disconnect the electrical connector from

20.3   Typical instrument cluster retaining screw locations (arrows) (later model Chevrolet shown)

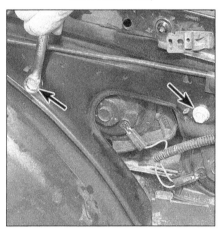

21.9   Disconnect the electrical connector, remove the bolt(s) (arrows) and detach the horn(s)

the horn. On 1994 and later models it will be necessary to remove the air filter housing to access the horns (see Chapter 5).
2   To test the horn(s), connect battery voltage to the horn terminal with a pair of jumper wires. If the horn doesn't sound, replace it.
3   If the horn does sound, check for voltage at the terminal when the horn button is depressed **(see illustration)**. If there's voltage at the terminal, check for a bad ground at the horn.
4   If there's no voltage at the horn, check the relay (see Section 6). Note that most horn relays are either the four-terminal or externally grounded three-terminal type.
5   If the relay is OK, check for voltage to the relay power and control circuits. If either of the circuits is not receiving voltage, inspect the wiring between the relay and the fuse panel.
6   If both relay circuits are receiving voltage, depress the horn button and check the circuit from the relay to the horn button for continuity to ground. If there's no continuity, check the circuit for an open. If there's no open circuit, replace the horn button.
7   If there's continuity to ground through the horn button, check for an open or short in the circuit from the relay to the horn.

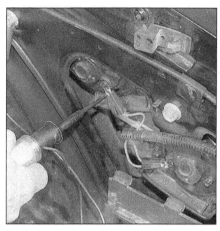

21.3   Connect a voltmeter to the horn wire and ground - test for voltage while the switch is depressed

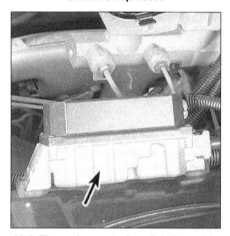

22.5   The cruise control module (arrow) is located in the engine compartment on the (driver's side) inner fenderwell - check for damage to the connectors

Replacement

Refer to illustration 21.9

8   To access the horns on 1994 and later vehicles the air filter housing must first be removed (see Chapter 4).
9   Disconnect the electrical connectors and remove the bracket bolt **(see illustration)**.
10   Installation is the reverse of removal.

22   Cruise control system - description and check

Refer to illustrations 22.5 and 22.7

1   The cruise control system maintains vehicle speed with an electronic servo motor located in the engine compartment, which is connected to the throttle linkage by a cable. The system consists of the electronic control module, brake switch, control switches, a relay, the vehicle speed sensor and associated wiring. Listed below are some general procedures that may be used to locate common cruise control problems.

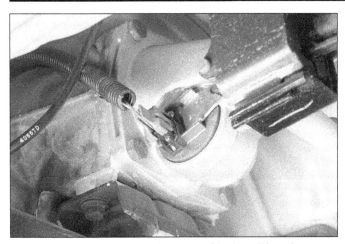

22.7 The speed sensor is mounted in the tail housing of the transmission

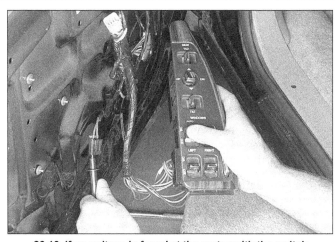

23.12 If no voltage is found at the motor with the switch depressed, check for voltage at the switch

2   Locate and check the fuse (see Section 3).

3   Have an assistant operate the brake lights while you check their operation (voltage from the brake light switch deactivates the cruise control).

4   If the brake lights don't come on or don't shut off, correct the problem and retest the cruise control.

5   Inspect the cable linkage between the cruise control module and the throttle linkage. The cruise control module is located on the left (drivers) inner fenderwell of the engine compartment **(see illustration)**.

6   Visually inspect the wires connected to the cruise control actuator and check for damage and broken wires.

7   The vehicle speed sensor is located on the transmission **(see illustration)**. Raise the front of the vehicle and support it on jack stands. Unplug the electrical connector and touch one probe of a digital voltmeter to the orange wire of the connector and the other to a good ground. With the vehicle in Neutral and key On, measure the voltage while rotating one wheel with the other one blocked. If the voltage doesn't vary as the wheel rotates, the sensor is defective.

8   Test drive the vehicle to determine if the cruise control is now working. If it isn't, take it to a dealer service department or an automotive electrical specialist for further diagnosis and repair.

23   Power window system - description and check

Refer to illustration 23.12

1   The power window system operates electric motors, mounted in the doors, which lower and raise the windows. The system consists of the control switches, the motors, regulators, glass mechanisms and associated wiring.

2   The power windows can be lowered and raised from the master control switch by the driver or by remote switches located at the individual windows. Each window has a separate motor which is reversible. The position of the control switch determines the polarity and therefore the direction of operation.

3   The circuit is protected by a fuse and a circuit breaker. Each motor is also equipped with an internal circuit breaker, this prevents one stuck window from disabling the whole system.

4   The power window system will only operate when the ignition switch is ON. In addition, many models have a window lockout switch at the master control switch which, when activated, disables the switches at the rear windows and, sometimes, the switch at the passenger's window also. Always check these items before troubleshooting a window problem.

5   These procedures are general in nature, so if you can't find the problem using them, take the vehicle to a dealer service department or other properly equipped repair facility.

6   If the power windows won't operate, always check the fuse and circuit breaker first.

7   If only the rear windows are inoperative, or if the windows only operate from the master control switch, check the rear window lockout switch for continuity in the unlocked position. Replace it if it doesn't have continuity.

8   Check the wiring between the switches and fuse panel for continuity. Repair the wiring, if necessary.

9   If only one window is inoperative from the master control switch, try the other control switch at the window. **Note:** *This doesn't apply to the drivers door window.*

10   If the same window works from one switch, but not the other, check the switch for continuity.

11   If the switch tests OK, check for a short or open in the circuit between the affected switch and the window motor.

12   If one window is inoperative from both switches, remove the trim panel from the affected door and check for voltage at the switch **(see illustration)** and at the motor while the switch is operated.

13   If voltage is reaching the motor, disconnect the glass from the regulator (see Chapter 11). Move the window up and down by hand while checking for binding and damage. Also check for binding and damage to the regulator. If the regulator is not damaged and the window moves up and down smoothly, replace the motor. If there's binding or damage, lubricate, repair or replace parts, as necessary.

14   If voltage isn't reaching the motor, check the wiring in the circuit for continuity between the switches and motors. You'll need to consult the wiring diagram for the vehicle. If the circuit is equipped with a relay, check that the relay is grounded properly and receiving voltage.

15   Test the windows after you are done to confirm proper repairs.

24   Power door lock system - description and check

Refer to illustration 24.6

The power door lock system operates the door lock actuators mounted in each door. The system consists of the switches, actuators, a control unit and associated wiring. Diagnosis can usually be limited to simple checks of the wiring connections and actuators for minor faults which can be easily repaired. Since this system uses an electronic control unit in-depth diagnosis should be left to a dealership service department. The door lock control unit is located behind the instrument panel, to the right of the fuse box.

Power door lock systems are operated by bi-directional solenoids located in the doors. The lock switches have two operating positions: Lock and Unlock. When activated, the switch sends a ground signal to the door lock control unit to lock or unlock the doors. Depending on which way the switch is activated, the control unit reverses polarity to the solenoids, allowing the two sides of the circuit to be used alternately as the feed (positive) and ground side.

24.6 Check for voltage at the lock solenoid while the switch is operated

26.3 The impact sensor(s) (arrow) are mounted at the front of the vehicle to detect an impact of sufficient G-force (1994 through 1996 sensor location shown)

Some vehicles may have an anti-theft systems incorporated into the power locks. If you are unable to locate the trouble using the following general Steps, consult your a dealer service department.

1   Always check the circuit protection first. Some vehicles use a combination of circuit breakers and fuses.

2   Operate the door lock switches in both directions (Lock and Unlock) with the engine off. Listen for the click of the solenoids operating.

3   Test the switches for continuity. Replace the switch if there's not continuity in both switch positions.

4   Check the wiring between the switches, control unit and solenoids for continuity. Repair the wiring if there's no continuity.

5   Check for a bad ground at the switches or the control unit.

6   If all but one lock solenoids operate, remove the trim panel from the affected door (see Chapter 11) and check for voltage at the solenoid while the lock switch is operated One of the wires should have voltage in the Lock position; the other should have voltage in the Unlock position **(see illustration)**.

7   If the inoperative solenoid is receiving voltage, replace the solenoid.

8   If the inoperative solenoid isn't receiving voltage, check for an open or short in the wire between the lock solenoid and the control unit.

Note: *It's common for wires to break in the portion of the harness between the body and door (opening and closing the door fatigues and eventually breaks the wires).*

25 Electric rear view mirrors - description and check

1   Electric rear view mirrors use two motors to move the glass; one for up-and-down adjustments and one for left-to-right adjustments.

2   The control switch has a selector portion which sends voltage to the left or right side mirror. With the ignition ON, engine OFF,

roll down the windows and operate the mirror control switch through all functions (left-right and up-down) for both the left and right side mirrors.

3   Listen carefully for the sound of the electric motors running in the mirrors.

4   If the motors can be heard but the mirror glass doesn't move, there's probably a problem with the drive mechanism inside the mirror. Remove and disassemble the mirror to locate the problem.

5   If the mirrors don't operate and no sound comes from the mirrors, check the fuse (see Section 3).

6   If the fuse is OK, remove the mirror control switch from its mounting without disconnecting the wires attached to it. Turn the ignition ON and check for voltage at the switch. There should be voltage at one terminal. If there's no voltage at the switch, check for an opening or short in the wiring between the fuse panel and the switch.

7   If there's voltage at the switch, disconnect it. Check the switch for continuity in all its operating positions. If the switch does not have continuity, replace it.

8   Re-connect the switch. Locate the wire going from the switch to ground. Leaving the switch connected, connect a jumper wire between this wire and ground. If the mirror works normally with this wire in place, repair the faulty ground connection.

9   If the mirror still doesn't work, remove the cover and check the wires at the mirror for voltage with a test light. Check with ignition ON and the mirror selector switch on the appropriate side. Operate the mirror switch in all its positions. There should be voltage at one of the switch-to-mirror wires in each switch position (except the neutral position).

10   If there's not voltage in each switch position, check the wiring between the mirror and control switch for opens and shorts.

11   If there's voltage, remove the mirror and test it off the vehicle with jumper wires. Replace the mirror if it fails this test (see Chapter 11).

26 Airbag - general information

Warning: *The models covered by this manual are equipped with airbags. Airbag system components are located in the steering wheel, steering column, instrument panel and center console. The airbag(s) could accidentally deploy if any of the system components or wiring harnesses are disturbed, so be extremely careful when working in these areas and don't disturb any airbag system components or wiring. You could be injured if an airbag accidentally deploys, or the airbag might not deploy correctly in a collision if any components or wiring in the system have been disturbed. The yellow wires and connectors routed through the instrument panel and center console are for this system. Do not use electrical test equipment on these yellow wires or tamper with them in any way while working in their vicinity.*

Caution: *On models equipped with a Delco Loc II or Theftlock audio system, be sure the lockout feature is turned off before performing any procedure which requires disconnecting the battery.*

Description

1   The models covered by this manual are equipped with a Supplemental Inflatable Restraint (SIR) system, more commonly known as an airbag system. The SIR system is designed to protect the driver and passenger from serious injury in the event of a head-on or frontal collision.

2   On 1991 through 1993 vehicles, the SIR system consists of a driver side airbag; located in the center of the steering wheel, two impact sensors; located behind the radiator support near the headlight opening; an arming sensor and a diagnostic/energy reserve module located under the instrument panel. On 1994 through 1996 vehicles, the SIR system consists of two airbags; located in the center of the steering wheel, and another located in the top of the dashboard above the glove box and one impact sensor located below the hood latch.

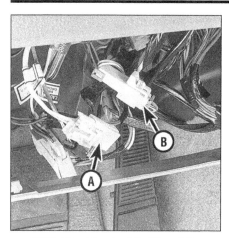

26.10 The driver's side (A) and the passenger's side (B) airbag connectors are located at the base of the steering column

The arming sensor and the diagnostic/energy reserve module have been incorporated into one unit which is located under the instrument panel.

Sensors

Refer to illustration 26.3

3   The sensors are basically pressure sensitive switches that complete an electrical circuit during an impact of sufficient G force. The electrical signal from the crash sensors is sent to the diagnostic module, that then completes circuit and inflates the airbags **(see illustration)**.

Diagnostic/energy reserve module

4   The diagnostic/energy reserve module contains an on-board microprocessor which monitors the operation of the system. It performs a diagnostic check of the system every time the vehicle is started. If the system is operating properly, the AIRBAG warning light will blink on and off seven times. If there is a fault in the system, the light will remain on and

the airbag control module will store fault codes indicating the nature of the fault. If the AIRBAG warning light remains on after staring, or comes on while driving, the vehicle should be taken to your dealer immediately for service. The diagnostic/energy reserve module also contains a back-up power supply to deploy the airbags in the event battery power is lost during a collision.

Operation

5   For the airbag(s) to deploy, an impact of sufficient G force must occur within 30-degrees of the vehicle centerline. When this condition occurs, the circuit to the airbag inflator is closed and the airbag inflates. If the battery is destroyed by the impact, or is too low to power the inflators, a back-up power supply inside the diagnostic/energy reserve module supplies current to the airbags.

Self-diagnosis system

6   A self-diagnosis circuit in the module displays a light when the ignition switch is turned to the On position. If the system is operating normally, the light should go out after seven flashes. If the light doesn't come on, or doesn't go out after seven flashes, or if it comes on while you're driving the vehicle, there's a malfunction in the SIR system. Have it inspected and repaired as soon as possible. Do not attempt to troubleshoot or service the SIR system yourself. Even a small mistake could cause the SIR system to malfunction when you need it.

Servicing components near the SIR system

7   Nevertheless, there are times when you need to remove the steering wheel, radio or service other components on or near the instrument panel. At these times, you'll be working around components and wiring harnesses for the SIR system. SIR system wiring is easy to identify; they're all covered by a bright yellow conduit. Do not unplug the connectors for the SIR system wiring, except to disable the system. And do not use elec-

trical test equipment on the SIR system wiring. **Always disable the SIR system before working near the SIR system components or related wiring.**

Disabling the SIR system

Refer to illustration 26.10

8   Turn the steering wheel to the straight ahead position, place the ignition switch in Lock and remove the key. Remove the airbag fuse from the fuse block (see Section 3).
9   Remove the steering column lower trim panel and sound insulator panel below the instrument panel (see Chapter 11).
10   Unplug the yellow Connector Position Assurance (CPA) steering column harness connectors **(see illustration)**.

Enabling the SIR system

11   After you've disabled the airbag and performed the necessary service, plug in the steering column (driver's side) and passenger side CPA connectors. Reinstall the steering column lower trim panel, and the sound insulator panel.
12   Install the airbag fuse.

27   Wiring diagrams - general information

Since it isn't possible to include all wiring diagrams for every year covered by this manual, the following diagrams are those that are typical and most commonly needed.

Prior to troubleshooting any circuit, check the fuse and circuit breakers (if equipped) to make sure they're in good condition. Make sure the battery is properly charged and check the cable connections (see Chapter 1).

When checking a circuit, make sure that all electrical connectors are clean, with no broken or loose terminals. When unplugging an electrical connector, do not pull on the wires. Pull only on the connector housings themselves.

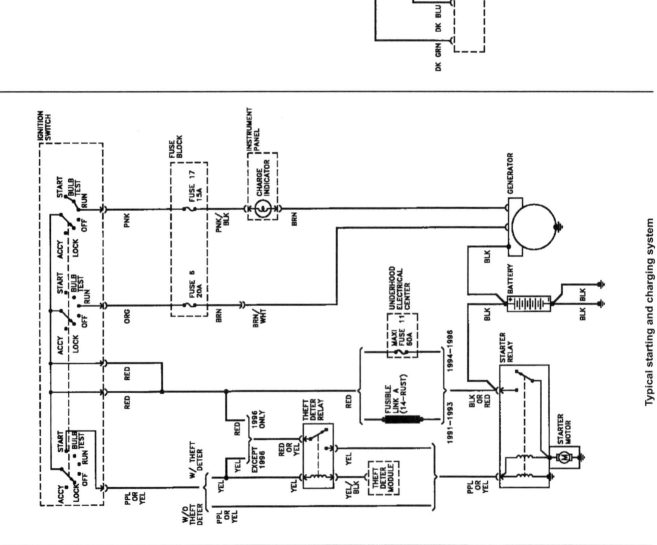

Typical 1994 and later engine cooling fan system

Typical starting and charging system

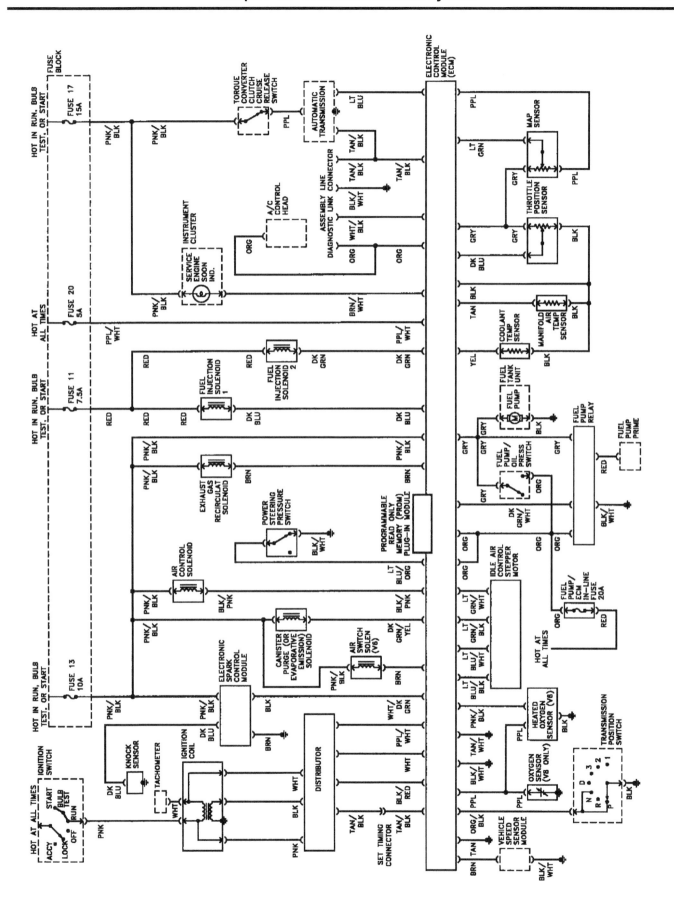

Typical 1993 and earlier engine control system

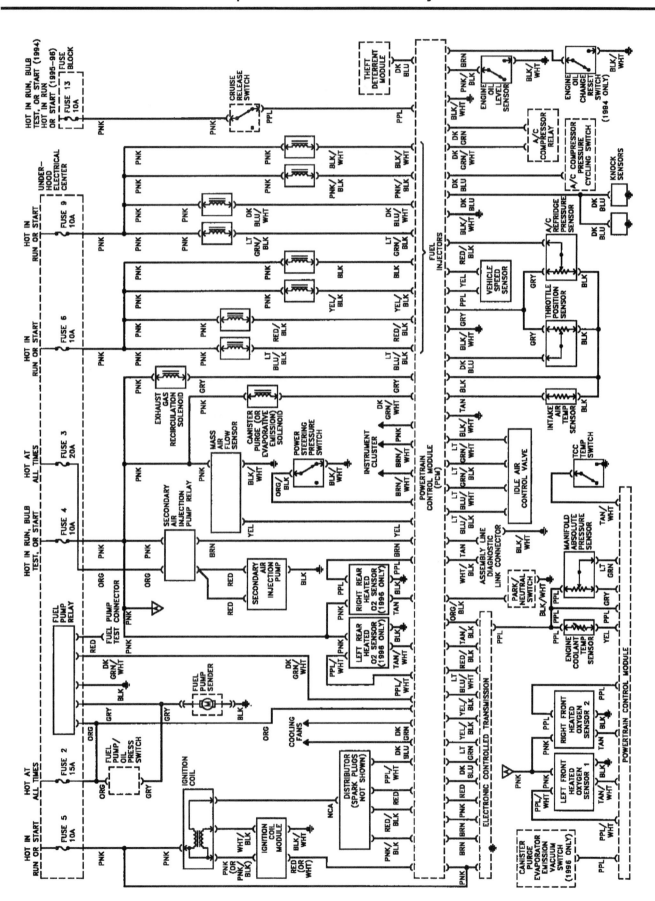

Typical 1994 and later engine control system

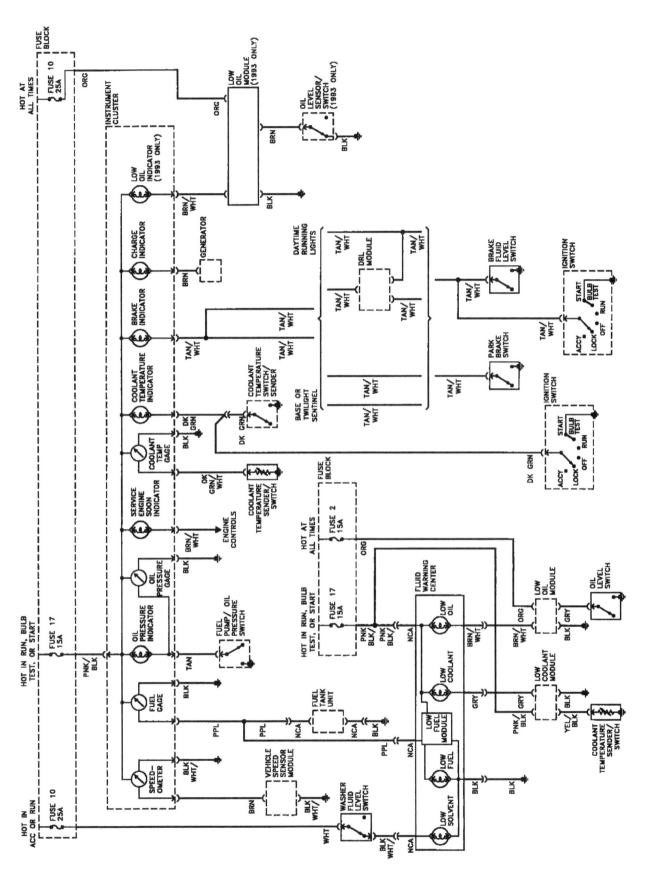

Typical 1993 and earlier engine warning system (Base system)

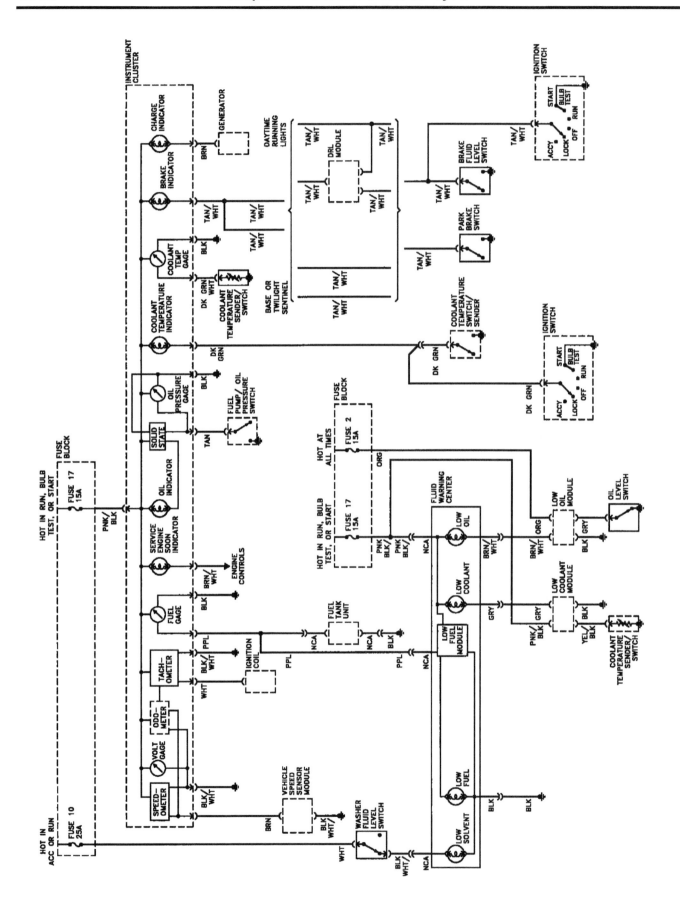

Typical 1993 and earlier engine warning system (Digital system)

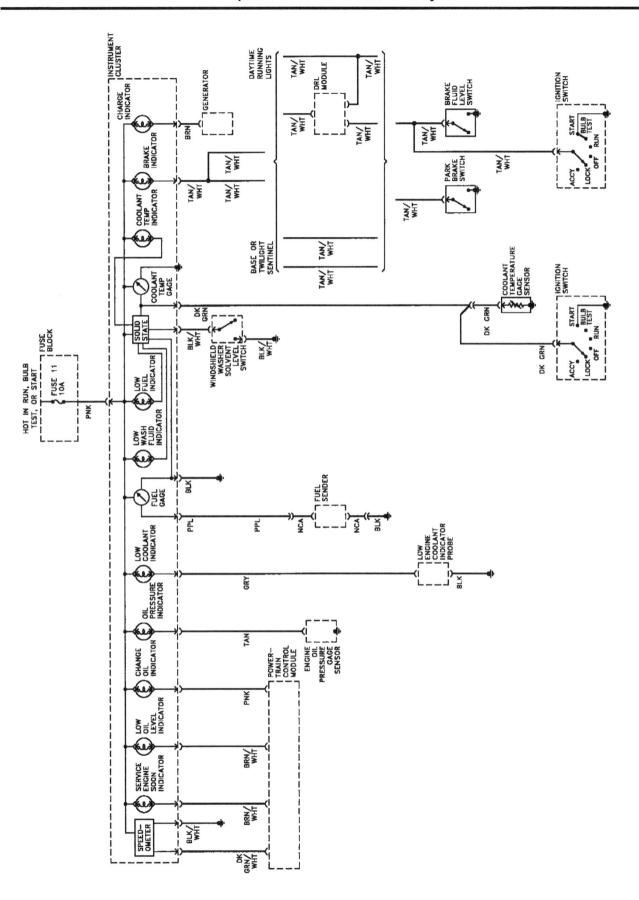

Typical 1994 and later engine warning system

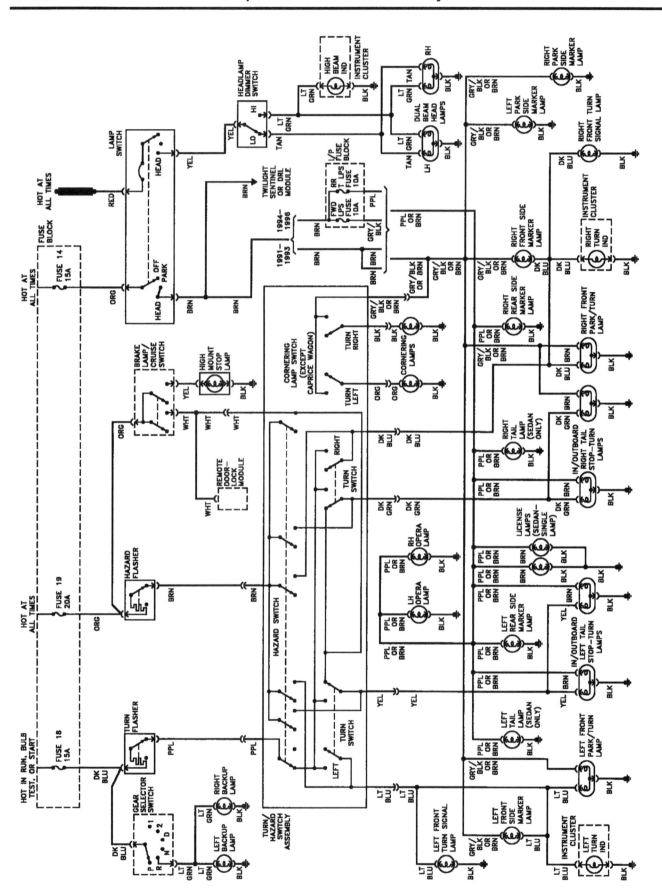

Typical exterior lighting system (with Base Headlights)

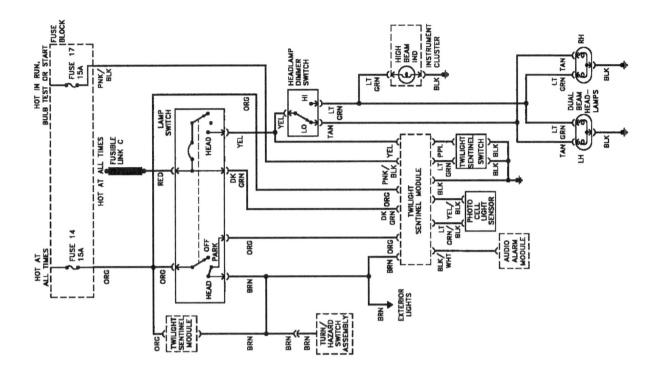

Typical exterior lighting system (with Twilight Sentinel)

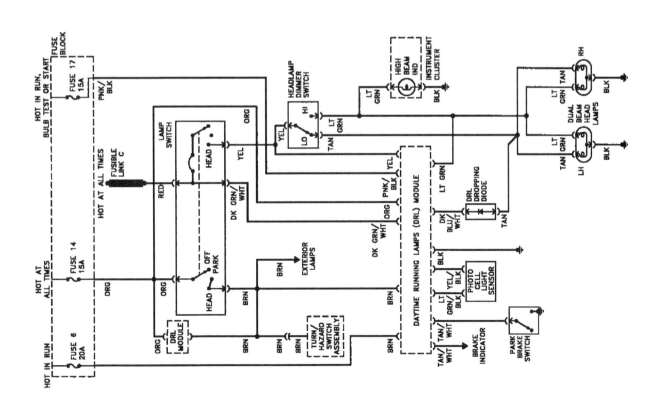

Typical exterior lighting system (with Daytime Running Lights)

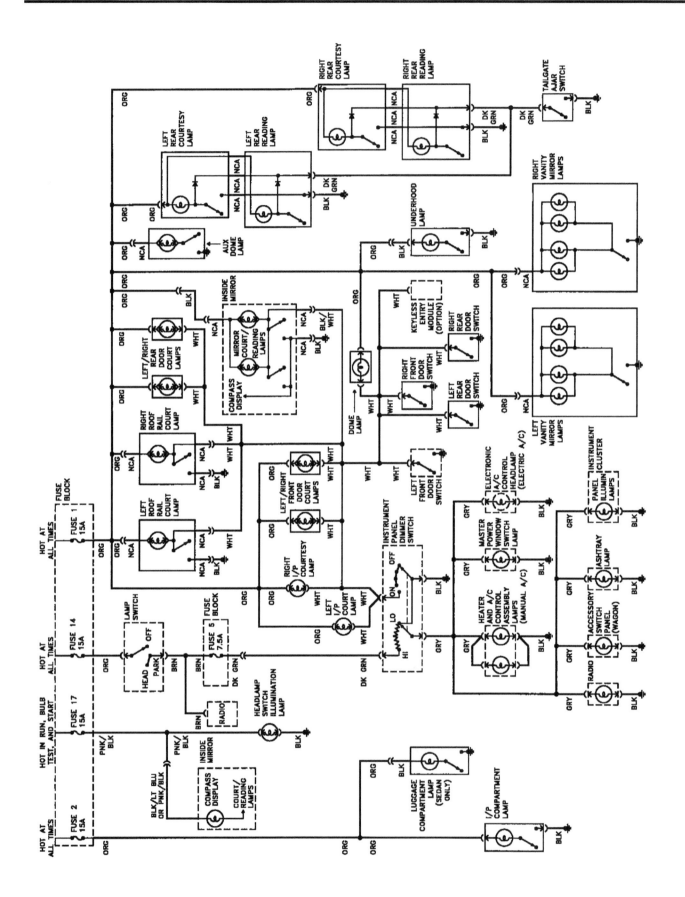

Typical 1993 and earlier interior lighting system

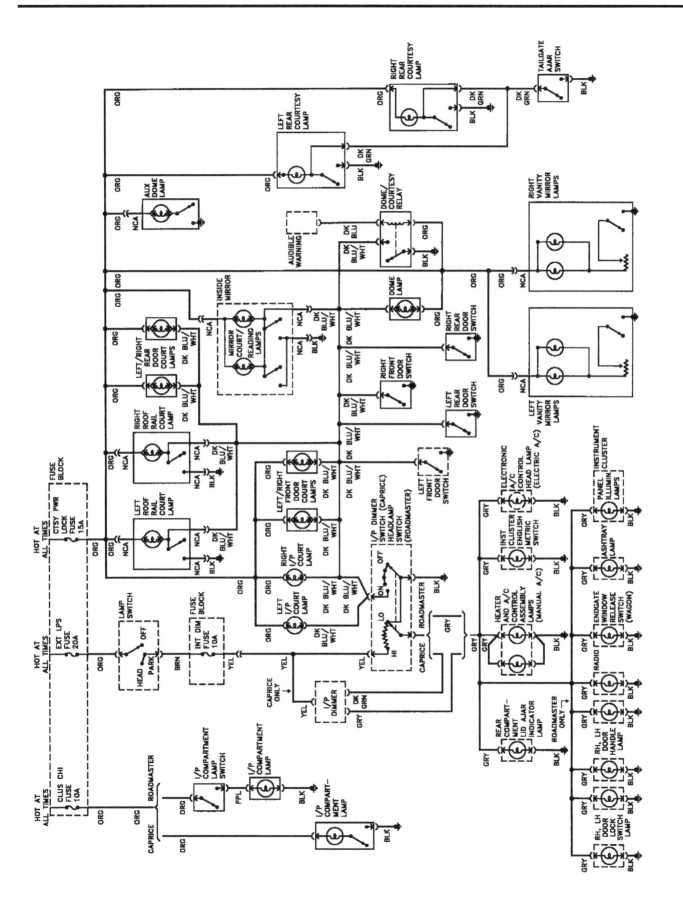

Typical 1994 and later interior lighting system

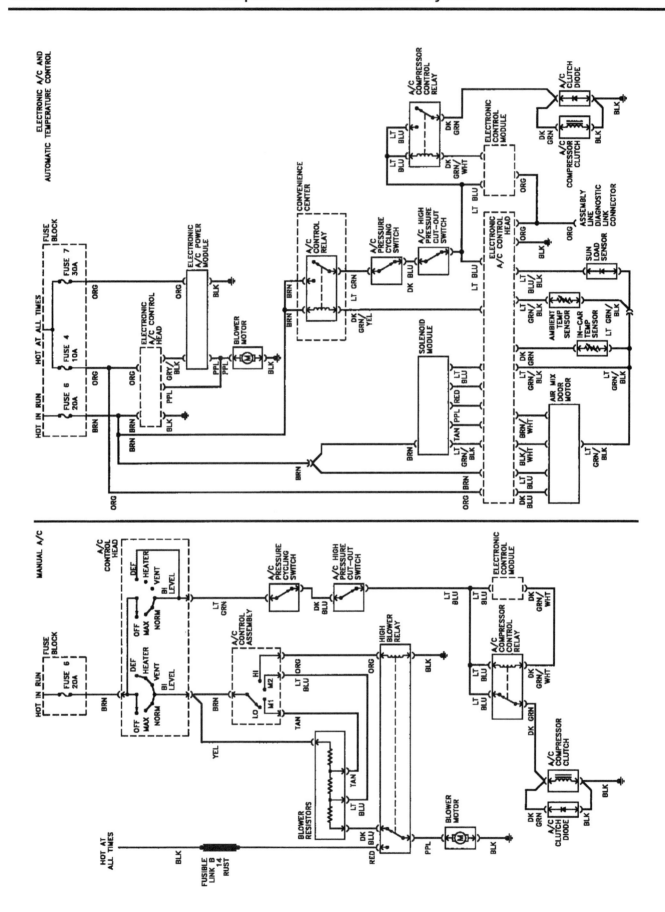

Typical 1993 and earlier heating and air conditioning system

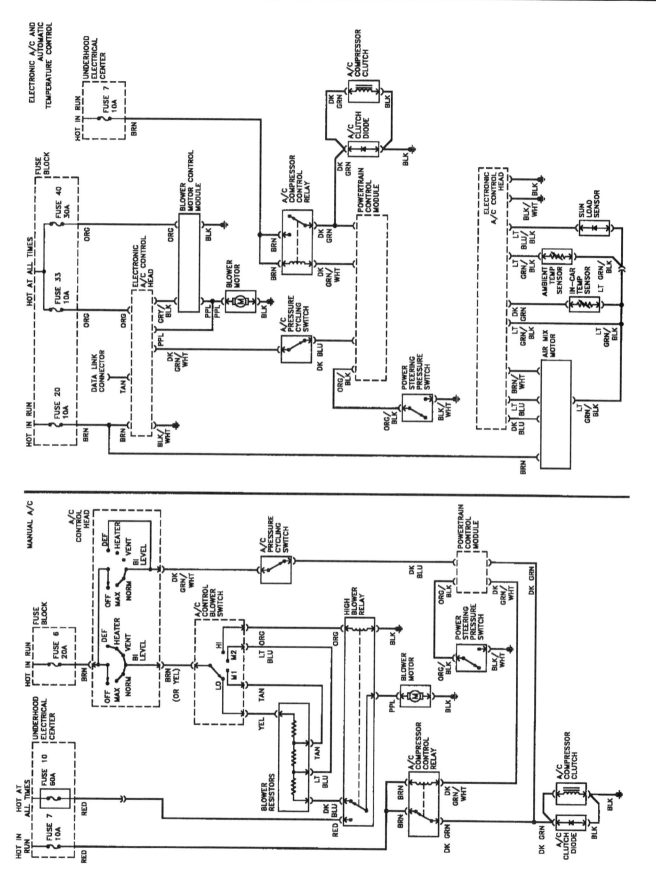

Typical 1994 and later heating and air conditioning system

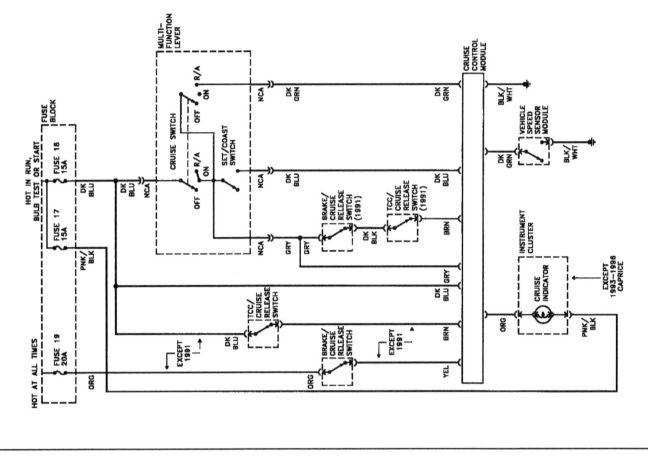

Typical power door lock system (without keyless entry)

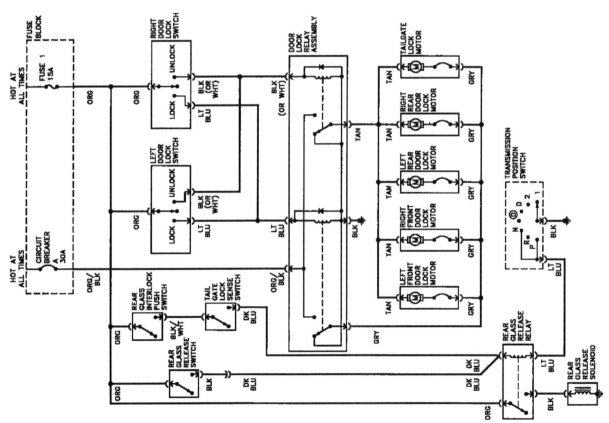

Typical cruise control system

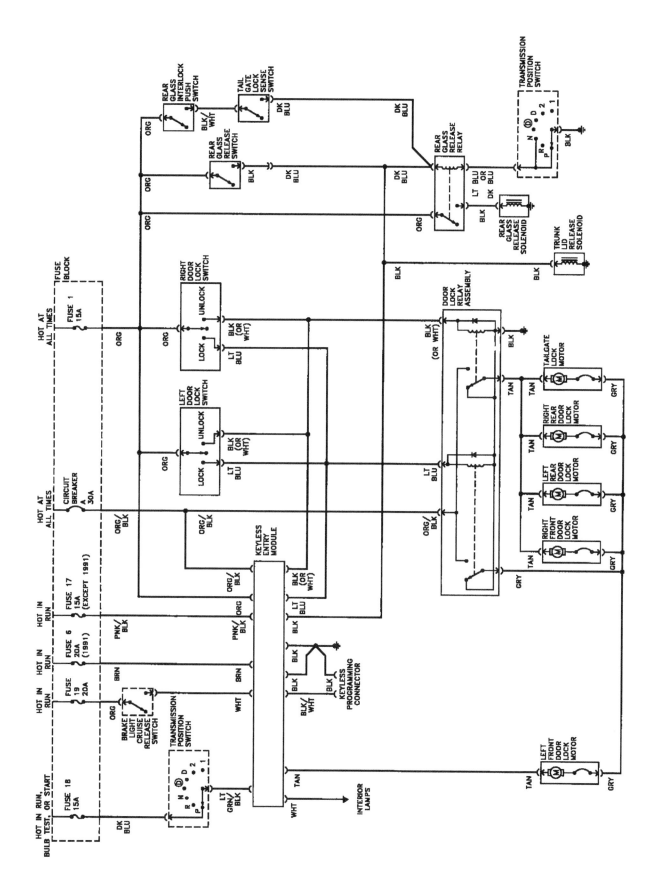

Typical 1993 and earlier power door lock system (with keyless entry)

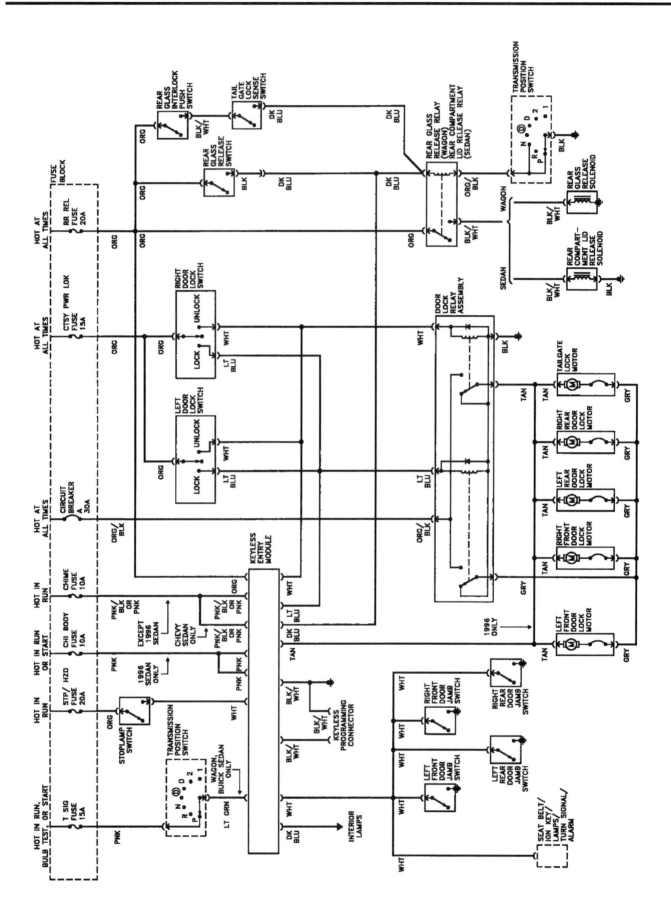

Typical 1994 and later power door lock system (with keyless entry)

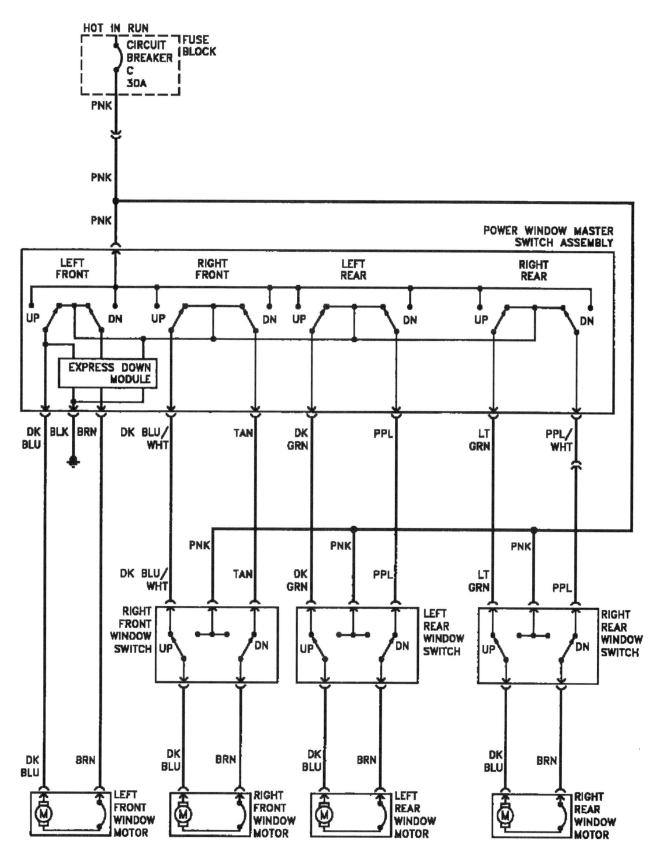

Typical 1993 and earlier power window system

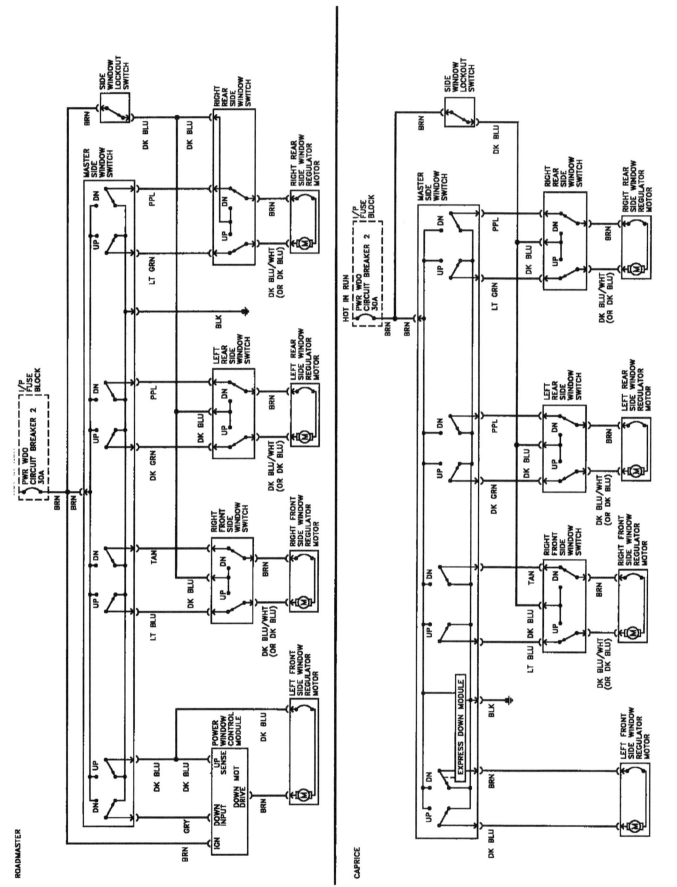

Typical 1994 and later power window system

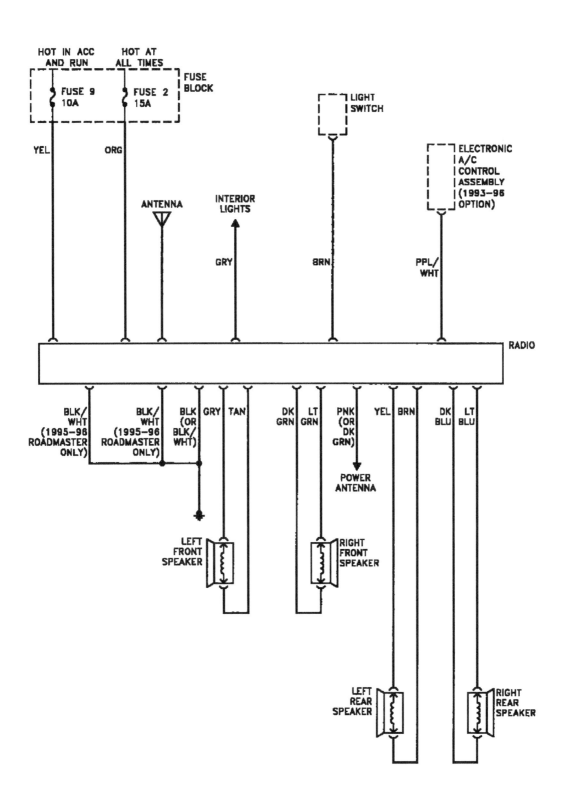

Typical audio system

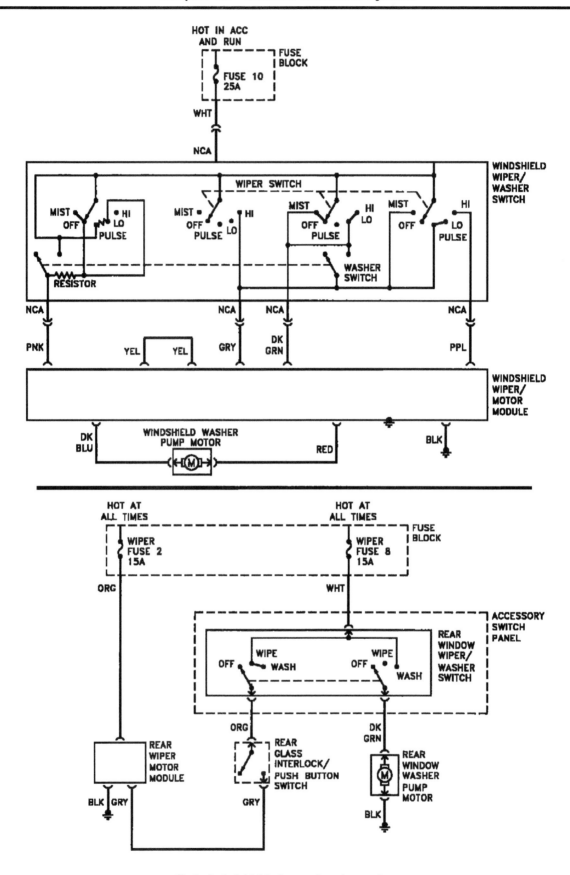

Typical windshield wiper and washer system

Index

Haynes Automotive Manuals

*NOTE: If you do not see a listing for your vehicle, please visit **haynes.com** for the latest product information and check out our **Online Manuals!***

ACURA
12020 Integra '86 thru '89 & Legend '86 thru '90
12021 Integra '90 thru '93 & Legend '91 thru '95
Integra '94 thru '00 - see HONDA Civic (42025)
MDX '01 thru '07 - see HONDA Pilot (42037)
12050 Acura TL all models '99 thru '08

AMC
14020 Concord/Hornet/Gremlin/Spirit '70 thru '83
14025 (Renault) Alliance & Encore '83 thru '87

AUDI
15020 4000 all models '80 thru '87
15025 5000 all models '77 thru '83
15026 5000 all models '84 thru '88
Audi A4 '96 thru '01 - see VW Passat (96023)
15030 Audi A4 '02 thru '08

AUSTIN-HEALEY
Sprite - see MG Midget (66015)

BMW
18020 3/5 Series '82 thru '92
18021 3-Series including Z3 models '92 thru '98
18022 3-Series including Z4 models '99 thru '05
18023 3-Series '06 thru '14
18025 320i all 4-cylinder models '75 thru '83
18050 1500 thru 2002 except Turbo '59 thru '77

BUICK
19010 Buick Century '97 thru '05
Century (front-wheel drive) - see GM (38005)
19020 Buick, Oldsmobile & Pontiac Full-size (Front wheel drive) '85 thru '05
19025 Buick, Oldsmobile & Pontiac Full-size (Rear wheel drive) '70 thru '90
19027 Buick LaCrosse '05 thru '13
Regal - see GENERAL MOTORS (38010)
Skyhawk - see GM (38015)
Skylark - see GM (38020, 38025)
Somerset - see GENERAL MOTORS (38025)

CADILLAC
21015 CTS & CTS-V '03 thru '14
21030 Cadillac Rear Wheel Drive '70 thru '93
Cimarron, Eldorado & Seville - see GM (38015, 38030, 38031)

CHEVROLET
10305 Chevrolet Engine Overhaul Manual
24010 Astro & GMC Safari Mini-vans '85 thru '05
24015 Camaro V8 all models '70 thru '81
24016 Camaro all models '82 thru '92
Cavalier - see GM (38016)
Celebrity - see GM (38005)
24017 Camaro & Firebird '93 thru '02
24018 Camaro '10 thru '15
24020 Chevelle, Malibu, El Camino '69 thru '87
24024 Chevette & Pontiac T1000 '76 thru '87
Citation - see GENERAL MOTORS (38020)
24027 Colorado & GMC Canyon '04 thru '12
24032 Corsica & Beretta all models '87 thru '96
24040 Corvette all V8 models '68 thru '82
24041 Corvette all models '84 thru '96
24042 Corvette all models '97 thru '13
24044 Cruze '11 thru '19
24045 Full-size Sedans Caprice, Impala, Biscayne, Bel Air & Wagons '69 thru '90
24046 Impala SS & Caprice and Buick Roadmaster '91 thru '96
Impala '00 thru '05 - see LUMINA (24048)
24047 Impala & Monte Carlo '06 thru '11
Lumina '90 thru '94 - see GM (38010)
24048 Lumina & Monte Carlo '95 thru '05
Lumina APV - see GM (38035)
24050 Luv Pick-up all 2WD & 4WD '72 thru '82
24051 Malibu '13 thru '19
24055 Monte Carlo all models '70 thru '88
Monte Carlo '95 thru '01 - see LUMINA (24048)
24059 Nova all V8 models '69 thru '79
24060 Nova/Geo Prizm '85 thru '92
24064 Pick-ups '67 thru '87 - Chevrolet & GMC
24065 Pick-ups '88 thru '98 - Chevrolet & GMC
24066 Pick-ups '99 thru '06 - Chevrolet & GMC
24067 Chevy Silverado & GMC Sierra '07 thru '14
24068 Chevy Silverado & GMC Sierra '14 thru '19
24070 S-10 & S-15 Pick-ups '82 thru '93
24071 S-10 & Sonoma Pick-ups '94 thru '04
24072 Chevrolet TrailBlazer, GMC Envoy & Oldsmobile Bravada '02 thru '09
24075 Sprint '85 thru '88, Geo Metro '89 thru '01
24080 Vans - Chevrolet & GMC '68 thru '96
24081 Chevrolet Express & GMC Savana Full-size Vans '96 thru '19

CHRYSLER
10310 Chrysler Engine Overhaul Manual
25015 Chrysler Cirrus, Dodge Stratus, Plymouth Breeze '95 thru '00
25020 Full-size Front-Wheel Drive '88 thru '93
K-Cars - see DODGE Aries (30008)
25025 Chrysler LHS, Concorde & New Yorker, Dodge Intrepid, Eagle Vision, '93 thru '97
25026 Chrysler LHS, Concorde, 300M, Dodge Intrepid '98 thru '04
25027 Chrysler 300 '05 thru '18, Dodge Charger '06 thru '18, Magnum '05 thru '08 & Challenger '08 thru '18
25030 Chrysler & Plymouth Mid-size '82 thru '95
Rear-wheel Drive - see DODGE (30050)
25035 PT Cruiser all models '01 thru '10
25040 Chrysler Sebring '95 thru '06, Dodge Stratus '01 thru '06 & Dodge Avenger '95 thru '00
25041 Chrysler Sebring '07 thru '10, 200 '11 thru '17 Dodge Avenger '08 thru '14

DATSUN
28005 200SX all models '80 thru '83
28012 240Z, 260Z & 280Z Coupe '70 thru '78
28014 280ZX Coupe & 2+2 '79 thru '83
300ZX - see NISSAN (72010)
28018 510 & PL521 Pick-up '68 thru '73
28020 510 all models '78 thru '81
28022 620 Series Pick-up all models '73 thru '79
720 Series Pick-up - see NISSAN (72030)

DODGE
30008 Aries & Plymouth Reliant '81 thru '89
30010 Caravan & Plymouth Voyager '84 thru '95
30011 Caravan & Plymouth Voyager '96 thru '02
30012 Challenger & Plymouth Sapporo '78 thru '83
30013 Caravan, Chrysler Voyager & Town & Country '03 thru '07
30014 Grand Caravan & Chrysler Town & Country '08 thru '18

DODGE (cont.)
30020 Dakota Pick-ups all models '87 thru '96
30021 Durango '98 & '99 & Dakota '97 thru '99
30022 Durango '00 thru '03 & Dakota '00 thru '04
30023 Durango '04 thru '09 & Dakota '05 thru '11
30025 Dart, Challenger/Plymouth Barracuda & Valiant 6-cylinder models '67 thru '76
30030 Daytona & Chrysler Laser '84 thru '89
Intrepid - see Chrysler (25025, 25026)
30034 Neon all models '95 thru '99
30035 Omni & Plymouth Horizon '78 thru '90
30036 Dodge & Plymouth Neon '00 thru '05
30040 Pick-ups full-size models '74 thru '93
30041 Pick-ups full-size models '94 thru '01
30042 Pick-ups full-size models '02 thru '08
30043 Pick-ups full-size models '09 thru '18
30045 Ram 50/D50 Pick-ups & Raider and Plymouth Arrow Pick-ups '79 thru '93
30050 Dodge/Plymouth/Chrysler RWD '71 thru '89
30055 Shadow & Plymouth Sundance '87 thru '94
30060 Spirit & Plymouth Acclaim '89 thru '95
30065 Vans - Dodge & Plymouth '71 thru '03

EAGLE
Talon - see MITSUBISHI (68030, 68031)
Vision - see CHRYSLER (25025)

FIAT
34010 124 Sport Coupe & Spider '68 thru '78
34025 X1/9 all models '74 thru '80

FORD
10320 Ford Engine Overhaul Manual
10355 Ford Automatic Transmission Overhaul
11500 Mustang '64-1/2 thru '70 Restoration Guide
36004 Aerostar Mini-vans '86 thru '97
36006 Contour & Mercury Mystique '95 thru '00
36008 Courier Pick-up all models '72 thru '82
36012 Crown Victoria & Mercury Grand Marquis '88 thru '11
36016 Escort & Mercury Lynx '81 thru '90
36020 Escort & Mercury Tracer '91 thru '02
36022 Escape '01 thru '17, Mazda Tribute '01 thru '11 & Mercury Mariner '05 thru '11
36024 Explorer & Mazda Navajo '91 thru '01
36025 Explorer & Mercury Mountaineer '02 thru '10
36026 Explorer '11 thru '17
36028 Fairmont & Mercury Zephyr '78 thru '83
36030 Festiva & Aspire '88 thru '97
36032 Fiesta all models '77 thru '80
36034 Focus all models '00 thru '11
36035 Focus '12 thru '14
36045 Ford Fusion '06 thru '14 & Mercury Milan '06 thru '10
36048 Mustang V8 all models '64-1/2 thru '73
36049 Mustang II 4-cylinder, V6 & V8 '74 thru '78
36050 Mustang & Mercury Capri '79 thru '93
36051 Mustang all models '94 thru '04
36052 Mustang '05 thru '14
36054 Pick-ups and Bronco '73 thru '79
36058 Pick-ups and Bronco '80 thru '96
36059 F-150 '97 thru '03, Expedition '97 thru '17, F-250 '97 thru '99, F-150 Heritage '04 & Lincoln Navigator '98 thru '17
36060 Super Duty Pick-up & Excursion '99 thru '10
36061 F-150 full-size '04 thru '14
36062 Pinto & Mercury Bobcat '75 thru '80
36063 F-150 full-size '15 thru '17
36064 Super Duty Pick-ups '11 thru '16
36066 Probe all models '89 thru '92
36070 Ranger & Bronco II gas models '83 thru '92
36071 Ranger '93 thru '11 & Mazda Pick-ups '94 thru '09
36074 Taurus & Mercury Sable '86 thru '95
36075 Taurus & Mercury Sable '96 thru '07
36076 Taurus '08 thru '14, Five Hundred '05 thru '07, Mercury Montego '05 thru '07 & Sable '08 thru '09
36078 Tempo & Mercury Topaz '84 thru '94
36082 Thunderbird & Mercury Cougar '83 thru '88
36086 Thunderbird & Mercury Cougar '89 thru '97
36090 Vans all V8 Econoline models '69 thru '91
36094 Vans full size '92 thru '14
36097 Windstar '95 thru '03, Freestar & Mercury Monterey Mini-van '04 thru '07

GENERAL MOTORS
10360 GM Automatic Transmission Overhaul
38005 Buick Century, Chevrolet Celebrity, Olds Cutlass Ciera & Pontiac 6000 '82 thru '96
38010 Buick Regal, Chevrolet Lumina, Oldsmobile Cutlass Supreme & Pontiac Grand Prix front wheel drive '88 thru '07
38015 Buick Skyhawk, Cadillac Cimarron, Chevrolet Cavalier, Oldsmobile Firenza Pontiac J-2000 & Sunbird '82 thru '94
38016 Chevrolet Cavalier/Pontiac Sunfire '95 thru '05
38017 Chevrolet Cobalt & Pontiac G5 '05 thru '11
38020 Buick Skylark, Chevrolet Citation, Olds Omega, Pontiac Phoenix '80 thru '85
38025 Buick Skylark & Somerset, Olds Achieva, Calais & Pontiac Grand Am '85 thru '98
38026 Chevrolet Malibu, Olds Alero & Cutlass, Pontiac Grand Am '97 thru '03
38027 Chevrolet Malibu '04 thru '12
38030 Cadillac Eldorado, Seville, Oldsmobile Toronado & Buick Riviera '71 thru '85
38031 Cadillac Eldorado, Seville, DeVille, Fleetwood, Oldsmobile Toronado & Buick Riviera '86 thru '93
38032 Cadillac DeVille '94 thru '05, Seville '92 thru '04 & Cadillac DTS '06 thru '10
38035 Chevrolet Lumina APV, Olds Silhouette & Pontiac Trans Sport all models '90 thru '96
38036 Chevrolet Venture, Olds Silhouette, Pontiac Trans Sport & Montana '97 thru '05
38040 Chevrolet Equinox '05 thru '17, GMC Terrain '10 thru '17 & Pontiac Torrent '06 thru '09

GEO
Metro - see CHEVROLET Sprint (24075)
Prizm - '85 thru '92 see CHEVY (24060), '93 thru '02 see TOYOTA Corolla (92036)
40030 Storm all models '90 thru '93
Tracker - see SUZUKI Samurai (90010)

GMC
Vans & Pick-ups - see CHEVROLET

HONDA
42010 Accord CVCC all models '76 thru '83
42011 Accord all models '84 thru '89
42012 Accord all models '90 thru '93
42013 Accord all models '94 thru '97
42014 Accord all models '98 thru '02
42015 Accord '03 thru '12 & Crosstour '10 thru '14
42016 Accord '13 thru '17

HONDA (cont.)
42020 Civic 1200 all models '73 thru '79
42021 Civic 1300 & 1500 CVCC '80 thru '83
42022 Civic 1500 CVCC all models '75 thru '79
42023 Civic all models '84 thru '91
42024 Civic & del Sol '92 thru '95
42025 Civic '96 thru '00 & CR-V '97 thru '01 & Acura Integra '94 thru '00
42026 Civic '01 thru '11 & CR-V '02 thru '11
42027 Civic '12 thru '15 & CR-V '12 thru '16
42035 Fit '07 thru '13
42035 Odyssey models '99 thru '10
Passport - see ISUZU Rodeo (47017)
42037 Honda Pilot '03 thru '08, Ridgeline '06 thru '14 & Acura MDX '01 thru '07
42040 Prelude CVCC all models '79 thru '89

HYUNDAI
43010 Elantra all models '96 thru '19
43015 Excel & Accent all models '86 thru '13
43050 Santa Fe all models '01 thru '12
43055 Sonata all models '99 thru '14

INFINITI
G35 '03 thru '08 - see NISSAN 350Z (72011)

ISUZU
Hombre - see CHEVROLET S-10 (24071)
47017 Rodeo '91 thru '02, Amigo '89 thru '94 & '98 thru '02 & Honda Passport '95 thru '02
47020 Trooper '84 thru '91 & Pick-up '81 thru '93

JAGUAR
49010 XJ6 all 6-cylinder models '68 thru '86
49011 XJ6 all models '88 thru '94
49015 XJ12 & XJS all 12-cylinder models '72 thru '85

JEEP
50010 Cherokee, Comanche & Wagoneer Limited all models '84 thru '01
50011 Cherokee '14 thru '19
50020 CJ all models '49 thru '86
50025 Grand Cherokee all models '93 thru '04
50026 Grand Cherokee '05 thru '19
50029 Grand Wagoneer & Pick-up '72 thru '91
50030 Wrangler all models '87 thru '17
50035 Liberty '02 thru '12 & Dodge Nitro '07 thru '11
50050 Patriot & Compass '07 thru '17

KIA
54050 Optima '01 thru '10
54060 Sedona '02 thru '14
54070 Sephia '94 thru '01, Spectra '00 thru '09, Sportage '05 thru '20
54077 Sorento '03 thru '13

LEXUS
ES 300/330 - see TOYOTA Camry (92007, 92008)
RX 300/330/350 - see TOYOTA Highlander (92095)

LINCOLN
Navigator - see FORD Pick-up (36059)
59010 Rear-Wheel Drive Continental '70 thru '87, Mark Series '70 thru '92 & Town Car '81 thru '10

MAZDA
61010 GLC (rear wheel drive) '77 thru '83
61011 GLC (front wheel drive) '81 thru '85
61012 Mazda3 '04 thru '11
61015 323 & Protegé '90 thru '03
61016 MX-5 Miata '90 thru '14
61020 MPV all models '89 thru '98
61030 Pick-ups '72 thru '93
Pick-ups '94 thru '09 - see Ford (36071)
61035 RX-7 all models '79 thru '85
61036 RX-7 all models '86 thru '91
61040 626 (rear-wheel drive) all models '79 thru '82
61041 626 & MX-6 (front-wheel drive) '83 thru '92
61042 626 '93 thru '01 & MX-6/Ford Probe '93 thru '02
61043 Mazda6 '03 thru '13

MERCEDES-BENZ
63012 123 Series Diesel '76 thru '85
63015 190 Series 4-cylinder gas models '84 thru '88
63020 230, 250 & 280 6-cylinder SOHC '68 thru '72
63025 280 123 Series gas models '77 thru '81
63030 350 & 450 all models '71 thru '80
63040 C-Class: C230/C240/C280/C320/C350 '01 thru '07

MERCURY
64200 Villager & Nissan Quest '93 thru '01
All other titles, see FORD listing.

MG
66010 MGB Roadster & GT Coupe '62 thru '80
66015 MG Midget & Austin Healey Sprite Roadster '58 thru '80

MINI
67020 Mini '02 thru '13

MITSUBISHI
68020 Cordia, Tredia, Galant, Precis & Mirage '83 thru '93
68030 Eclipse, Eagle Talon & Plymouth Laser '90 thru '94
68031 Eclipse '95 thru '05 & Eagle Talon '95 thru '98
68035 Galant '94 thru '12
68040 Pick-up '83 thru '96 & Montero '83 thru '93

NISSAN
72010 300ZX all models incl. Turbo '84 thru '89
72011 350Z & Infiniti G35 '03 thru '08
72015 Altima all models '93 thru '06
72016 Altima '07 thru '12
72020 Maxima all models '85 thru '92
72021 Maxima all models '93 thru '08
72025 Murano '03 thru '14
72030 Pick-ups '80 thru '97 & Pathfinder '87 thru '95
72031 Frontier, Xterra & Pathfinder '96 thru '04
72032 Frontier & Xterra '05 thru '14
72037 Pathfinder '05 thru '14
72040 Pulsar all models '83 thru '86
72042 Rogue all models '08 thru '20
72050 Sentra all models '82 thru '94
72051 Sentra & 200SX all models '95 thru '06
72060 Stanza all models '82 thru '90
72070 Titan pick-ups '04 thru '10 & Armada '05 thru '14
72080 Versa all models '07 thru '19

OLDSMOBILE
73015 Cutlass V6 & V8 gas models '74 thru '88
For other OLDSMOBILE titles, see BUICK, CHEVROLET or GM listings.

PLYMOUTH
For PLYMOUTH titles, see DODGE.

PONTIAC
79008 Fiero all models '84 thru '88
79018 Firebird V8 models except Turbo '70 thru '81
79019 Firebird all models '82 thru '92
79025 G6 all models '05 thru '09
79040 Mid-size Rear-wheel Drive '70 thru '87
Vibe '03 thru '10 - see TOYOTA Corolla (92037)
For other PONTIAC titles, see BUICK, CHEVROLET or GM listings.

PORSCHE
80020 911 Coupe & Targa models '65 thru '89
80025 914 all 4-cylinder models '69 thru '76
80030 924 all models including Turbo '76 thru '82
80035 944 all models including Turbo '83 thru '89

RENAULT
Alliance & Encore - see AMC (14025)

SAAB
84010 900 all models including Turbo '79 thru '88

SATURN
87010 Saturn all S-series models '91 thru '02
Saturn Ion '03 thru '07- see GM (38017)
87020 Saturn L-series all models '00 thru '04
87040 Saturn VUE '02 thru '09

SUBARU
89002 1100, 1300, 1400 & 1600 '71 thru '79
89003 1600 & 1800 2WD & 4WD '80 thru '94
89080 Impreza '02 thru '11, WRX '02 thru '14 & WRX STI '04 thru '14
89100 Legacy all models '90 thru '99
89101 Legacy & Forester '00 thru '09
89102 Legacy '10 thru '16 & Forester '12 thru '16

SUZUKI
90010 Samurai/Sidekick & Geo Tracker '86 thru '01

TOYOTA
92005 Camry all models '83 thru '91
92006 Camry '92 thru '96 & Avalon '95 thru '96
92007 Camry, Avalon, Solara, Lexus ES 300 '97 thru '01
92008 Camry, Avalon, Lexus ES 300/330 '02 thru '06 & Solara '02 thru '08
92009 Camry, Avalon & Lexus ES 350 '07 thru '17
92015 Celica Rear-wheel Drive '71 thru '85
92020 Celica Front-wheel Drive '86 thru '99
92025 Celica Supra all models '79 thru '92
92030 Corolla all models '75 thru '79
92032 Corolla rear-wheel drive models '80 thru '87
92035 Corolla front-wheel drive models '84 thru '92
92036 Corolla & Geo/Chevrolet Prizm '93 thru '02
92037 Corolla '03 thru '19, Matrix '03 thru '14, & Pontiac Vibe '03 thru '10
92040 Corolla Tercel all models '80 thru '82
92045 Corona all models '74 thru '82
92050 Cressida all models '78 thru '82
92055 Land Cruiser FJ40/45/55 '68 thru '82
92056 Land Cruiser FJ60/62/80/FZJ80 '80 thru '96
92060 Matrix '03 thru '11 & Pontiac Vibe '03 thru '10
92065 MR2 all models '85 thru '87
92070 Pick-up all models '69 thru '78
92075 Pick-up all models '79 thru '95
92076 Tacoma '95 thru '04, 4Runner '96 thru '02 & T100 '93 thru '98
92077 Tacoma all models '05 thru '18
92078 Tundra '00 thru '06, Sequoia '01 thru '07
92079 4Runner all models '03 thru '09
92080 Previa all models '91 thru '95
92081 Prius all models '01 thru '12
92082 RAV4 all models '96 thru '12
92085 Tercel all models '87 thru '94
92090 Sienna all models '98 thru '10
92095 Highlander '01 thru '19 & Lexus RX330/330/350 '99 thru '19
92179 Tundra '07 thru '19 & Sequoia '08 thru '19

TRIUMPH
94007 Spitfire all models '62 thru '81
94010 TR7 all models '75 thru '81

VW
96008 Beetle & Karmann Ghia '54 thru '79
96009 New Beetle '98 thru '10
96016 Rabbit, Jetta, Scirocco & Pick-up gas models '75 thru '92
96017 Golf, GTI & Jetta '93 thru '98, Cabrio '95 thru '02
96018 Golf, GTI & Jetta '99 thru '05
96019 Jetta, Rabbit, GLI, GTI & Golf '05 thru '11
96020 Rabbit, Jetta, Pick-up diesel '77 thru '84
96021 Jetta '11 thru '18 & Golf '15 thru '19
96023 Passat '98 thru '05, Audi A4 '96 thru '01
96030 Transporter 1600 all models '68 thru '79
96035 Transporter 1700, 1800, 2000 '72 thru '79
96040 Type 3 1500 & 1600 '63 thru '73
96045 Vanagon Air-Cooled all models '80 thru '83

VOLVO
97010 120, 130 Series & 1800 Sports '61 thru '73
97015 140 Series all models '66 thru '74
97020 240 Series all models '76 thru '93
97040 740 & 760 Series all models '82 thru '88
97050 850 Series all models '93 thru '97

TECHBOOK MANUALS
10205 Automotive Computer Codes
10206 OBD-II & Electronic Engine Management
10210 Automotive Emissions Control Manual
10215 Fuel Injection Manual '78 thru '85
10225 Holley Carburetor Manual
10230 Rochester Carburetor Manual
10305 Chevrolet Engine Overhaul Manual
10320 Ford Engine Overhaul Manual
10330 GM and Ford Diesel Engine Repair Manual
10331 Duramax Diesel Engine '01 thru '19
10332 Cummins Diesel Engine Performance
10333 GM, Ford & Chrysler Engine Performance
10334 GM Engine Performance
10340 Small Engine Repair Manual, 5 HP & Less
10341 Small Engine Repair Manual, 5.5 thru 20 HP
10345 Suspension, Steering & Driveline Manual
10355 Ford Automatic Transmission Overhaul
10360 GM Automatic Transmission Overhaul
10405 Automotive Body Repair & Painting
10410 Automotive Brake Manual
10411 Automotive Anti-lock Brake (ABS) Systems
10425 Automotive Heating & Air Conditioning
10435 Automotive Tools Manual
10445 Welding Manual

Over a 100 Haynes motorcycle manuals also available

10/22

Haynes North America, Inc. • (805) 498-6703 • www.haynes.com